Numerical Methods for Differential Equations and Applications

Liviu Gr. Ixaru

Institute of Physics and Nuclear Engineering
Division of Fundamental Physics, Bucharest, Romania

Numerical Methods

for Differential Equations and Applications

EDITURA ACADEMIEI
Bucureşti, Romania

D. Reidel Publishing Company
A MEMBER OF THE KLUWER ACADEMIC PUBLISHERS GROUP
Dordrecht / Boston / Lancaster

Library of Congress Cataloging in Publication Data

Ixaru. Liviu Gr.
Numerical methods for differential equations and applications.

(Mathematics and its applications. East European series)
Translation of: Metode numerice pentru ecuaţii diferenţiale cu aplicaţii.
Includes bibliographical references and index.
1. Differential equations—Numerical solutions.
2. Differential equations, Partial—Numerical solutions.
I. Title. II. Title: Numerical methods. III. Series: Mathematics and its applications (D. Reidel Publishing Company). East European series.
QA372.1913 1984 515.3'5 83-26948
ISBN 978-90-481-8383-8

Distributors for the U.S.A. and Canada
Kluwer Academic Publishers
190 Old Derby Street, Hingham, MA 02043, U.S.A.

Distributors for Albania, Bulgaria, Chinese People's Republic, Cuba, Czechoslovakia, German Democratic Republic, Hungary, Korean People's Republic, Mongolia, Poland, Romania, the U.S.S.R., Vietnam and Yugoslavia
Editura Academiei, Bucharest, Romania

Distributors for all remaining countries
Kluwer Academic Publishers Group,
P.O. Box 322, 3300 AH Dordrecht, Holland

Original title: *Metode numerice pentru ecuaţii diferenţiale cu aplicaţii*
First edition published in 1979 by Editura Academiei.
First English edition published in 1984 by Editura Academiei and D. Reidel Publ. Co.

To the memory of my father

Contents

Editor's preface

Growing specialization and diversification have brought a host of monographs and textbooks on increasingly specialized topics. However, the "tree" of knowledge of mathematics and related fields does not grow only by putting forth new branches. It also happens, quite often in fact, that branches which were thought to be completely disparate are suddenly seen to be related.

Further, the kind and level of sophistication of mathematics applied in various sciences has changed drastically in recent years: measure theory is used (non-trivially) in regional and theoretical economics; algebraic geometry interacts with physics; the Minkowsky lemma, coding theory and the structure of water meet one another in packing and covering theory; quantum fields, crystal defects and mathematical programming profit from homotopy theory; Lie algebras are relevant to filtering; and prediction and electrical engineering can use Stein spaces. And in addition to this there are such new emerging subdisciplines as "completely integrable systems", "chaos, synergetics and large-scale order", which are almost impossible to fit into the existing classification schemes. They draw upon widely different sections of mathematics.

This program, *Mathematics and Its Applications*, is devoted to such (new) interrelations as *exempli gratia:*

— a central concept which plays an important role in different mathematical and/or scientific specialized areas;

— new applications of the results and ideas from one area of scientific endeavor into another;

— influences which the results, problems and concepts of one field of enquiry have and have had on the development of another.

The *Mathematics and Its Applications* programme tries to make available a careful selection of books which fit the philosophy outlined above. With such books, which are stimulating rather than definitive, intriguing rather than encyclopaedic, we hope to contribute something towards better communication among the practitioners in diversified fields.

Because of the wealth of scholarly research being undertaken in the Soviet Union, Eastern Europe, and Japan, it was decided to devote special attention to the work emanating from these particular regions. Thus it was decided to start three regional series under the umbrella of the main MIA programme.

This book in the Eastern Europe series concerns numerical analysis and, more particularly, the numerical analysis needed to deal with the ordinary differential equations one meets in quantum physics, chemistry and engineering. The book was especially written with such potential active users of numerical methods in mind. The first Romanian edition (now sold out) was particularly well received, so it seemed a good idea to also make this work available to the English reading scientific community.

The unreasonable effectiveness of mathematics in science...

Eugene Wigner

Well, if you knows of a better 'ole, go to it.

Bruce Bairnsfather

What is now proved was once only imagined.

William Blake

As long as algebra and geometry proceeded along separate paths, their advance was slow and their applications limited.
But when these sciences joined company they drew from each other fresh vitality and thenceforward marched on at a rapid pace towards perfection.

Joseph Louis Lagrange

Amsterdam, March 1983

MICHIEL HAZEWINKEL

Preface

This book is merely addressed to persons who, without being professionals in applied mathematics, are often faced with the problem of numerically solving differential equations.

Graduate students and persons who are active in the research in theoretical physics, chemistry or engineering are perhaps the most suited readers.

In writing this book, special attention was paid to keep some balance between rigour and accessibility. Thus, use is made of only a background level of knowledge in mathematics which is typically achieved during training in the faculties of physics, chemistry and technical branches. At the same time, although the presentation is generally rigorous, on some occasions we deliberately sacrificed rigour in favour of accessibility.

The presentation of various methods includes all aspects of interest: theory -- which is necessary for those who are rather deeply engaged in the field — , algorithms and programming advices — for hurried readers. The book also contains a lot of numerical examples treated up to the very end, sometimes just with programs included.

The book is organized into five chapters. In each of the first three chapters a definite class of methods is discussed for the solution of the initial value problem for ordinary differential equations. It should be mentioned that, while the first two classes, that is of multistep and of one-step methods, are largely presented in the literature, we also consider a new class of methods (the so-called piecewise perturbation methods) which was developed only recently. As a matter of fact, it is now the first time, to our knowledge, that such methods are treated in a self-standing chapter of a book of numerical methods.

The fourth chapter is mainly focussed on the boundary value problems for *linear* second-order equations. A special section is here devoted to the Schrödinger equation, the basic equation of quantum mechanics.

In the fifth chapter a numerical example is taken and solved completely. Algorithms are given for six different methods, with the corresponding programs in FORTRAN-4. Special emphasis is here laid on some of the typical programming tricks that make computation faster.

This version of the book is substantially modified and enlarged in comparison with the original Romanian edition in a way which is expected to better serve its purpose.

In writing this book I benefited from help and encouragement from many colleagues and collaborators. Special thanks are addressed to Dr. Gh. Adam and Dr. M. Rizea for fruitful discussions, especially in the early stages of my work, and to Dr. I. Brânduş and Dr. M. I. Cristu for permanent support and encouragement. Finally I wish to thank Silvia Mihăilescu for beneficial stylistic adjustments and Mariana Pripas for carefully typing the manuscript.

Bucharest, March 1983 LIVIU IXARU

Preliminaries

The first two chapters of this book are merely devoted to the numerical methods of solving the initial value problem associated to first-order ordinary differential equations.

GENERALITIES, CONVENTIONS AND NOTATION

Let $f(x, y)$ be a real-valued function of two variables defined for the real $x \in [a, b]$ and all real y. Suppose that y is a real valued function defined on $[a, b]$. The equation

$$y' = f(x, y) \tag{1}$$

is called an ordinary differential equation of the first order. Any real valued function $y(x)$ which is differentiable and satisfies (1) for $a \leqslant x \leqslant b$ is said to be a solution of this differential equation.

In general, such a solution, if it exists, will not be unique because an arbitrary constant is introduced on integrating (1). To make the solution unique we must impose an extra condition on the solution. This condition usually takes the form of requiring $y(x)$ to have a specified value at a particular value c of the argument x, that is

$$y(c) = s, \tag{2}$$

where $c \in [a, b]$ is the particular value of x and s is the preset value of $y(c)$.

Equations (1) and (2) are collectively called an initial value (or Cauchy) problem. Equation (2) is called the initial condition and s is the initial value of y.

For practical reasons we will further assume that the point c is just the left-hand end of the domain of the equation, i.e., $c = a$. The initial value will be denoted by y_0 so that the initial condition reads

$$y(a) = y_0. \tag{3}$$

This assumption is not restrictive at all for our purpose which is to derive algorithms to numerically solve the initial value problem (1)—(3). In fact, once we have such algorithms, we can equally apply them to solve the (theoretically more general) problem (1)—(2).

To illustrate this, consider the problem

$$y' = \frac{1}{1 + y^2}$$

with $a = 0$, $b = 10$, $c = 5$ and $y(5) = s$. To solve it numerically by an algorithm for problem (1)—(3) we divide the domain [0, 10] into two subdomains $I_1 \equiv [a = 0, b_1 = 5]$ and $I_2 \equiv [a_1 = 5, b = 10]$. The numerical integration of it on I_2 is straight forward because the problem is manifestly of the very form (1)—(3). To solve the problem on I_1 we must first transform the equation by introducing a new variable $X = 5 - x$ and a new function $Y(X) = y(x)$. Now, since $Y'(X) = - y'(x)$ the problem to be solved becomes

$$Y' = -\frac{1}{1 + Y^2}$$

for $X \in [0, 5]$, with the initial condition $Y(0) = y(5) = s$. The latter is also of the form (1)—(3) and thus it can be integrated numerically by means of the same algorithm.

Another mathematical problem of interest is the one of the systems of first-order differential equations, formulated as follows.

A system of N ordinary differential equations of the first order is a system of equations of the form

$$y^{i\prime} = f^i(x, y^1, y^2, \ldots, y^N),\ i = 1, 2, \ldots, N,\ x \in [a, b], \tag{4}$$

in which $f^1, f^2, \ldots, f^N$ are known real functions of their $N + 1$ real arguments $x, y^1, \ldots, y^N$.

The initial value (or Cauchy) problem for system (4) consists of finding a set of functions $y^1(x), y^2(x), \ldots, y^N(x)$ defined and differentiable on $[a, b]$, which assume some preset initial conditions, viz.:

$$y^i(c) = s^i,\ i = 1, 2, \ldots, N, \tag{5}$$

for given $c \in [a, b]$ and real numbers $s^1, s^2, \ldots, s^N$, and satisfy system (4) identically in x.

Problem (4)—(5) can be more compactly written in vector form,

$$\mathbf{y}'(x) = \mathbf{f}(x, \mathbf{y}), \tag{6}$$

$$\mathbf{y}(c) = \mathbf{s}, \tag{7}$$

where $\mathbf{y}, \mathbf{f}$ and $\mathbf{s}$ are real N dimensional vectors,

$$\mathbf{y} = \begin{pmatrix} y^1 \\ y^2 \\ \vdots \\ y^N \end{pmatrix}, \quad \mathbf{f}(x, \mathbf{y}) = \begin{pmatrix} f^1(x, y^1, y^2, \ldots, y^N) \\ f^2(x, y^1, y^2, \ldots, y^N) \\ \vdots \\ f^m(x, y^1, y^2, \ldots, y^N) \end{pmatrix}, \quad \mathbf{s} = \begin{pmatrix} s^1 \\ s^2 \\ \vdots \\ s^N \end{pmatrix}. \tag{8}$$

Once again we will focus on the particular case when $c = a$, that is

$$\mathbf{y}' = \mathbf{f}(x, \mathbf{y}), \quad x \in [a, b], \tag{9}$$

$$\mathbf{y}(a) = \mathbf{y}_0. \tag{10}$$

If eqs (1)—(3) are compared with eqs (9)—(10) it clearly appears that the initial value problems for single equations and for systems of equations are mathematically identical except for the fact that f, y, and s are scalars in the first case, but N-dimensional vectors in the second case. The same holds also true for the mathematical approaches to the two problems in the sense that the mathematical methods of investigation for the second problem, resulting theorems, numerical algorithms, etc. are simply generalizations in vector language of the ones in the scalar case.

In connection with each of the problems (1)—(3) and (9)—(10) mathematicians usually say that the solution, *if it exists*, is unique. The question is *when* does the solution actually exist, or in other words, under which conditions to be satisfied by function f (or the vector of functions f) does the solution exist?

The question is answered by a well-known theorem, the so-called existence theorem. For the scalar case this theorem says that the solution of problem (1)—(3) exists and is unique provided the real function (x, y) satisfies the following two conditions:

(AS) $f(x, y)$ is defined and continuous in the strip $a \leqslant x \leqslant b$, $-\infty < y < +\infty$, where a and b are finite;

(BS) There exists a constant L such as for any $x \in [a, b]$ and any two numbers y and y^* one has

$$|f(x, y) - f(x, y^*)| \leqslant L|y - y^*|.$$

The constant L and the written inequality are called the Lipschitz constant and the Lipschitz condition, respectively; $|z|$ is the standard notation for the absolute value of a scalar z. A detailed proof of this theorem is given in [1].

This theorem can be generalized at once to accommodate the vector case, problem (9)—(10), provided a vector analog of the absolute value of a scalar is defined. This analog is called *the norm* of an N dimensional vector

$$\mathbf{v} = \begin{pmatrix} v^1 \\ v^2 \\ \vdots \\ v^N \end{pmatrix}, \tag{11}$$

and it is defined as the sum of the absolute values of each component

$$||\mathbf{v}|| = |v^1| + |v^2| + \ldots + |v^N|. \tag{12}$$

With this definition *, the existence theorem (which is proved separately in [1]) says that the solution of problem (9)—(10) exists and is unique provided the real vector $\mathbf{f}(x, \mathbf{y})$ satisfies the two conditions which are the direct generalizations of the previous conditions AS and BS, namely:
(AV) $\mathbf{f}(x, \mathbf{y})$ is defined and continuous (in the sense that each component

---

* This is not the only possible definition of a vector norm. Some other possible definitions are for instance

$$||\mathbf{v}|| = (|v^1|^2 + |v^2|^2 + \ldots + |v^N|^2)^{1/2},$$

called the Euclidean norm (or length), or

$$||\mathbf{v}|| = \max_{1 \leqslant i \leqslant N} |v^i|.$$

However, in order not to complicate things we will always think of the norm as given by definition (12). As a matter of fact, most of the theorems involving vector norms remain valid, no matter which is the actual definition of this quantity. This is true, in particular, also for the existence theorem which follows in the main text.

$f^i(x, y^1, y^2, \ldots, y^N)$, $i = 1, 2, \ldots, N$, is defined and continuous) in the region $a \leqslant x \leqslant b$ and $-\infty < y^i < \infty$, $i = 1, 2, \ldots, N$;

(BV) There exists a constant L such that for arbitrary $x \in [a, b]$ and any two N dimensional vectors $\mathbf{y}$ and $\mathbf{y}^*$ one has

$$\|\mathbf{f}(x, \mathbf{y}) - \mathbf{f}(x, \mathbf{y}^*)\| \leqslant L\|\mathbf{y} - \mathbf{y}^*\|.$$

It is important to note in connection with the existence theorem that sometimes differential equations might occur in practice for which the conditions of the theorem are violated. Yet, this does not exclude that the solution still exists. For instance, in the scalar problem

$$y' = -y^2, \quad x \in [1, 10], \quad y(1) = 1,$$

condition **BS** is clearly violated since one can never find a constant L such that the inequality

$$|-y^2 + y^{*2}| \leqslant L|y - y^*|$$

holds for *arbitrary* y and y^*. Nevertheless, the solution exists and it is $y(x) = x^{-1}$.

Even in such cases of doubt the existence theorem is of practical help as in the following possible experiment aimed at settling whether the solution exists. Suppose that for problem (1)—(3) condition (BS) is violated (as in the previous example). First modify the definition of $f(x, y)$ to bring it into a form $\bar{f}(x, y)$ that automatically satisfies condition **BS**. The simplest way to do it is

$$\bar{f}(x, y) = \begin{cases} f(x, Y) & \text{for} \quad y > Y \\ f(x, y) & \text{for} \quad -Y \leqslant y \leqslant Y \\ f(x, -Y) & \text{for} \quad y < Y \end{cases}$$

where Y is some suitable large number. The existence theorem ensures that the solution of the problem

$$y' = \bar{f}(x, y), \qquad x \in [a, b] \quad y(a) = y_0,$$

exists. Now compute it numerically and check if the curve $y(x)$ lies inside the rectangle $a \leqslant x \leqslant b$, $-Y \leqslant y \leqslant Y$. If this happens, then the solution of the original problem exists, is unique and coincides with the solution of the latter problem because in this range $f(x, y)$ and $\bar{f}(x, y)$ coincide. If not, the experiment can be repeated with new larger values of Y until one possibly reaches the previous situation.

From now on we shall always assume that the solution of problem (1)—(3) or (9)—(10) exists.

Before going into the very matter of devising algorithms to solve these problems, we pause awhile to answer several questions concerning the applicability of such algorithms. All the questions are various facets of the following basic question: to what extent can the algorithms namely devised for the initial value problem for first-order equations (or systems of equations) be used to solve more general problems?

Remember, for instance, that problems (1)—(3) and (9)—(10) are defined in the real field. Can the corresponding numerical algorithms be used to solve the same problems in the complex field (except for x)? The answer is positive and is based on the fact that any complex number is a set of two real numbers. As an illustration we take the scalar problem

$$y' = (\lambda_1 + i\lambda_2)\, y, \quad x \in [0, 1], \quad y(0) = 1,$$

with real λ_1 and λ_2.

Write

$$y(x) = y^1(x) + iy^2(x),$$

where y^1 and y^2 are real functions. Introduce it in the equation and in the initial condition, respectively, and finally identify the real and imaginary components separately. One gets

$$y^{1'} = \lambda_1 y^1 - \lambda_2 y^2$$

$$y^{2'} = \lambda_2 y^1 + \lambda_1 y^2,$$

$$y^1(0) = 1, \quad y^2(0) = 0,$$

which is a system of two equations in the real field, i.e. a problem of the type (9)—(10).

Another question of practical interest is: can such algorithms be used to solve initial value problems for equations (or systems) of higher order? The answer is also positive provided the derivative of the highest order can be separated from the other elements of the equation.

To illustrate this we take the following tth order differential equation:

$$F(x, y, y', y'', \dots, y^{(t)}) = 0, \quad x \in [a, b], \tag{13}$$

with the initial conditions

$$y(a) = y_0,\ y'(a) = y_0',\ y''(a) = y_0'', \dots, y^{(t-1)}(a) = y_0^{(t-1)}. \tag{14}$$

If eq. (13) can be converted to the form

$$y^{(t)} = f(x, y, y', y'', \dots, y^{(t-1)}), \tag{15}$$

then, by introducing the new functions $y^1(x) = y(x)$, $y^2(x) = y'(x), \dots$ $\dots, y^t(x) = y^{(t-1)}(x)$, eq. (15) becomes a system of t first-order equations, viz.:

$$\begin{aligned} y^{1'} &= y^2, \\ y^{2'} &= y^3, \\ &\vdots \\ y^{t-1'} &= y^t, \\ y^{t'} &= f(x, y^1, y^2, \dots, y^t), \end{aligned} \tag{16}$$

with the initial conditions

$$y^1(a) = y_0, \quad y^2(a) = y_0', \dots, y^t(a) = y_0^{(t-1)}. \tag{17}$$

Problem (16)—(17) is manifestly of the form (9)—(10) and therefore it can be solved numerically by any algorithm devised for the latter problem.

There are also problems in which the additional conditions are no more imposed at one and the same point c as before, but at different points. Such problems are coined as boundary value problems and they typically occur in connection with second-order linear differential equations. The question is whether such problems are tractable by means of algorithms for the initial value problem. Once again the answer is positive and an outline of the procedure will be given in Chapter 4 of the book.

The cases mentioned above show that the numerical algorithms for the initial value problem (1)—(3) or (9)—(10) enjoy the salient capability of being applicable for many more problems than the ones for which they have been namely devised.

When deriving algorithms in the first two chapters of the book we will always have in mind only the scalar problem (1)—(3). We hope that in doing so the procedure gains in clarity and can be easier followed by the reader. As mentioned before, the algorithms for systems of equations result by direct generalization of the previous ones. They will be written explicitly only on some occasions, particularly in connection with practical issues.

Any numerical method starts with the division of the domain $[a, b]$ in subdomains, by means of the points

$$x_0 = a,\ x_1, x_2, \ldots, x_{n-1}, x_n = b \tag{18}$$

and solves the problem progressively outwards: by using the initial data (3) (or (10)) and the expression of f (or $\mathbf{f}$) the solution is evaluated at x_1. Next, this is used as the initial condition to find the solution at x_2 by the same procedure, and so on.

There are actually two classes of numerical methods. In the first class (multistep methods) the algorithm yields the solution at x_p in terms of the data from *several* previous mesh points, for instance $x_{p-3}, x_{p-2}, x_{p-1}$. The second class (one-step methods) consists of methods which require information from only the previous mesh point x_{p-1}. Chapter 1 will be devoted to multistep methods, and Chapter 2 to one-step methods.

REFERENCES

1. P. Henrici, *Discrete Variable Methods in Ordinary Differential Equations*, John Wiley and Sons, New York, 1964.

There are also many other books on numerical analysis dealing partly or in full with numerical solution of first-order differential equations as, for instance:

L. Fox, *Numerical Solution of Ordinary and Partial Differential Equations*, Pergamon Press, New York, 1962.

L. Lapidus and J. H. Seinfeld, *Numerical Solution of Ordinary Differentia Equations*, in the series "Mathematics in Science and Engineering", vol. 74, Academic Press, New York, 1971.

J. W. Daniel and R. E. Moore, *Computation and Theory in Ordinary Differential Equations*, W. H. Freeman and Co., San Francisco, 1970.

G. M. Phillips and P. J. Taylor, *Theory and Applications of Numerical Analysis*, Academic Press, London, 1973.

J. D. Lambert, *Computational Methods in Ordinary Differential Equations*, Wiley, London, 1973.

K. Moszynski, *Metode numerice de rezolvare a ecuaţiilor diferenţiale* (in Romanian), Editura tehnică, Bucharest, 1974.

Gh. Marinescu, *Analiza numerică* (in Romanian), Ed. Academiei, Bucharest, 1974.

N. Bakhvalov, *Méthodes Numériques*, Editions Mir, Moscow, 1976.

Chapter 1

Multistep methods

1.1. STRATEGIES TO DERIVE ALGORITHMS

There are several strategies to derive algorithms for problem (1)—(3):
(i) Integrating the identity

$$y'(x) = f(x, y(x))$$

between some limits X and $X + \Delta X$ inside $[a, b]$, we obtain:

$$y(X + \Delta X) - y(X) = \int_{X}^{X+\Delta X} f(x, y(x))\, dx. \tag{1.1.1}$$

Now, replacing f through various interpolating polynomials, we can obtain various algorithms.
(ii) Approximating the first derivative appearing in the differential equation by a finite difference formula also yields algorithms. For instance, the approximation

$$y'(X) \simeq \frac{1}{\Delta X}(y(X + \Delta X) - y(X)) \tag{1.1.2}$$

yields the algorithm

$$y(X + \Delta X) \simeq y(X) + \Delta X f(X, y(X)). \tag{1.1.3}$$

(iii) We may simply start from the possibility of expanding the solution in a Taylor series around X

$$y(X + \Delta X) = y(X) + \Delta X f(X, y(X)) + \frac{1}{2}(\Delta X)^2 y''(X) + \ldots \tag{1.1.4}$$

and retain as many terms as necessary.

Strategy (iii) forms the basis for generating one-step methods while strategy (i) is the most natural way to generate multistep methods.

To put the latter into practice first we need some elements of the theory of backward differences.

1.2. ELEMENTS OF THE THEORY OF BACKWARD DIFFERENCES

Consider the function $g(x)$, $x \in [a, b]$ and denote by g_k its values at the equidistant points x_k, viz.: $g_k \equiv g(x_k)$. We define the backward difference operator of order p, denoted as ∇^p, by the following recurrence formula

$$\nabla^p g_k = \nabla^{p-1} g_k - \nabla^{p-1} g_{k-1}, \quad p = 1, 2, \ldots \tag{1.2.1}$$

with ∇° taken as the identity operator. Thus we have:

$$\nabla^\circ g_k = g_k, \tag{1.2.2}$$

$$\nabla^1 g_k = \nabla^\circ g_k - \nabla^\circ g_{k-1} = g_k - g_{k-1}, \tag{1.2.3}$$

$$\nabla^2 g_k = \nabla^1 g_k - \nabla^1 g_{k-1} = g_k - 2g_{k-1} + g_{k-2}, \tag{1.2.4}$$

and, in general,

$$\nabla^p g_k = \sum_{j=0}^{p} (-1)^j \binom{p}{j} g_{k-j}, \tag{1.2.5}$$

in which

$$\binom{x}{j} = \begin{cases} 1 & \text{if } j = 0 \\ 0 & \text{if } x \text{ is an integer and } x < j \neq 0 \\ \dfrac{x(x-1)(x-2)\ldots(x-j+1)}{j!} & \text{otherwise.} \end{cases} \tag{1.2.6}$$

Conversely, the original values of g at successively decreasing points read:

$$g_k = \nabla^{\circ} g_k, \tag{1.2.7}$$

$$g_{k-1} = (\nabla^{\circ} - \nabla^1)\, g_k, \tag{1.2.8}$$

$$g_{k-2} = (\nabla^{\circ} - 2\nabla^1 + \nabla^2)\, g_k, \tag{1.2.9}$$

and, in general,

$$g_{k-p} = \sum_{j=0}^{p} (-1)^j \binom{p}{j} \nabla^j g_k. \tag{1.2.10}$$

Another important mathematical object is the interpolating polynomial. The *interpolating polynomial* for the function g relative to the $r+1$ knots $x_k, x_{k-1}, \ldots, x_{k-r}$ is defined as the polynomial $P(x)$ whose degree is less than, or equal to r, and whose value at each knot coincides with the value of g, viz.:

$$P(x_{k-i}) = g_{k-i}, \quad i = 0, 1, 2, \ldots, r. \tag{1.2.11}$$

For given g, r and positions of the knots the interpolating polynomial is unique.

In this definition the distance between the knots is arbitrary. However, from now on we will assume that the points x_{k-i} are equidistant with the step size h, i.e. $x_{k-i} = x_k - ih$. The reason is that in this case we can make a direct connection with the backward difference operators introduced before.

In fact the interpolating polynomial with r knots for the function g admits a very simple expression in terms of the backward difference operators, viz.:

$$P(x) = \sum_{j=0}^{r} (-1)^j \binom{-s}{j} \nabla^j g_k, \quad s = \frac{x - x_k}{h}. \tag{1.2.12}$$

Eq. (1.2.12) can be easily verified. Indeed $P(x)$ is a polynomial whose degree is less than or equal to r and at the point x_{k-i} we have

$$P(x_{k-i}) = \sum_{j=0}^{r} (-1)^j \binom{i}{j} \nabla^j g_k = g_{k-i}, \tag{1.2.13}$$

as resulting from eq. (1.2.10).

1.3. THE GENERAL ALGORITHM OF THE MULTISTEP METHODS

To solve the initial value problem

$$y' = f(x, y), \quad x \in [a, b], \quad y(a) = y_0,$$

we first divide the domain $[a, b]$ in *equidistant* intervals through the mesh points $x_0 = a, x_1, x_2, \ldots, x_{n-1}, x_n = b$, with the step size h, that is $x_i = x_0 + ih$. Now use formula (1.1.1) in which we identify X and $X + \Delta X$ as mesh points of the partition, namely $X = x_{k-q}$ and $X + \Delta X = x_{k+p}$, where k, q and p are nonnegative integers. Formula (1.1.1) becomes

$$y_{k+p} = y_{k-q} + \int_{x_{k-q}}^{x_{k+p}} f(x, y(x))\, dx, \tag{1.3.1}$$

and it yields a relationship between the values of the solution at the points x_{k+p} and x_{k-q}.

The general algorithm of the multistep methods is obtained by replacing function f by its r-knotted interpolating polynomial (1.2.12),

$$\bar{y}_{k+p} = \bar{y}_{k-q} + h \sum_{j=0}^{r} \alpha_j \nabla^j f_k, \tag{1.3.2}$$

in which

$$\alpha_j = (-1)^j \int_{-q}^{p} \binom{-s}{j} ds, \quad f_k \equiv f(x_k, \bar{y}_k). \tag{1.3.3}$$

From now on the exact solution of problem (1)—(3) will remain denoted by y, while the solution computed by a numerical algorithm will be denoted by $\bar{y}$.

Of course, the solution computed by the numerical algorithm will not coincide in general with the exact solution, because any numerical algorithm relies on some specific approximations. Algorithm (1.3.2), for instance, embeds two approximations, namely the use of an interpolating polynomial for f (not of f itself) and the implicit supposition that the values of f at the mesh points are given by the second of the formulas in eq. (1.3.3) i.e., with approximate values for the solution y, not by the value of $f(x_k, y(x_k))$ as it would be correct. The errors associated to these approximations will be discussed in sections 1.6 and 1.7. As for the round-off errors, see section 1.10.

By assigning various values to the nonnegative integers p, q, and r we obtain different formulas for the algorithm. According to the value of p these formulas (or methods) are divided into two classes: (i) open (or explicit) methods when $p = 1$, and (ii) closed (or implicit) methods when $p = 0$. (Larger values for p are out of discussion since the very goal of the algorithm is to compute the solution at each new mesh point in terms of the known values of it at the very preceding mesh points.)

At first sight only the open methods seem practicable since they compute the solution at x_{k+1} in terms of the data from the previous points $x_{k-r}, x_{k-r+1}, \ldots, x_k$, if $r \geqslant q$, or $x_{k-q}, x_{k-r}, x_{k-r+1}, \ldots, x_k$, if $r < q$, and all of these are known. For the closed methods the new point is x_k and then such methods seem simply unapplicable since the computation of the solution at this point requires in input the value of $f_k = f(x_k, \bar{y}_k)$ which is of course as unknown as is the very value of $\bar{y}_k$. Yet, the situation is not so dramatic and, as it will be seen in section 1.9, the closed methods are equally applicable provided a predictor-corrector procedure is added. In addition, it will be seen that the closed methods are more accurate than their open counterparts.

The first point at which algorithm (1.3.2) can be used to solve the problem (1)—(3) is x_{i+1} if the algorithm is open and x_i if it is closed, where i is the largest from q and r. Then to put the algorithm in practice the user must first furnish the values of the solution at the points $x_0, x_1, \ldots, x_i$ in the first case and $x_0, x_1, \ldots, x_{i-1}$ in the second. These can be calculated by some extra procedure, for instance by using Taylor series for the solution, strategy (iii) in section 1.1.

Coming back to algorithm (1.3.2), it is worth expressing it in a form in which the operator ∇^i is explicited. One gets

$$\bar{y}_{k+p} = \bar{y}_{k-q} + h \sum_{j=0}^{r} \beta_j f_{k-j} \tag{1.3.4}$$

in which

$$\beta_j = (-1)^j \left[\binom{j}{j} \alpha_j + \binom{j+1}{j} \alpha_{j+1} + \dots + \binom{r}{j} \alpha_r \right]. \tag{1.3.5}$$

By assigning various values to p, q, and r, various particular algorithms can be obtained. Some of these, the ones which are most often met in applications, are listed in the next two sections. The reader interested in more details can also consult the book of Henrici [1].

1.4. THE OPEN METHODS EULER, ADAMS-BASHFORTH, NYSTRÖM

Euler's method is the particular case of eq. (1.3.2) or (1.3.4) with $p = 1$, $q = 0$, $r = 0$. Its algorithm is

$$\bar{y}_{k+1} = \bar{y}_k + hf(x_k, \bar{y}_k). \tag{1.4.1}$$

This method is also called the method (formula, algorithm) of polygonal lines. It is interesting to note that, on one hand, the method needs information from only the previous point x_k and therefore it is a one-step method; on the other hand it resulted as a particular case of the general multistep algorithm (1.3.2) or (1.3.4). The latter is the very reason for which it was inserted in this chapter. Moreover, the reader should realize that, once this method is known, the problem of finding the starting values for a genuine multistep method can be solved at once, see also section 1.8.

The Adams-Bashforth method (originally discussed in [2]) results from eq. (1.3.2) in the particular case $p = 1$, $q = 0$ and without specifying r. Hence the algorithm is

$$\bar{y}_{k+1} = \bar{y}_k + h\left[\nabla^0 f_k + \frac{1}{2} \nabla^1 f_k + \frac{5}{12} \nabla^2 f_k + \frac{3}{8} \nabla^3 f_k + \right.$$
$$\left. + \frac{251}{720} \nabla^4 f_k + \frac{95}{288} \nabla^5 f_k + \frac{19087}{60480} \nabla^6 f_k + \dots \right]. \tag{1.4.2}$$

Table 1 collects the values of the coefficients β_j of this method for several values of r. (For $r = 0$ one gets the Euler method.)

Table 1

Coefficients β_j of the Adams-Bashforth method

r \ j	0	1	2	3	4	5
1	$\frac{3}{2}$	$-\frac{1}{2}$				
2	$\frac{23}{12}$	$-\frac{16}{12}$	$\frac{5}{12}$			
3	$\frac{55}{24}$	$-\frac{59}{24}$	$\frac{37}{24}$	$-\frac{9}{24}$		
4	$\frac{1901}{720}$	$-\frac{2774}{720}$	$\frac{2616}{720}$	$-\frac{1274}{720}$	$\frac{251}{720}$	
5	$\frac{4227}{1440}$	$-\frac{7673}{1440}$	$\frac{9482}{1440}$	$-\frac{6798}{1440}$	$\frac{2627}{1440}$	$-\frac{425}{1440}$

To illustrate the use of this table we write the Adams-Bashforth algorithm for $r = 1$

$$\bar{y}_{k+1} = \bar{y}_k + h\left(\frac{3}{2} f_k - \frac{1}{2} f_{k-1}\right). \tag{1.4.3}$$

The Nyström method [3] results from eq. (1.3.2) in the particular case $p = 1$, $q = 1$ and arbitrary r,

$$\bar{y}_{k+1} = \bar{y}_{k-1} + h\left[2\nabla^0 f_k + \frac{1}{3}(\nabla^2 f_k + \nabla^3 f_k) + \frac{29}{90}\nabla^4 f_k + \right.$$

$$\left. + \frac{14}{45}\nabla^5 f_k + \frac{1139}{3780}\nabla^6 f_k + \dots\right]. \tag{1.4.4}$$

The values of the coefficients β_j vs. r for this method are collected in table 2. The special case $r = 0$ of Nyström's method,

$$\bar{y}_{k+1} = \bar{y}_k + 2hf_k, \tag{1.4.5}$$

is alternatively called the *mid-point rule* or *central difference formula*.

Table 2

Coefficients β_j of the Nyström method

$r \backslash j$	0	1	2	3	4
0	2				
1	2	0			
2	$\frac{7}{3}$	$-\frac{2}{3}$	$\frac{1}{3}$		
3	$\frac{8}{3}$	$-\frac{5}{3}$	$\frac{4}{3}$	$-\frac{1}{3}$	
4	$\frac{269}{90}$	$-\frac{266}{90}$	$\frac{294}{90}$	$-\frac{146}{90}$	$\frac{29}{90}$

The same formula results also when $r = 1$. This is not surprising at all; it is actually due to the fact that $\alpha_1 = 0$ in eq. (1.4.4).

1.5. THE CLOSED METHODS ADAMS–MOULTON, MILNE–SIMPSON, AND MILNE

The Adams-Moulton method [4] is the particular case of the general algorithm (1.3.2) with $p = 0$, $q = 1$ and arbitrary r, viz.

$$\bar{y}_k = \bar{y}_{k-1} + h\left[\nabla^0 f_k - \frac{1}{2}\nabla^1 f_k - \frac{1}{12}\nabla^2 f_k - \frac{1}{24}\nabla^3 f_k - \right.$$

$$\left. - \frac{19}{720}\nabla^4 f_k - \frac{3}{160}\nabla^5 f_k - \frac{863}{60480}\nabla^6 f_k - \dots\right]. \tag{1.5.1}$$

The special case $r = 1$ of this method,

$$\bar{y}_k = \bar{y}_{k-1} + \frac{1}{2} h(f_k + f_{k-1}), \tag{1.5.2}$$

is known as the *trapezoidal* rule.

For completeness we give in table 3 the coefficients β_j for various relevant values of r. Using this table we can write at once the algorithm for any $r \leqslant 5$. In particular, for $r = 2$ the algorithm is

$$\bar{y}_k = \bar{y}_{k-1} + h\left(\frac{5}{12} f_k + \frac{2}{3} f_{k-1} - \frac{1}{12} f_{k-2}\right). \tag{1.5.3}$$

Table 3

Coefficients β_j for the Adams-Moulton method

r \ j	0	1	2	3	4	5
1	$\frac{1}{2}$	$\frac{1}{2}$				
2	$\frac{5}{12}$	$\frac{8}{12}$	$-\frac{1}{12}$			
3	$\frac{9}{24}$	$\frac{19}{24}$	$-\frac{5}{24}$	$\frac{1}{24}$		
4	$\frac{251}{720}$	$\frac{646}{720}$	$-\frac{264}{720}$	$\frac{106}{720}$	$-\frac{19}{720}$	
5	$\frac{475}{1440}$	$\frac{1427}{1440}$	$-\frac{798}{1440}$	$\frac{482}{1440}$	$-\frac{173}{1440}$	$\frac{27}{1440}$

The Milne-Simpson method results from the general algorithm (1.3.2) in the particular case $p = 0$, $q = 2$ and arbitrary r,

$$\bar{y}_k = \bar{y}_{k-2} + h\left[2\nabla^0 f_k - 2\nabla^1 f_k + \frac{1}{3}\nabla^2 f_k - \frac{1}{90}(\nabla^4 + \nabla^5) f_k - \right.$$

$$\left. - \frac{37}{3780}\nabla^6 f_k + \ldots\right] \tag{1.5.4}$$

The coefficients β_j of this method are collected in table 4. One sees that they are identical for $r = 2$ and $r = 3$, which is a normal result in view of the fact that there is no term with $\nabla^3 f_k$ in eq. (1.5.4).

Table 4

Coefficients β_j for the Milne-Simpson method

$r \backslash j$	0	1	2	3	4
0	2				
1	0	2			
2	$\frac{1}{3}$	$\frac{4}{3}$	$\frac{1}{3}$		
3	$\frac{1}{3}$	$\frac{4}{3}$	$\frac{1}{3}$	0	
4	$\frac{29}{90}$	$\frac{124}{90}$	$\frac{24}{90}$	$\frac{4}{90}$	$-\frac{1}{90}$

The corresponding algorithm,

$$\bar{y}_k = \bar{y}_{k-2} + \frac{h}{3}(f_k + 4f_{k-1} + f_{k-2}), \tag{1.5.5}$$

is also known as the *method of Milne* [5]. Note in passing that if f depends on x only, eq. (1.5.5) reduces to the familiar Simpson formula in the theory of numerical integration.

1.6. ERROR ANALYSIS: LOCAL TRUNCATION ERROR AND CONSISTENCY

All the numerical methods discussed in the previous two sections have an algorithm of the form

$$a_i\bar{y}_{k+i} + a_{i-1}\bar{y}_{k+i-1} + \ldots + a_0\bar{y}_k = h(b_if_{k+i} + b_{i-1}f_{k+i-1} + \\ + \ldots + b_0f_k), \tag{1.6.1}$$

where $a_0, a_1, \ldots, a_i$ and $b_0, b_1, \ldots, b_i$ are some given numbers and $|a_0| + |b_0| \neq 0$. i is the number of steps corresponding to the quoted algorithm. If $b_i \neq 0$ the method is called a closed method while if $b_i = 0$ it is an open method. (Recall that the coefficient a_i must always be different from zero; otherwise the algorithm (1.6.1), whose aim is to compute $\bar{y}_{k+i}$ in terms of the other quantities entering there, would be simply meaningless.) Algorithm (1.6.1) is called an i-step algorithm.

To illustrate that any of the previous algorithms is a special case of eq. (1.6.1), consider the Adams-Bashforth method for $r = 1$, eq. (1.4.3). By a suitable label translation (i.e. new k = old $k + 1$) eq. (1.4.3) becomes

$$\bar{y}_{k+2} - \bar{y}_{k+1} = h\left(\frac{3}{2} f_{k+1} - \frac{1}{2} f_k\right), \tag{1.6.2}$$

which is manifestly of the form (1.6.1) with $i = 2$, $a_2 = 1$, $a_1 = -1$, $a_0 = 0$, $b_2 = 0$, $b_1 = 3/2$ and $b_0 = -1/2$.

The immediate purpose of the error analysis is to find out a measure of the accuracy to be expected in the results when algorithms of the previous two sections are used to numerically solve the initial value problem (1)—(3).

Of course this can be done by examination of each algorithm in part. Yet, a global treatment of multistep methods is equally possible and it leads to results which can be particularized at once for any algorithm of interest. The mathematical apparatus necessary for such a general approach is, however, too long to be inserted here. We will consider instead only a brief sketch of it by merely insisting upon elements which are of primary importance for the users of such algorithms. The reader who wants details can consult, for instance, the book of Henrici [1].

Thus we will consider the general algorithm (1.6.1) having in mind that, as already mentioned, all the algorithms derived in the previous two sections by means of the interpolating polynomial procedure are particular cases of this algorithm. As a matter of fact, the algorithm (1.6.1) is even more general in the sense that it potentially includes also formulas which may be generated by any other procedure that can be imagined. This feature is however less important for a typical user and thus it will be disregarded.

There are two types of errors occurring when solving problem (1)—(3) by a numerical algorithm *:

(i) *Errors in the starting values.* The first application of algorithm (1.6.1) requires the knowledge of the values of $\bar{y}_0, \bar{y}_1, \bar{y}_2, \ldots, \bar{y}_{i-1}$ **. Only the first value, y_0, is known exactly while the others must be furnished by some extra computation in which another algorithm in used. It follows that the resultant values of $\bar{y}_1, \bar{y}_2, \ldots, \bar{y}_{i-1}$ will normally differ (hopefully slightly) from the exact $y(x_1), y(x_2), \ldots, y(x_{i-1})$.
The deviations

$$\delta_s \equiv y(x_s) - \bar{y}_s, \quad s = 1, 2, \ldots i - 1, \tag{1.6.3}$$

are called errors in the starting values of algorithm (1.6.1).

(ii) *Local truncation errors.* By its very construction, any algorithm represents an *approximate* way to compute the value of the solution. For instance, at the kth application of eq. (1.6.1), even if the input values $\bar{y}_{k+i-1}, \ldots, \bar{y}_k, f_{k+i}, \ldots, f_k$ happen to coincide with the exact values $y(x_{k+i-1}), \ldots, y(x_k), f(x_{k+i}, y(x_{k+i})), \ldots, f(x_k, y(x_k))$, respectively, the resultant value of $\bar{y}_{k+i}$ does not coincide generally with the exact solution $y(x_{k+i})$.
The deviation

$$e_{k+i} \equiv y(x_{k+i}) - \bar{y}_{k+i} \tag{1.6.4}$$

corresponding to this situation is called local truncation error and its magnitude naturally depends not only on the algorithm and step size used, but also on the differential equation to be solved.

When, as in practice, apply the algorithm recursively (i.e. first with $k = 0$ to compute $\bar{y}_i$, next with $k = 1$, to compute $\bar{y}_{i+1}$, and so on) these errors typically accumulate in such a way that the larger the value of k, the more distanced the value of $\bar{y}_{k+i}$ from the desired $y(x_{k+i})$, and a good numerical method is only the one in which the local error and the accumulated errors are kept under control, or as mathematicians say, the method must be convergent. We give below the rigorous definition of this concept.

---

* In this section the round-off errors are thoroughly disregarded. Their influence will be examined in section 1.10.

** This is rigorously true only when algorithm (1.6.1) is open. When it is closed it requires in addition the knowledge of a trial value for y_i.

We take some arbitrarily fixed value of $x \in (a, b]$, $x = X$, say, and consider problem (1)—(3) on $[a, X]$ only. We next consider *equidistant* partitions of $[a, X]$ with preset value for the total number of steps n, and finally make n arbitrarily large. Clearly, each n defines uniquely the step size, namely $h = (X - a)/n$. For instance, if $a = 1$ and $X = 2$, then the step size corresponding to $n = 10$ is $h = 1/10$, to $n = 20$ corresponds $h = 1/20$, etc. When n increases the mesh points $x_s = x_0 + sh$, $s = 1, 2, \ldots, i - 1$, come closer to the initial point x_0. In turn the starting values for the algorithm, $\bar{y}_1, \bar{y}_2, \ldots, \bar{y}_{i-1}$, thought as functions of h, must approach just the very initial value y_0, that is

$$\lim_{h \to 0} \bar{y}_s(h) = y_0, \quad s = 1, 2, \ldots, i - 1. \tag{1.6.5}$$

Algorithm (1.6.1) to integrate problem (1)—(3), for which the conditions of the existence theorem are satisfied, will be called *convergent* if for any $X \in [a, b]$ and any value for the initial condition y_0 one has

$$\lim_{h \to 0} \bar{y}_n = y(X), \tag{1.6.6}$$

provided the starting values in the algorithm also tend to y_0, as indicated in eq. (1.6.5). Here $y(X)$ is the exact solution at $x = X$ of problem (1)—(3). It is also implied that all operators occurring in algorithm (1.6.1) and in the evaluation of starting values are computed with *absolute accuracy* i.e. without round-off errors.

The convergence theory for multistep methods consists of finding the conditions to be satisfied by the coefficients a_s and b_s, $s = 0, 1, \ldots, i$, such that algorithm (1.6.1) is convergent. (We shall see later on that all special algorithms of the last two sections satisfy these conditions.)

This theory proceeds in two stages. In the first stage the behaviour of the local error is investigated and it leads to the concept of *consistency*. It further searches for relationships between the coefficients a_s and b_s to secure the consistency of algorithm (1.6.1). The second stage consists of investigating the error *propagation* and it leads to the concept of *stability*. Also properties of the coefficients are found out to secure that the method is stable. Finally the two features merge together by virtue of a celebrated theorem which says that algorithm (1.6.1) is convergent if and only if it is both consistent and stable. (This theorem, due to Dalquist [6], is also proved in [1].) In this section we cope with only the first stage, and leave the second stage to the next section.

To investigate the behaviour of the local truncation error let us first introduce some auxiliary definitions and concepts.

Thus we associate to algorithm (1.6.1) the following functional

$$L(z(x); h) \equiv a_i z(x + ih) + a_{i-1} z(x + (i-1)h) + \dots + a_0 z(x)$$

$$- h(b_i z'(x + ih) + b_{i-1} z'(x + (i-1)h) + \dots + b_0 z'(x)), \qquad (1.6.7)$$

where h and coefficients a_s, b_s are the same as in eq. (1.6.1). $z(x)$ is any function which has continuous derivatives of sufficiently high order and all the points x, $x + h, \dots, x + ih$ are supposed to lie within $[a, X]$. L is called an i-step difference operator.

We now expand $L(z(x); h)$ in powers of h. In view of the formulas

$$z(x + \alpha h) = z(x) + \alpha h\, z'(x) + \frac{1}{2} \alpha^2 h^2 z''(x) + \dots,$$

$$z'(x + \alpha h) = z'(x) + \alpha h z''(x) + \dots,$$

one finds out, after a little algebra, that $L(z(x); h)$ results of the form

$$L(z(x); h) = C_0 z(x) + C_1 h z'(x) + \dots + C_q h^q z^{(q)}(x) + \dots \qquad (1.6.8)$$

where the coefficients C_q, $q = 0, 1, 2, \dots$ are constants which depend only on the coefficients $a_0, a_1, \dots, a_i, b_0, b_1, \dots, b_i$, as follows

$$\begin{aligned} C_0 &= a_0 + a_1 + a_2 + \dots + a_i, \\ C_1 &= a_1 + 2a_2 + \dots + ia_i - (b_0 + b_1 + \dots + b_i), \\ &\vdots \\ C_q &= \frac{1}{q!}(a_1 + 2^q a_2 + \dots + i^q a_i) - \\ &\quad - \frac{1}{(q-1)!}(b_1 + 2^{q-1} b_2 + \dots + i^{q-1} b_i),\ q = 2, 3 \dots \end{aligned} \qquad (1.6.9)$$

Now a definition follows. Operator L is said to be of *order m* if the first nonvanishing coefficient is C_{m+1}, i.e., $C_0 = C_1 = \ldots = C_m = 0$ but $C_{m+1} \neq 0$. Thus if L is of order m then

$$L(z(x); h) = C_{m+1}h^{m+1}z^{(m+1)}(x) + 0(h^{m+2}). \tag{1.6.10}$$

What may be the importance of the concept of a difference operator for our very goal which is to use algorithms of form (1.6.1) for the initial value problem (1)—(3) on $[a, X]$?

To make the answer transparent we first rewrite eq. (1.6.7) at $x = x_k$ for $z(x) = y(x)$ where y is the solution of the equation to be integrated, $y' = f(x, y)$:

$$L(y(x_k); h) = a_i y(x_{k+i}) + a_{i-1}y(x_{k+i-1}) + \ldots + a_0 y(x_k)$$
$$- h(b_i f(x_{k+i}, y(x_{k+i})) + \ldots + b_0 f(x_k, y(x_k))). \tag{1.6.11}$$

Next rewrite algorithm (1.6.1) under a convenient form, viz.

$$\mathscr{L}(\bar{y}_k, h) \equiv a_i\bar{y}_{k+i} + a_{i-1}\bar{y}_{k+i-1} + \ldots + a_0\bar{y}_k -$$
$$- h(b_{k+i}f_{k+i} + \ldots + b_0 f_k) = 0. \tag{1.6.12}$$

(Remember that algorithm (1.6.1) calculates the value of $\bar{y}_{k+i}$ in terms of the other quantities supposed as known.)

Now substract eq. (1.6.12) from eq. (1.6.11) and use eq. (1.6.10). One gets

$$a_i E_{k+i} + a_{i-1}E_{k+i-1} + \ldots + a_0 E_k - h(b_i \Delta f_{k+i} + \ldots + b_0 \Delta f_k)$$
$$= C_{m+1}h^{m+1}y_k^{(m+1)} + 0(h^{m+2}), \tag{1.6.13}$$

where

$$E_s = y(x_s) - \bar{y}_s, \quad \Delta f_s = f(x_s, y(x_s)) - f(x_s, \bar{y}_s),$$
$$s = k,\ k+1, \ldots, k+i. \tag{1.6.14}$$

Finally suppose that all data taken as known in algorithm (1.6.1) coincide with the exact values,

$$E_s = 0,\ s = k,\ k+1, \ldots, k+i-1,\ \Delta f_s = 0,\ s = k, k+1, \ldots, k+i. \tag{1.6.15}$$

Under these circumstances, E_{k+i} in just the local truncation error e_{k+i}. Then eq. (1.6.13) reduces to

$$e_{k+i} = \frac{1}{a_i} C_{m+1} h^{m+1} y_k^{(m+1)} + 0(h^{m+2}). \tag{1.6.16}$$

This formula answers the question; it shows the direct connection between the local truncation error of algorithm (1.6.1) and the accuracy (1.6.10) of its associated difference operator L.

In practice algorithm (1.6.1) requires small values for h. We will actually assume that h is so small that the correction term $0(h^{m+2})$ is numerically negligible. If so, eq. (1.6.16) reads

$$e_{k+i} \simeq \frac{C_{m+1}}{a_i} h^{m+1} y_k^{(m+1)}, \tag{1.6.17}$$

a formula which represents a possible measure of the local accuracy of the algorithm (1.6.1). It shows that the local truncation error e_{k+i} depends on the algorithm (through the values of m and of $C \equiv C_{m+1}/a_i$), on the step size h, and also on the very equation to be solved (through $y_k^{(m+1)}$). The number $C = C_{m+1}/a_i$ is called the *error constant* *. Originally the integer m was called the order of the associated operator (1.6.7). Yet, since we have seen meantime that it governs also the behaviour of the local truncation error for algorithm (1.6.1) we will equally call m the order of the algorithm.

Formula (1.6.17) enables comparison of various particular algorithms in terms of efficiency. To this aim we list in Table 5 the values of m and of C for, the special algorithms of the previous two sections. For all of them one has $a_i = 1$ so that $C = C_{m+1}$.

Let us compare for instance the algorithms Adams-Bashforth (AB) and Adams-Moulton (AM). As seen in Table 5, for fixed r they are of the same order $m = r + 1$. On the other hand, by inspecting the values of the coefficients α_j in eqs (1.4.2) and (1.5.1) one finds out that $|\alpha_{r+1}^{AB}| \geqslant |\alpha_{r+1}^{AM}|$. In conclusion, the closed version AM should be preferred. The price to be paid when putting this decision in practice is

---

* In [1] the error constant is defined as

$$\overline{C} = C_{m+1}/(b_0 + b_1 + \ldots + b_i)$$

This value is however more transparent to describe accumulated errors than local truncation error, see section 1.10.

Table 5

Number of steps, order and error constant for some special algorithms

Algorithm	Number of steps i	Order m	Error constant C
Adams-Bashforth	$r+1$	$r+1$	α_{r+1}
Adams-Moulton, $r \neq 0$	r	$r+1$	α_{r+1}
Adams-Moulton, $r=0$	1	1	$-\dfrac{1}{2}$
Nyström, $r \neq 0$	$r+1$	$r+1$	α_{r+1}
Nyström, $r=0$	2	2	$\dfrac{1}{3}$
Milne-Simpson, $r>2$	r	$r+1$	α_{r+1}
Milne-Simpson, $r=2$	2	4	$-\dfrac{1}{90}$
Milne-Simpson, $r=1$	2	2	$\dfrac{1}{3}$
Milne-Simpson, $r=0$	2	1	-2

that, because AM algorithm is closed, an extra procedure must be added to make it applicable. This is the predictor-corrector procedure discussed in section 1.9.

The concept of consistency of a multistep method is closely related to the elements discussed above. Algorithm (1.6.1) is called *consistent* if its order is at least equal to unity. An equivalent definition is that algorithm (1.6.1) is consistent if the numbers C_0 and C_1 defined in (1.6.9) are equal to zero. Thus the two relations to be satisfied by the coefficients a_s and b_s for a consistent method are $C_0 = 0$ and $C_1 = 0$. Table 5 shows that *all* algorithms of the previous two sections are consistent.

The consistency is a necessary condition for an algorithm to be convergent. An intuitive argument to understand this statement follows.

Actually, when computing the solution from a up to $X = x_n$ (we have in mind the procedure explained in connection with the definition of convergence of algorithm (1.6.1)) we have to apply the algorithm successively, namely first take $k = 0$ to compute $\bar{y}_i$, next put $k = 1$ to compute $\bar{y}_{i+1}$, and so on, up to $k = n - i$, which yields $\bar{y}_n$. All together we repeat the algorithm $n - i + 1$ times. As the local error behaves as h^{m+1}, one should expect that the error accumulated after $n - i + 1$ successive applications should roughly behave as $(n - i + + 1)h^{m+1}$. Now remember that $n = (X - a)/h$. Then, if n is large $(n \gg i - 1)$ the error in $\bar{y}_n$ is approximately $nh^{m+1} = (X - a)h^{-1} \times \times h^{m+1} \sim h^m$.

The condition of convergence requires that this must tend to zero when h tends to zero, and the smallest integer m for which this happens is $m = 1$. This is equivalent to saying that the consistency of the algorithm (1.6.1) is a necessary condition for its convergence.

1.7. ERROR ANALYSIS: ERROR PROPAGATION AND STABILITY

In the previous section we have seen that the order of the algorithm is an important parameter to characterize the local truncation error. Indeed, we have seen that the higher the order, the more efficient the algorithm.

Such a statement is however true only if the algorithm is also stable, which means that it is of a form that prevents erratic error propagation during its recurrent application to solve the differential equation.

To illustrate this restriction we consider here a very simple example. We will solve a differential equation by two similar methods, the second of which is of higher order. It will be seen however that the second algorithm is simply useless because the error propagated by it cannot be controlled, to finally swamp the desired solution. In contrast, the propagated error will be found negligible for the first method. We will say that the first algorithm is stable but the second is unstable.

Our example is

$$y' = f(x), \quad x \in [a, b], \quad y(a) = y_0. \tag{1.7.1}$$

(Note that f depends *only* on x.) Its solution is simply

$$y(x) = y_0 + \int_a^x f(x')dx'$$

so that to compute $y(x)$ we may use any standard procedure for numerical integration of functions, such as Simpson or Gauss formulas, see for instance [7]. Alternatively, we can use just algorithms to solve the initial value problem (1)—(3), and we shall actually do so by taking the Adams-Bashforth method with $r = 1$,

$$\bar{y}_{k+2} = \bar{y}_{k+1} + h\left(\frac{3}{2} f_{k+1} - \frac{1}{2} f_k\right), \quad k = 0, 1, \ldots, n-2, \tag{1.7.2}$$

as the first method, and the algorithm

$$\bar{y}_{k+2} = (1-\alpha)\,\bar{y}_{k+1} + \alpha\bar{y}_k + h\left[\frac{1}{2}(3+\alpha) f_{k+1} - \frac{1}{2}(1-\alpha) f_k\right],$$
$$k = 0, 1, \ldots, n-2, \tag{1.7.3}$$

where α is an arbitrary parameter, as the second method. Algorithm (1.7.2) is the special case $\alpha = 0$ of algorithm (1.7.3). For convenience we assume that the final point for the numerical solution of problem (1.7.1) is just the r.h. side of the domain, that is $X \equiv x_n = b$.

Both algorithms are of second order and their local truncation errors are

$$e^1_{k+2} = \frac{5}{12} h^3 y'''(x_k), \tag{1.7.4}$$

and

$$e^2_{k+2} = \frac{5-\alpha}{12} h^3 y'''(x_k), \tag{1.7.5}$$

respectively, as it can be verified easily by means of eqs (1.6.17) and (1.6.9). So, for the moment, we are tempted to think that for $\alpha \in (0, 10)$ the latter algorithm is more efficient than the former because its error constant is, in absolute value, smaller than 5/12. Moreover, when $\alpha = 5$ the second algorithm seems to work at its best because in this case it becomes a third-order algorithm, with the local error

$$e^2_{k+2} = \frac{1}{6} h^4 y^{(4)}(x_k). \tag{1.7.6}$$

We now examine the error propagation for the algorithm (1.7.3) with arbitrary α and finally particularize the result for the two relevant values of α, namely $\alpha = 5$, the presumably best value, and also $\alpha = 0$, to accommodate the Adams-Bashforth algorithm (1.7.2).

In our analysis we shall use two assumptions. The first is that the starting values y_0 and $\bar{y}_1$ coincide with the exact values of the solution we search for. The second assumption is that we have at our disposal an almost ideal computer, i.e. a computer which works without round-off errors except once, namely at the first application ($k = 0$) of algorithm (1.7.3). Thus apply algorithm (1.7.3) to $k = 0$,

$$\bar{y}_2 = (1 - \alpha)\,\bar{y}_1 + \alpha y_0 + h\left(\frac{1}{2}(3 + \alpha) f_1 - \frac{1}{2}(1 - \alpha) f_0\right), \tag{1.7.7}$$

and admit that this is the only sequence in which the computer works with a finite number of decimal figures, so that it yields a value $\tilde{y}_2$ which is displaced from the correct value $\bar{y}_2$ by the quantity $\varepsilon_2 = 10^{-t}$, with integer $t > 0$

$$\tilde{y}_2 = y_2 + \varepsilon_2. \tag{1.7.8}$$

All subsequent computations are assumed absolutely accurate. At the next application, which means $k = 1$, algorithm (1.7.3) produces $\tilde{y}_3$ in terms of the input data $\tilde{y}_2$ (affected by the error ε_2), $\bar{y}_1, f_1, f_2$, all these three being exact,

$$\tilde{y}_3 = (1 - \alpha)(\bar{y}_2 + \varepsilon_2) + \alpha \bar{y}_1 + h\left[\frac{1}{2}(3 + \alpha) f_2 - \frac{1}{2}(1 - \alpha) f_1\right]. \tag{1.7.9}$$

It should be stressed once more that the deviation of $\tilde{y}_3$ from the correct $\bar{y}_3$ has nothing to do with the accuracy of the computer (remember that now the whole computation is supposed as absolutely accurate); it is only due to the fact that the very input value of $\bar{y}_2$ is now the slightly erroneous number $\tilde{y}_2$. Thus we write $\tilde{y}_3 = \bar{y}_3 + \varepsilon_3$ where ε_3 is the error coming out solely from ε_2. (Naturally if ε_2 were zero, then ε_3 is also zero.) Eq. (1.7.9) can be written in a more suggestive way as a set of two equations:

$$\bar{y}_3 = (1 - \alpha)\,\bar{y}_2 + \alpha \bar{y}_1 + h\left[\frac{1}{2}(3+\alpha) f_2 - \frac{1}{2}(1 - \alpha) f_1\right], \tag{1.7.10 a}$$

$$\varepsilon_3 = (1 - \alpha)\varepsilon_2 + \alpha\varepsilon_1, \ (\varepsilon_1 = 0). \tag{1.7.10 b}$$

This is equivalent to saying that the computation of r.h.s. of eq. (1.7.9) proceeds as a sequence of three stages: the first two stages consist of computation of the r.h. sides of eq. (1.7.10 a) and of eq. (1.7.10 b) and the third is the addition of these two results to produce the value of $\tilde{y}_3$. As stated, all these computations are performed with absolute accuracy.

Next, $\tilde{y}_3$ and $\tilde{y}_2$ are used as input values in the algorithm with $k = 2$. Once again the algorithm can be alternatively written as two separate equations, one for the genuine solution of the problem,

$$\bar{y}_4 = (1 - \alpha)\,\bar{y}_3 + \alpha\bar{y}_2 + h\left[\frac{1}{2}(3 + \alpha)f_3 - \frac{1}{2}(1 - \alpha)f_2\right], \quad (1.7.11\text{ a})$$

and the other for the error propagation,

$$\varepsilon_4 = (1 - \alpha)\,\varepsilon_3 + \alpha\varepsilon_2. \quad (1.7.11\text{ b})$$

Altogether, it can be said that everything happens such as the computer were running progressively over the mesh points of the interval to calculate separately two recurrence relations:

$$\bar{y}_{k+2} = (1 - \alpha)\,\bar{y}_{k+1} + \alpha\bar{y}_k + h\left[\frac{1}{2}(3 + \alpha)\,f_{k+1} - \frac{1}{2}(1 - \alpha)\,f_k\right],$$
$$k = 0, 1, \ldots, n - 2, \quad (1.7.12\text{ a})$$

$$\varepsilon_{k+2} = (1 - \alpha)\,\varepsilon_{k+1} + \alpha\varepsilon_k,\ k = 1, 2, \ldots, n - 2 \quad (1.7.12\text{ b})$$

and finally adds the partial results to yield

$$\tilde{y}_{k+2} = \bar{y}_{k+2} + \varepsilon_{k+2}. \quad (1.7.13)$$

The solution of the error propagation equation (1.7.12 b) can be solved easily by hand. Thus we search for a solution of the form $\varepsilon_k = d^k (d \neq 0)$ which is next introduced in the equation to get

$$d^k(d^2 - (1 - \alpha)d - \alpha) = 0, \quad (1.7.14)$$

with the nonvanishing roots $d_1 = 1$ and $d_2 = -\alpha$. Thus, the general solution of eq. (1.7.12 b) is a linear combination of the two particular solutions d_1^k and d_2^k, viz.

$$\varepsilon_k = C_1 d_1^k + C_2 d_2^k = C_1 + (-1)^k\,C_2\alpha^k. \quad (1.7.15)$$

To evaluate the constants C_1 and C_2 we insert the initial conditions ε_1 and ε_2 in eq. (1.7.15),

$$\varepsilon_1 = C_1 - C_2\alpha, \quad \varepsilon_2 = C_1 + C_2\alpha^2.$$

As in our case $\varepsilon_1 = 0$ the result is

$$C_1 = \frac{\varepsilon_2}{1+\alpha}, \quad C_2 = \frac{\varepsilon_2}{\alpha(1+\alpha)},$$

and it gives the following expression for the propagated error ε_k:

$$\varepsilon_k = \frac{\varepsilon_2}{1+\alpha}(1 + (-1)^k \alpha^{k-1}). \tag{1.7.16}$$

Therefore the solution at $b = x_n$ furnished by this almost ideal computer is

$$\tilde{y}_n = \bar{y}_n + \varepsilon_n = \bar{y}_n + \frac{\varepsilon_2}{1+\alpha}(1 + (-1)^n \alpha^{n-1}). \tag{1.7.17}$$

The second term represents the contribution of the propagation of the only admitted error $\varepsilon_2 = 10^{-t}$.

Let us now examine the practical consequences of the law (1.7.16) of error propagation for the problem $y' = 3x^2 + 1$, $a = 0$, $b = 10$, $y(0) = 1$. The exact solution is $y(x) = x^3 + x + 1$ so that $y(X = b = x_n) = 1011$.

Case $\alpha = 5$. The local truncation error (1.7.6) is strictly zero because $y^{(4)}(x) = 0$ for any x in the quoted interval and the same holds true for any derivative of higher order. Then for any n (and h such that $nh = b - a$), one has $\bar{y}_n = y(10) = 1011$. On the opposite, the magnitude of the propagated error ε_n strongly depends on n. To be specific we take $t = 20$ i.e. $\varepsilon_2 = 10^{-20}$. If $n = 50$ the propagated error is $2.96\ 10^{13}$ while if $n = 51$ (that is we use only slightly narrower intervals) it is $-1.48\ 10^{14}$. One sees that in general the propagated error is positive for even n and negative for odd n and grows up in absolute value to swamp out the very solution of the problem.

Such a phenomenon is called instability and we say that algorithm (1.7.3) is unstable for $\alpha = 5$.

Formula (1.7.16) also shows that the propagated error can be kept under control only if $\alpha \in (-1, 1]$. In other words all algorithms of the form (1.7.3) must be rejected unless α belongs to the domain of stability $(-1, 1]$.

As a matter of fact, it is instructive to insist that algorithm (1.7.3) is unstable for $\alpha = -1$ but this result is not based on the formula (1.7.16) which is not correct in this case. The reason is that for $\alpha = -1$ the two roots of eq. (1.7.14) coincide, $d_1 = d_2 = 1$, and therefore the ansatz of a linear combination as in eq. (1.4.15) becomes senseless.

A simple technique to derive a correct formula for this case is as follows: Consider that α is slightly displaced from -1, that is $\alpha = -1 + \Delta\alpha$. Then $d_1 = 1$ and $d_2 = 1 - \Delta\alpha$. The general solution ε_k is again a linear combination as in eq. (1.7.15), but we prefer to write it in a slightly modified form, viz.

$$\varepsilon_k = C_1' d_1^k + C_2' \left[\frac{1}{\Delta\alpha} (d_1^k - d_2^k) \right].$$

Now evaluate the term in brackets for $\Delta\alpha \to 0$

$$\lim_{\Delta\alpha \to 0} \frac{1}{\Delta\alpha} (d_1^k - d_2^k) = \lim_{\Delta\alpha \to 0} \frac{1}{\Delta\alpha} (1 - (1 - \Delta\alpha)^k) = k,$$

so that

$$\varepsilon_k = C_1' + C_2' k. \tag{1.7.18}$$

The values of C_1' and C_2' result from the initial conditions $\varepsilon_1 = 0 = C_1' + C_2'$ and $\varepsilon_2 = C_1' + 2C_2'$. The result is $C_1' = -C_2' = -\varepsilon_2$ so that the formula of the propagated error is

$$\varepsilon_k = \varepsilon_2 (k - 1).$$

This formula proves that algorithm (1.7.3) is unstable for $\alpha = -1$.

Case $\alpha = 0$. This value belongs to the stability range $(-1, 1]$. Formula (1.7.16) becomes $\varepsilon_k = \varepsilon_2$ thus showing that the propagated error remains ε_2 no matter how many steps n we use. [To be rigorous, the straightforward particularization $\alpha = 0$ in eq. (1.7.16) is illegitimate because eq. (1.7.14) now has only one nonvanishing root $d_1 = 1$. Yet, the expression of ε_n given by this particularization happens to coincide with the one that results from a rigorous derivation. In fact, the general solution is in this case $\varepsilon_k = C''d_1^k = C''$, and to fix C'' we need only one condition, namely the one at $k = 2$ which gives $C'' = \varepsilon_2$, thus leading to $\varepsilon_k = \varepsilon_2$.] Nevertheless, the value of $\tilde{y}_n$ computed by means of the Adams-Bashforth algorithm (1.7.2) differs from the desired value $y(b)$ by more than the initial error ε_2, and the reason is that the local truncation error (1.7.4) is nonvanishing for our sample equation $y' = 3x^2 + 1$.

It is instructive to evaluate $\tilde{y}_n$ for this case and to do it we repeat the previous procedure. We also preserve the previous assumptions that the initial values y_0 and $\bar{y}_1$ are exact and that the computer is almost perfect.

By applying the algorithm for $k = 0$ we obtain some $\tilde{y}_2$ which differs from the exact $y(x_2)$ by the local truncation error and round-off error,

$$\tilde{y}_2 = \bar{y}_2 + \varepsilon_2 = y(x_2) - e_2 + \varepsilon_2.$$

(See definition (1.6.4) of the local truncation error.)
At the next application we have

$$\tilde{y}_3 = \tilde{y}_2 + h\left(\frac{3}{2} f_2 - \frac{1}{2} f_1\right) = \left[y(x_2) + h\left(\frac{3}{2} f_2 - \frac{1}{2} f_1\right)\right] - e_2 + \varepsilon_2.$$

The term in brackets is just $\bar{y}_3$ and this differs from $y(x_3)$ by the local truncation error e_3 because at this stage, as also at all subsequent ones, the computation is supposed absolutely exact. Thus

$$\tilde{y}_3 = \bar{y}_3 - e_2 + \varepsilon_2 = y(x_3) - (e_3 + e_2) + \varepsilon_2.$$

By repeating the same arguments one finds out that after k steps we have

$$\tilde{y}_{k+2} = y(x_{k+2}) - (e_{k+2} + e_{k+1} + \ldots + e_3 + e_2) + \varepsilon_2 \tag{1.7.19}$$

For our sample equation the third derivative of the solution is $y'''(x) = 6$, so that the sum E_n of the local truncation errors can be calculated at once,

$$E_n = e_n + e_{n-1} + \ldots + e_2 = \frac{5}{12} \cdot h^3 \cdot 6(\underbrace{1 + 1 + \ldots + 1}_{n-1 \text{ times}}) = \frac{5}{2} h^3(n-1). \tag{1.7.20}$$

When n is large enough we can make the approximation $n - 1 \simeq n$. We can also use the connection between n and h, viz. $nh = b - a = 10$. Thus we can express E_n in terms of either n or h, and thus in the approximation $n - 1 \simeq n$ we have $E_n = 25h^2$ or $E_n = 2500/n^2$.
In conclusion the deviation of the computed $\tilde{y}_n$ from the exact $y(x_n = b)$ reads

$$y(b) - \tilde{y}_n = E_n - \varepsilon_2, \tag{1.7.21}$$

which in the limit $n \to \infty$ becomes $\lim\limits_{n\to\infty} \tilde{y}_n = y(b) + \varepsilon_2$. We thus have found out that for this sample problem the Adams-Bashforth algorithm yields results which, except for the round-off errors, tend to the exact value of the solution, in other words that it is a convergent algorithm.

Four remarks follow:

REMARK 1. The reader must not remain with the impression that the instability of a numerical method leads to disastrous effects only because of the inherent round-off errors of a computer. Consider for instance the above problem in which the computer is entirely perfect, that the function $f(x)$ is a polynomial of degree higher than two, and that the algorithm used is (1.7.3) with $\alpha = 5$. The value $\tilde{y}_n$ never coincides with $y(b)$ because now the very algorithm has nonvanishing local truncation errors. For instance, at the first application, eq. (1.7.7), it produces $\tilde{y}_2 = \bar{y}_2$ (because $\varepsilon_2 = 0$ in (1.7.8)) but this $\bar{y}_2$ differs from $y(x_2)$ by the local truncation error e_2,

$$\bar{y}_2 = y(x_2) - e_2,$$

and the point is that this $-e_2$ propagates at the next iterations according to the same disastrous law (1.7.12 b) as it were for ε_2 in the discussed example.

REMARK 2. As we have seen, algorithm (1.7.3) is stable only if $\alpha \in (-1, 1]$ and a natural question is to find out the value of α which ensures the most efficient algorithm of this form. The answer is given by the examination of the local truncation error (1.7.5) which shows that the smallest error constant is that corresponding to the algorithm with $\alpha = 1$. This is nothing else than a formula already mentioned before, namely the Nyström algorithm for $r = 0$ or 1, eq. (1.4.5).

REMARK 3. The real computers are far from being almost perfect as we supposed before. For instance the computers IBM of the series

360 or 370 work with about seven significant figures in simple precision arithmetic. This implies that the maximal coincidence expected between the numerical values of the solution at the mesh points and its exact values is limited at the first seven significant figures. In practice the range of coincidence is much narrower because of the large number of repeated applications of the algorithm, each new use adding a new round-off error. A possible strategy to minimize the influence of the round-off errors is to use algorithms of orders as high as possible since these are expected to produce reasonable accurate results at larger step size, i.e. by using a smaller number of repetitions of the algorithm. It happens however that the higher the order of an algorithm, the longer its expression, so that the number of arithmetic operations in a single step is larger than for an algorithm of lower order and thus it might lead to larger values for the round-off error/step. Altogether, it is impossible to make a general statement of what algorithm is optimal for practical use. This also depends very much on the equation to be solved. A detailed investigation of the round-off errors is presented in [1]. See also section 1.10 of this book.

REMARK 4. Formula (1.7.20) shows that the accumulated error is given by the sum over the local truncation errors. However this simple relation takes place only in the case when the function f depends only on x as in our example. When f is of the general form $f(x, y)$ the relation between the individual local truncation errors and the accumulated error is significantly more complicated. Again, see section 1.10.

The example discussed above provides a realistic image upon the importance of the stability of a numerical algorithm. Our treatment was close to the main lines of the general theory of stability and we give below a brief sketch of this theory.

We have seen that the error propagation was governed by equation (1.7.12 b) which led to solution (1.7.15). One can verify at once that eq. (1.7.12 b) is nothing else than the equation

$$a_2\varepsilon_{k+2} + a_1\varepsilon_{k+1} + a_0\varepsilon_k = 0, \ k = 0, 1, \ldots, n-2, \tag{1.7.22}$$

where a_0, a_1, and a_2 are the coefficients of the two-step algorithm (1.7.3) written as (1.6.1).

In general, for the i-step algorithm (1.6.1) the error propagation is governed by the direct extension of eq. (1.7.22) viz.:

$$a_i\varepsilon_{k+i} + a_{i-1}\varepsilon_{k+i-1} + \ldots + a_0\varepsilon_k = 0, \; k = 0, 1, \ldots, n - i, \qquad (1.7.23)$$

the characteristic polynomial of which is

$$P(d) = a_i d^i + a_{i-1} d^{i-1} + \ldots + a_0. \qquad (1.7.24)$$

The general solution of eq. (1.7.23) is then a linear combination of the form

$$\varepsilon_k = \sum_{j=1}^{i} C_j d_j^k, \qquad (1.7.25)$$

where d_j, $j = 1, 2, \ldots, i$, are the i roots of the characteristic equation

$$P(d) = 0. \qquad (1.7.26)$$

One sees that the algorithm is stable only if the roots are smaller or at most equal to unity in modulus and also if the roots of modulus one are simple. Indeed, if one of the roots, d_1, say, is in modulus larger than 1 then the component $C_1 d_1^k$ will blow up with increasing k, which means that the error propagation can not be kept under control. If two roots of modulus 1 would be equal then, in eq. (1.7.25), instead of powers of these roots, sums, as in eq. (1.7.18), will occur which equally blow up with increasing k, and the same phenomenon takes place also when more than two equal roots of modulus 1 are present. Finally, there is no problem if $|d_j| < 1$ for all $j = 1, 2, \ldots, i$ because expression (1.7.25) shows that in such a case the propagated error not only that can be controlled, but simply tends to zero with increasing k. Expressed in other words, when seen solely from the point of view of stability requirements, the most attractive algorithms seem to be the ones for which all the roots of eq. (1.7.26) are of modulus smaller than unity. Nevertheless this is a false impression, since it contradicts the condition of consistency. In fact, such an algorithm must be also consistent and the first condition for consistency ($C_0 = 0$, see eq. (1.6.9)) can be alternatively written as $P(1) = 0$, i.e. the characteristic polynomial of each applicable algorithm has always a root equal to unity.

It can be verified that for all the algorithms discussed in sections 1.4 and 1.5 this root is simple so that all of them are stable. Now, by virtue of the theorem:

Algorithm (1.6.1) is convergent if and only if it is both consistent and stable,

one can conclude that all algorithms discussed in those two sections are convergent.

There exists, however, a finer delimitation in the class of convergent multistep algorithms in terms of the *number* of nonequal roots of modulus 1 of the characteristic polynomial, eq. (1.7.26). The algorithms for which only a single such root exists (and this is of course $d_1 = 1$) are called *absolutely* or *strongly stable.* The methods for which more such roots exist are called *weakly stable.*

In both kinds of methods the numerical results tend, when h is decreased, to the exact values of the solution at the mesh points but the manner in which this takes place is markedly different. In fact, in case of absolutely stable methods the errors tend smoothly to zero while for weakly stable methods they tend to zero in a way which, for some particular forms of the function f, might seem confusing at first sight.

To illustrate the difference we consider two algorithms, namely Adams-Bashforth and Nyström.

The Adams-Bashforth algorithm is absolutely stable for any r since its characteristic polynomial

$$P(d) = d^2 - d \tag{1.7.27}$$

admits a unique root of this type, $d_1 = 1$; the other root is $d_2 = 0$.

In contrast, the Nyström algorithm is weakly stable (also for any r) since its characteristic polynomial

$$P(d) = d^2 - 1 \tag{1.7.28}$$

has *two* roots of modulus 1, namely $d_1 = 1$ and $d_2 = -1$.

To be specific we consider two particular forms of these algorithms,
(i) the Adams-Bashforth algorithm with $r = 1$, eq. (1.7.2)
(ii) the Nyström algorithm with $r = 0$ or $r = 1$,

$$\bar{y}_{k+2} = \bar{y}_k + 2hf_k, \tag{1.7.29}$$

and use them to solve the simple initial value problem

$$y' = \lambda y, \ x \in [0, b], \ y_0 = y(0) = 1, \ \lambda < 0, \tag{1.7.30}$$

with the exact solution $y(x) = \exp(\lambda x)$.

In table 6 we collect some relevant numerical results for $\lambda = -1$ and $b = 2.2$, computed by means of a Texas Instruments 59 pocket calculator. The two initial data to start the integration were just the values of the exact solution at $x_0 = 0$ and at $x_1 = h$, i.e. $y_0 = 1$, $\bar{y}_1 = y(h) = \exp(-h)$.

Table 6 consists of two blocks of data. In the first block we give the values of the deviation of the computed solution from the exact one $\Delta_k = y(x_k) - \bar{y}_k$ at several points*, for $n = 11$, that is $h = 2.2/11 =$

Table 6

Numerical results for the test problem $y' = -y$, $x \in [0, 2.2]$, $y(0) = 1$.

Deviation of the computed solution from the exact one for $n = 11$ in units of 10^{-5}

x_k	Δ_k	
	Adams-Bashforth	Nyström
0	0	0
0.2	0	0
0.4	−279	−219
0.6	−424	−916
0.8	−512	−329
.	.	.
.	.	.
.	.	.
1.6	−528	−459
1.8	−494	127
2.0	−455	−555
2.2	−414	313

Values of Δ_n for various values of n, in units of 10^{-5}

n	Δ_n	
	Adams-Bashforth	Nyström
11	−414	313
12	−347	−500
13	−296	185
14	−255	−338
.	.	.
.	.	.
.	.	.
37	−36.2	0.7
38	−34.3	− 27
39	−32.6	− 0.05
40	−30.9	− 24
.	.	.
.	.	.
.	.	.
66	−11.3	− 7.2
67	−11.0	− 1.8
68	−10.7	− 6.7
69	−10.4	− 1.7

$= 0.2$. In the second block we collect the values of the same deviation at $x_n = b = 2.2$ for various values of the number of intervals n.

---

* Since the round-off errors are found negligible for this test problem, Δ_k represents simply the accumulated truncation error at x_k.

The data in the first block show that for both methods the deviations are more or less uniform at the first several mesh points. They preserve the same behaviour also at the outer end of the integration domain in the case of the first method but become oscillating in sign for the second method. If data of the second block are analysed, it is actually suggested that two distinct ways of extinction of Δ_n vs. n exist in the case of the Nyström method, one for even values of n and the other for odd values. When n increases each of these two sets of data tend rather smoothly to zero but at the very beginning, when n is still too small, it happens that the values of the deviation at successive values of n oscillate in sign, as it were in the example discussed before in connection with the phenomenon of instability.

To see the reasons of these different behaviours of the error we will consider the two algorithms in more detail by taking advantage of the fact that the numerical solution of problem (1.7.30) can be expressed in compact form for both algorithms.

The Adams-Bashforth algorithm

For eq. $y' = \lambda y$ we obviously have $f_{k+1} = \lambda \bar{y}_{k+1}$ and $f_k = \lambda \bar{y}_k$. Then algorithm (1.7.2) is

$$\bar{y}_{k+2} = \left(1 + \frac{3}{2}\lambda h\right)\bar{y}_{k+1} - \frac{1}{2}h\lambda\bar{y}_k,\ h = b/n,\ k = 0, 1, \ldots, n-2. \tag{1.7.31}$$

This is a finite difference equation which can be solved by standard procedure, that is we search for a solution of the form $\bar{y}_k = D^k$, which we next introduce into eq. (1.7.31) to obtain

$$D^k\left[D^2 - \left(1 + \frac{3}{2}\lambda h\right)D + \frac{1}{2}\lambda h\right] = 0. \tag{1.7.32}$$

This algebraic equation has the following two nonvanishing roots:

$$D_1 = \frac{1}{2} + \frac{3}{4}\lambda h + \sqrt{\frac{1}{4} + \frac{1}{4}\lambda h + \frac{9}{16}(\lambda h)^2},\ D_2 = \frac{\lambda h}{2D_1}. \tag{1.7.33}$$

A remark which will be useful later on is that when h tends to zero D_1 tends to d_1 and D_2 tends to d_2 where d_1 et d_2 are the two roots of the characteristic polynomial (1.7.27).
The general solution of eq. (1.7.31) is

$$\bar{y}_k = C_1 D_1^k + C_2 D_2^k. \tag{1.7.34}$$

Constants C_1 and C_2 result from the initial conditions $y_0 = 1 = C_1 + C_2$, $\bar{y}_1 = C_1 D_1 + C_2 D_2$, as follows

$$C_1 = 1 - C_2,\ C_2 = \frac{\bar{y}_1 - D_1}{D_2 - D_1}. \tag{1.7.35}$$

Now assume that h is sufficiently small, more rigorously that $|\lambda h| \ll 1$, and use the truncated series in powers of λh for the square root in D_1. We get

$$D_1 = 1 + \lambda h + \frac{1}{2}(\lambda h)^2 - \frac{1}{4}(\lambda h)^3 + 0((\lambda h)^4) \tag{1.7.36}$$

Next use the power series representation of $\exp(\lambda h)$,

$$\exp(\lambda h) = 1 + \lambda h + \frac{1}{2}(\lambda h)^2 + \frac{1}{6}(\lambda h)^3 + 0((\lambda h)^4)$$

and compare it with expansion for D_1 to get

$$D_1 = \exp(\lambda h) - \frac{5}{12}(\lambda h)^3 + 0((\lambda h)^4).$$

Now evaluate D_1^k by means of the familiar Newton formula

$$(a + b)^k = a^k + ka^{k-1}b + \frac{k(k-1)}{2}a^{k-2}b^2 + \dots$$

in which we take $a = \exp(\lambda h)$, $b = -\frac{5}{12}(\lambda h)^3 + 0((\lambda h)^4)$ and retain only two terms. (We assume that h is so small that this truncation is also justified.) One gets

$$D_1^k = \exp(\lambda k h) + k\left[-\frac{5}{12}(\lambda h)^2 + O((\lambda h)^4)\right]\exp[\lambda(k-1)h]. \quad (1.7.37)$$

If we recall that $\exp[\lambda(k-1)h] = \exp(k\lambda h)\exp(-\lambda h) = \exp(\lambda x_k)(1 + O(\lambda h))$ we obtain

$$D_1^k = \exp(\lambda x_k)\left(1 - \frac{5}{12}\lambda^3 x_k h^2 + O(h^3)\right). \quad (1.7.38)$$

We can equally apply a similar procedure to evaluate D_2^k but because $D_2 = O(\lambda h)$ we will have $D_2^k = O((\lambda h)^k)$ and therefore the second term in eq. (1.7.34) can be safely neglected for large k.

Thus the solution given by the Adams-Bashforth algorithm is

$$\bar{y}_k = (1 - C_2)\, D_1^k + O((\lambda h)^k) =$$

$$= \exp(\lambda x_k)\left(1 - \frac{5}{12}\lambda^3 x_k h^2 + O(h^3)\right)(1 - C_2). \quad (1.7.39)$$

To evaluate C_2 we introduce $\bar{y}_1$, D_1 and D_2 in the second of eqs (1.7.35), to find out

$$C_2 = -\frac{5}{12}(\lambda h)^3 + O((\lambda h)^4), \quad (1.7.40)$$

so that finally

$$\bar{y}_k = \exp(\lambda x_k)\left(1 - \frac{5}{12}\lambda^3 x_k h^2 + O(h^3)\right). \quad (1.7.41)$$

The first term in the parentheses stands for the exact solution while the next ones account for the accumulated error.

It is important to note at this moment that the exact solution is a common factor here so that the deviation

$$\Delta_k = \exp(\lambda x_k) - \bar{y}_k = \exp(\lambda x_k)\left(\frac{5}{12}\lambda^3 x_k h^2 + O(h^3)\right) \quad (1.7.42)$$

smoothly tends to zero with increasing k for constant h and fixed $\lambda < 0$. This behaviour is well illustrated in table 6 by the last four values presented in the first block of data. As a matter of fact, the values of Δ_k furnished by eq. (1.7.42), in which the term $0(h^3)$ is disregarded, are -538, -496, -451 and -406 (in the same units of 10^{-5}), in reasonable agreement with the experimental values presented in Table 6.

When we fix the value of x, say $x = x_n = b$ and vary the total number of intervals n to integrate the equation from zero to b we can express the value of Δ_n (i.e. the accumulated error at the point $x_n = b$) in terms of $n = b/h$,

$$\Delta_n = \exp(\lambda b)\left(\frac{5}{12}\lambda^3 b^3 n^{-2} + 0(n^{-3})\right) \tag{1.7.43}$$

This also exhibits a smooth decrease with increasing n, thus supporting the experimental evidence given in the second block.

To summarize, the results given by the Adams-Bashforth method tend smoothly to the values of the exact solution and the reason for it is the fact that the second root of eq. (1.7.32) tends to zero when h tends to zero. The latter circumstance is simply the reflection of the fact that the second root of the characteristic polynomial (1.7.27) is $d_2 = 0$, or, in other words, that the Adams-Bashforth algorithm is absolutely stable.

The Nyström algorithm

For eq. (1.7.30) this algorithm is

$$\bar{y}_{k+2} = 2\lambda h \bar{y}_{k+1} + \bar{y}_k$$

so that the analog of eq. (1.7.32) is

$$D^k(D^2 - 2\lambda h D - 1) = 0 \tag{1.7.44}$$

with the roots

$$D_1 = \lambda h + \sqrt{1 + (\lambda h)^2},\ D_2 = \lambda h - \sqrt{1 + (\lambda h)^2},\ D_2 = -\frac{1}{D_1}. \tag{1.7.45}$$

Note that $\lim_{h\to 0} D_1 = d_1$ and $\lim_{h\to 0} D_2 = d_2$ where d_1 and d_2 are the roots of the characteristic polynomial (1.7.28).

We repeat the procedure developed in connection with D_1 of eq. (1.7.33) to find out that

$$D_1^k = \exp(\lambda x_k)\left(1 - \frac{1}{6}\lambda^3 x_k h^2 + O(h^3)\right). \tag{1.7.46}$$

Of course we can use the same procedure also to derive the expression of D_2^k. Yet, the last of eqs (1.7.45) suggests a simpler way, viz.:

$$D_2^k = \left(-\frac{1}{D_1}\right)^k = (-1)^k \exp(-\lambda x_k)\left(1 + \frac{1}{6}\lambda^3 x_k h^2 + O(h^3)\right). \tag{1.7.47}$$

A major difference from the previous method occurs here, namely that the second root is no longer negligible so that it must be taken explicitly in the linear combination (1.7.34). By doing so, after a little algebra we find out that

$$\bar{y}_k = \exp(\lambda x_k)\left(1 - \frac{1}{6}\lambda^3 x_k h^2 + O(h^3)\right)$$

$$-\frac{1}{12}(-1)^k \exp(-\lambda x_k)((\lambda h)^3 + O(h^4)), \tag{1.7.48}$$

in which we also used $C_2 = -\dfrac{1}{12}(\lambda h)^3 + O((\lambda h)^4)$, as resulting from the second of eqs (1.7.35).

The accumulated error Δ_k then is

$$\Delta_k = \exp(\lambda x_k)\left(\frac{1}{6}\lambda^3 x_k h^2 + O(h^3)\right)$$

$$+\frac{1}{12}\exp(-\lambda x_k)((\lambda h)^3 + O(h^4))(-1)^k \tag{1.7.49}$$

and it has two components which are proportional to the exact solution $\exp(\lambda x_k)$ and to the spurious function $\exp(-\lambda x_k)$, respectively.

To examine the practical consequences of this structure we place again the discussion along the two lines represented by the two data blocks in Table 6.

First assume that h is fixed and k increases. The behaviour of the two components is governed by the exponential functions $\exp(\pm\lambda x_k)$. Now, since λ is negative, the first component typically decreases with increasing k while the second component, which oscillates in sign for successive values of k, increases in absolute value. The latter begins to dominate at larger and larger k thus explaining the oscillating behaviour of the errors in the first block.

Now assume that x_n is fixed ($x_n = b$) and we vary n. We use $h = b/n$ and rewrite eq. (1.7.49) for the error at $x_n = b$,

$$\Delta_n = \exp(\lambda b)\left(\frac{1}{6}\ \lambda^3 b^3 n^{-2} + O(n^{-3})\right)$$

$$+\frac{1}{12}(-1)^n \exp(-\lambda b)(\lambda^3 b^3 n^{-3} + O(n^{-4})). \tag{1.7.50}$$

This formula explains the behaviour of the errors in the second block of Table 6. Indeed, for small values of n the second component is numerically dominating so that, at the beginning, the errors exhibit alternating signs at successive values of n. However, when n increases the second component damps out as n^{-3}, i.e. faster than the first component so that a critical value n_c exists at which the two components compensate each other. For values of n above this n_c the values of Δ_n will no longer change the sign, although they will still exhibit some extincting jumps at successive values of n on the account that Δ_n is the sum of the two components for even n but their difference for odd n.

The occurrence of the second component comes from the fact that the second root of eq. (1.7.44) is never negligible. When h tends to zero this root tends to -1 i.e. it reproduces the second root $d_2 = -1$ of the characteristic polynomial (1.7.28) which is just the supplementary root of modúlus 1 which places the Nyström algorithm in the class of weakly stable methods.

1.8. PRACTICAL ISSUES CONCERNING MULTISTEP ALGORITHMS

We are here concerned with two practical issues namely how to start the integration procedure and how to solve systems of differential equations.

Starting the integration

As mentioned before on several occasions, this is one of the most delicate points raised in connection to the multistep algorithms. The difficulty consists of that, while the very problem to be solved, eqs (1)—(3), specifies only the value at x_0 of the solution, the iteration process (1.6.1) requires also the values of the solution at $x_1, x_2, \ldots, x_{i-1}$. As an example, the first application, $k = 0$, of the Adams-Bashforth algorithm with $r = 1$,

$$\bar{y}_{k+2} = \bar{y}_{k+1} + h\left(\frac{3}{2} f_{k+1} - \frac{1}{2} f_k\right), \; k = 0, 1, 2, \ldots \tag{1.8.1}$$

requires $\bar{y}_0$, which occurs as one of arguments of f_0, *and also* $\bar{y}_1$ which occurs both separately and as one of arguments of f_1. As mentioned, $\bar{y}_0$ is known since it is just the given initial condition, but $\bar{y}_1$ must be furnished by some extra procedure.

In general, to compute the other values $\bar{y}_1, \bar{y}_2, \ldots, \bar{y}_{i-1}$ which are necessary to start the iteration by means of algorithm (1.6.1), we can use any one-step numerical method, as for instance the Euler method discussed in section 1.4. It is however very important that the resultant values of $\bar{y}_1, \bar{y}_2, \ldots \bar{y}_{i-1}$ are as accurate as possible, anyway more accurate than is algorithm (1.6.1) used to further propagate the solution.

Specifically, if the chosen algorithm (1.6.1) is of order m, the mathematicians will generally recommend a one-step method of order $m' > m$ to compute starting values. As a matter of fact, the Euler method, which is of the *first* order, is never recommended to furnish $\bar{y}_1$ which enters the *second*-order algorithm (1.8.1).

Yet, there is one escape from such an interdiction. In fact, the interdiction refers actually to the use of the Euler method with *the same h* as in eq. (1.8.1), but it does not rule out the possibility to use the same method at smaller values of the step size. For instance, to evaluate

$\bar{y}_1$ we can integrate problem (1)—(3) on $[x_0, x_0 + h]$ by the Euler method with the step $h_E = h/N$ where N is a suitable (large) positive integer, and $\bar{y}_N$ resulted from such a run is finally assigned to the entrance value $\bar{y}_1$ in eq. (1.8.1). Clearly, the larger N the more accurate $\bar{y}_1$.

In general, such a procedure can be used in all cases when the one step algorithm chosen to compute starting values is less accurate than multistep algorithm (1.6.1) which propagates the solution. The only shortcoming is that some extra tests are required to settle the suitable value of N.

There exist also procedures which enable to compute $\bar{y}_1, \bar{y}_2, \ldots, \bar{y}_{i-1}$ with arbitrary accuracy. The simplest of them is based on Taylor expansion.

To be specific, suppose that the chosen algorithm of the form (1.6.1) is of order m, so that we have to compute the starting values with some better accuracy, of order $m' = m + 1$, say.

To this aim we consider the Taylor series of $y(x)$ around x_0

$$y(x_0 + \delta) = y(x_0) + \delta y'(x_0) + \frac{\delta^2}{2!} y''(x_0) + \ldots + \frac{\delta^m}{m!} y^{(m)}(x_0) + \\ + \frac{\delta^{m+1}}{(m+1)!} y^{m+1}(\zeta), \tag{1.8.2}$$

$\delta > 0$, where ζ is some point in the open interval $(x_0, x_0 + \delta)$. We see that to compute $y(x_0 + \delta)$ with accuracy of order $m + 1$ we need the values at x_0 of the solution and of its derivatives up to the order m. The value of $y(x_0)$ is known, as being just the initial condition. The first derivative results directly from the very equation: $y'(x_0) = f(x_0, y_0)$. The higher order derivatives are evaluated by successive total differentiation with respect to x of the known function $f(x, y(x))$, for instance

$$y''(x) = [f(x, y(x))]' = f_x + f_y y' = f_x + f_y f,$$

where f_x and f_y are the partial derivatives of f with respect to x and y, respectively.

As an illustration, we apply this procedure for $f = x^2 + y^2$, and $m = 3$.

We get

$$y'(x) = x^2 + y^2$$

$$y''(x) = 2x + 2y(x^2 + y^2)$$

$$y'''(x) = 2 + 4xy + (2x^2 + 6y^2)(x^2 + y^2)$$

in which we further introduce the given values of x_0 and y_0. For instance, when $x_0 = 0$ and $y_0 = 1$ the corresponding weights in eq. (1.8.2) are $y(0) = 1$, $y'(0) = 1$, $y''(0) = 2$, $y'''(0) = 8$. On taking $\delta = h$ in formula (1.8.2), we obtain that

$$\bar{y}_1 = 1 + h + h^2 + \frac{4}{3} h^3 \tag{1.8.3}$$

approximates $y(x_0 + h)$ with an error of order h^4.

The above method can also be taken as a genuine numerical method and it is sometimes used as such, see chapter 2. The only difficulty is that if f is of complicated form, then some time and patience is needed to calculate analytically the corresponding total derivatives.

Integration of systems of first-order differential equations

As mentioned in "Preliminaries", such systems occur in practice either independently, or via a suitable transcription of equations (or systems) of higher order. We have also mentioned that the algorithms for single equations of the first order can be easily extended to accommodate systems as well.

Here we give an example. We consider the following initial value problem for a second-order differential equation

$$y'' = F(x, y, y'), \ x \in [a, b], \ y(a) = y_0, \ y'(a) = y'_0. \tag{1.8.4}$$

On introducing the new functions $y^1(x) \equiv y(x)$ and $y^2(x) \equiv y'(x)$, eq. (1.8.4) becomes an initial value problem for a system of two first-order differential equations,

$$y^{1\prime} = y^2(x)$$

$$y^{2\prime} = F(x, y^1, y^2)$$

with the initial conditions $y^1(a) = y_0$, and $y^2(a) = y_0'$. This system can be written compactly in vector form

$$\mathbf{y}' = \mathbf{f}(x, \mathbf{y}), \quad \mathbf{y}(a) = \mathbf{y}_0 \tag{1.8.5}$$

with the vectors

$$\mathbf{y} = \begin{pmatrix} y^1 \\ y^2 \end{pmatrix}, \quad \mathbf{f}(x, \mathbf{y}) = \begin{pmatrix} y^2 \\ F(x, y^1, y^2) \end{pmatrix}. \tag{1.8.6}$$

Thus any algorithm of the form (1.6.1) can be applied provided we accordingly think of $\bar{y}_k, \bar{y}_{k+1}, \ldots, \bar{y}_{k+i}, f_k, f_{k+1}, \ldots, f_{k+i}$ as two-dimensional vectors, not as scalar as it was for single equations. For instance when the Adams-Bashforth algorithm with $\imath = 1$ is chosen to integrate the system, eq. (1.8.1) becomes a set of two equations, viz.

$$\bar{y}^1_{k+2} = \bar{y}^1_{k+1} + h\left(\frac{3}{2}\,\bar{y}^2_{k+1} - \frac{1}{2}\,\bar{y}^2_k\right), \tag{1.8.7 a}$$

$$\bar{y}^2_{k+2} = \bar{y}^2_{k+1} + h\left(\frac{3}{2}\,F(x_{k+1}, \bar{y}^1_{k+1}, \bar{y}^2_{k+1}) - \frac{1}{2}\,F(x_k, \bar{y}^1_k, \bar{y}^2_k)\right), \tag{1.8.7 b}$$

which gives the components of the vector $\bar{\mathbf{y}}_{k+2}$ in terms of the input vectors $\bar{\mathbf{y}}_k$ and $\bar{\mathbf{y}}_{k+1}$.

All features of algorithm (1.8.1) remain valid. For instance it remains a second-order method, with the same error constant $C = 5/12$, and the local truncation error is given by the same eq. (1.7.4) but in vector form, viz.

$$e^s_{k+2} = \frac{5}{12}\,h^3 y^{s(3)}(x_k), \quad s = 1, 2. \tag{1.8.8}$$

As for the starting vector $\mathbf{y}_1$, its components can be calculated by appropriately applying any of the procedures discussed at the beginning of this section.

1.9. PREDICTOR-CORRECTOR PROCEDURE FOR CLOSED MULTISTEP ALGORITHMS

The objective of formula (1.6.1) is to yield $\bar{y}_{k+i}$ in terms of the known input values $\bar{y}_k, \bar{y}_{k+1}, \ldots, \bar{y}_{k+i-1}$. This task is not trivial at all when algorithm (1.6.1) is closed, that is when $b_i \neq 0$. In fact, in this case the unknown $\bar{y}_{k+i}$ also occurs in r.h.s. of this formula, namely as one of the arguments of f_{k+i}. Thus eq. (1.6.1) is an implicit algebraic equation for $\bar{y}_{k+i}$.

It is then worth remembering how such equations can be solved numerically. We consider the implicit algebraic equation

$$Z = F(Z) \tag{1.9.1}$$

where F is some given function of Z. Suppose that eq. (1.9.1) has a solution $\bar{Z}$ (and only one) in some interval $[Z_{\min}, Z_{\max}]$.

To locate $\bar{Z}$ we first assign some starting value $Z^0 \in [Z_{\min}, Z_{\max}]$ and calculate a first approximation Z^1 by

$$Z^1 = F(Z^0).$$

This Z^1 is next taken as an argument of F to obtain a new, hopefully better approximation Z^2, viz.

$$Z^2 = F(Z^1),$$

and so on. Thus the process is

$$Z^n = F(Z^{n-1}), \; n = 1, 2, 3, \ldots \tag{1.9.2}$$

and it is called convergent if $\lim_{n \to \infty} Z^n = \bar{Z}$.

The following theorem concerning the convergence exists:

The iteration process (1.9.2) is convergent if F satisfies two conditions

(i) $Z \in [Z_{\min}, Z_{\max}]$ implies that $F(Z) \in [Z_{\min}, Z_{\max}]$, a condition sometimes referred to as the *closure condition*, and
(ii) $F(Z)$ satisfies a Lipschitz condition on $[Z_{\min}, Z_{\max}]$ with a Lipschitz constant $L_F < 1$. Thus

$$|F(Z^*) - F(Z^{**})| \leqslant L_F|Z^* - Z^{**}|, \quad L_F < 1 \tag{1.9.3}$$

for any $Z^*, Z^{**} \in [Z_{\min}, Z_{\max}]$.

Note in passing that a function F satisfying the two conditions is said to be a *contraction mapping* on $[Z_{\min}, Z_{\max}]$ and the above theorem is known as the contraction mapping theorem. For its proof see section 7.7 in ref. [7].

We now show that the function F corresponding to algorithm (1.6.1) is a contraction mapping indeed. To this aim we re-write eq. (1.6.1) in a convenient form, viz.

$$Z = ha_i^{-1}b_i f(x_{k+i}, Z) + a_i^{-1}[h(b_{i-1}f_{k+i-1} + \dots + b_0 f_k) -$$

$$- a_{i-1}\bar{y}_{k+i-1} - \dots - a_0\bar{y}_k] \tag{1.9.4}$$

and remark that the second term in its right member is independent of Z. Then if this term is denoted by K, eq. (1.6.1) is of the form (1.9.1) with

$$F(Z) = ha_i^{-1}b_i f(x_{k+i}, Z) + K. \tag{1.9.5}$$

and its solution is the desired $\bar{y}_{k+i}$.

If we allow Z to take any real value, that is $Z_{\min} = -\infty$ and $Z_{\max} = +\infty$, then clearly the closure is satisfied. To check condition (ii) we use the fact that the function f satisfies the Lipschitz condition with respect to the second variable (see "Preliminaries"). We thus have in order

$$|F(Z^*) - F(Z^{**})| = h|a_i^{-1}b_i| \cdot |f(x_{k+i}, Z^*) - f(x_{k+i}, Z^{**})|$$

$$\leqslant Lh|a_i^{-1}b_i| \cdot |Z^* - Z^{**}|$$

and $L_F < 1$ implies

$$h < L^{-1}|a_i b_i^{-1}|. \tag{1.9.6}$$

Thus the conclusion is that the quoted F is a contraction mapping if h is satisfying condition (1.9.6). As a matter of fact, the values of h which are typically employed in practical applications are well within this range and therefore the iteration process (1.9.2) *is* appropriate to solve eq. (1.6.1). It is also obvious that the number of iterations necessary to compute the solution $\bar{y}_{k+i}$ with the desired numerical accuracy depends on the value of the starting number Z^0. Thus the procedure definitely gains in efficiency if we are able to supplement the iteration process (1.9.2) by an extra step whose aim is to furnish an adequate starting value Z^0. This extra step may consist of any other numerical method to solve the same initial value problem, eq. (1)—(3). Specifically, such a numerical method, called predictor, will yield some $\bar{y}^p_{k+i}$ which is an approximate representation of the exact $y(x_{k+i})$. Because of this, $\bar{y}^p_{k+i}$ is also close to $\bar{y}_{k+i}$ we search for, or in other words, $\bar{y}^p_{k+i}$ is an adequate starting value Z^0 in the iteration process.

Thus the solution $\bar{y}_{k+i}$ of eq. (1.6.1) results by successively correcting the "predicted" $Z^0 = \bar{y}^p_{k+i}$ via the iteration process (1.9.2). For this reason eq. (1.6.1) is called corrector, and the whole procedure to get $\bar{y}_{k+i}$ is known as the predictor-corrector procedure.

Naturally, the predictor-corrector procedure makes sense only if the predictor is less accurate than the corrector, with the meaning that $\bar{y}^p_{k+i}$ is a more crude approximation of $y(x_{k+i})$ than $\bar{y}_{k+i}$. Yet, the predictor must not be much too less accurate than the corrector since this would imply a larger number of corrections.

A reasonable compromise is to use an open algorithm of order $m^p = m - 1$ or $m^p = m$ for predictor where m is the order of corrector (1.6.1). In the latter case (that is $m^p = m$) only *one* iteration is sufficient; see [1] for the proof.

As an illustration, if corrector (1.6.1) is the Adams-Moulton algorithm with $r = 1$,

$$\bar{y}_{k+1} = \bar{y}_k + \frac{1}{2} h(f_{k+1} + f_k) \tag{1.9.7}$$

(this is a second-order algorithm, see table 5), reasonable candidates for predictor are either Adams-Bashforth algorithm with $r = 0$,

$$\bar{y}^p_{k+1} = \bar{y}_k + hf_k, \tag{1.9.8}$$

which is a first-order algorithm, or the same open algorithm with $r = 1$,

$$\bar{y}^p_{k+1} = \bar{y}_k + h\left(\frac{3}{2} f_k - \frac{1}{2} f_{k-1}\right). \tag{1.9.9}$$

which is a second-order method. (Recall that this $\bar{y}^p_{k+1}$ is less accurate than $\bar{y}_{k+1}$ of eq. (1.9.7), as resulting from the general comparison between closed and open multistep methods, an issue discussed in section 1.6.)

Moreover, if we choose the second, only one iteration is sufficient, that is

$$Z^1 = \bar{y}_k + \frac{1}{2} h(f(x_{k+1}, \bar{y}^p_{k+1}) + f_k), \tag{1.9.10}$$

so that we can further assign simply $\bar{y}_{k+1} = Z^1$. This does not mean at all that the exact solution of eq. (1.9.7) coincides with this Z^1. It only indicates that the deviation $|\bar{y}_{k+1} - Z^1|$ is of higher order than the deviation $|y(x_{k+1}) - \bar{y}_{k+1}|$ which measures the accuracy of the very numerical algorithm. So, if we theoretically go on with the iteration process to get $Z^2, Z^3, \ldots$, these will be naturally better approximations to $\bar{y}_{k+1}$ *but not* to $y(x_{k+1})$.

One of the most popular uses of the predictor-corrector pair consists in the numerical evaluation of the value of $y^{(m+1)}(x_k)$ which is further necessary to measure the local truncation error, eq. (1.6.17), or the accumulated error, see eq. (1.10.41). The associated procedure is often referred to as the Milne procedure.

Specifically, we show below that if the predictor associated to a corrector of order m is the open algorithm of the same order m and with the same structure of the coefficients multiplying the values of the solution, then $y^{(m+1)}(x_k)$ is given by formula

$$y^{(m+1)}(x_k) = h^{-(m+1)}\left[\frac{a_i}{C^p_{m+1} - C_{m+1}}(\bar{y}_{k+i} - \bar{y}^p_{k+i}) + O(h)\right]. \tag{1.9.11}$$

There are two typical examples of such suitable pairs. The first is the Adams-Bashforth predictor associated to the Adams-Moulton corrector, each taken at the same r. Both are of equal order $m = r + 1$ and have the same structure of the terms relating the solution, viz.: $\bar{y}_{k+1} = \bar{y}_k +$ + terms involving f. The second is the Nyström predictor associated to the Milne-Simpson corrector, again for the same r. These are also of equal order $m = r + 1$ and have the same structure $\bar{y}_k = \bar{y}_{k-2}$ + terms involving f.

Notice however that each of these pairs, if written in general form (1.6.1), correspond to different values of the number of steps i, and the value of i for the predictor is larger by one unit than for the corrector. (As an example, eqs. (1.9.7) and (1.9.9) are one-step and two-step algorithms, respectively.) Nevertheless we can formally restore the equality between the number of steps in corrector and predictor by appropriately translating the coefficient labels of the former. For instance, eq. (1.9.7) can be equally viewed as a two-step algorithm with $a_2 = 1$, $a_1 = -1$, $a_0 = 0$, $b_2 = 1/2$, $b_1 = 1/2$, and $b_0 = 0$; notice that the coefficients a_s in this translated form coincide with the corresponding coefficients of the associated predictor (1.9.9) written in orthodox form (1.6.1).

True, rigorously speaking, the translated form violates the very condition $|a_0| + |b_0| \neq 0$ imposed initially on eq. (1.6.1) but this fact is however immaterial for the arguments which follow in this section.

These points in mind, we can write formally both algorithms of the suitable pair as i-step algorithms, with identical coefficients $a_j, j = 0, 1, 2, \ldots, i$, viz.

Predictor

$$a_i \bar{y}^p_{k+i} + a_{i-1}\bar{y}_{k+i-1} + \ldots + a_0 \bar{y}_k = h(b^p_{i-1} f_{k+i-1} + \ldots + b^p_0 f_k). \tag{1.9.12}$$

Corrector

$$a_i \bar{y}_{k+i} + a_{i-1}\bar{y}_{k+i-1} + \ldots + a_0 \bar{y}_k = h(b_i f_{k+i} + b_{i-1} f_{k+i-1} + \\ + \ldots + b_0 f_k). \tag{1.9.13}$$

As announced, both are supposed to be of the same order m. Notice also that the term $b^p_i f_{k+i}$ is absent in the predictor, as it must actually be, since $b^p_i = 0$. It is also important to remark that the entrance values $\bar{y}_{k+i-1}, \bar{y}_{k+i-2}, \ldots, \bar{y}_k$ in both algorithms are the same and they are actually the ones resulting from the *corrector* equation at previous stages.

We also assume that $\bar{y}_{k+i}$ satisfies exactly eq. (1.9.13), that is it results from the predictor-corrector procedure which starts with $Z^0 = \bar{y}^p_{k+i}$ and iterates infinitely many times. (The round-off errors are thoroughly excluded.) As for the starting values $\bar{y}_s$, these are assumed to be highly accurate, viz.:

$$y(x_s) - \bar{y}_s = O(h^{m'}), \quad s = 1, 2, \ldots, i-1, \; m' > m \quad \text{and} \quad \bar{y}_0 = y(x_0).$$

Now we proceed to the derivation of formula (1.9.11). Since the corrector (1.9.13) is a method of order m, eq. (1.6.13), is valid, thus

$$a_i E_{k+i} + a_{i-1} E_{k+i-1} + \ldots + a_0 E_k = h(b_i \Delta f_{k+i} + \ldots + b_0 \Delta f_k) +$$

$$+ h^{m+1} C_{m+1} y^{(m+1)}(x_k) + O(h^{m+2}), \tag{1.9.14}$$

where

$$E_s = y(x_s) - \bar{y}_s, \; \Delta f_s = f(x_s, y(x_s)) - f(x_s, \bar{y}_s)), \; s = 0, 1, 2, \ldots, n. \tag{1.9.15}$$

On using the Taylor series against y_s for $f(x, y)$ we get

$$\Delta f_s = g_s E_s + O(E_s^2), \quad s = 0, 1, 2, \ldots, n, \tag{1.9.16}$$

where

$$g_s = \left. \frac{\partial f(x, y)}{\partial y} \right|_{x = x_s, \; y = y(x_s)} .$$

A necessary ingredient here is the knowledge of the fact that, if h is sufficiently small, then $E_s = h^m \bar{E}_s$ where $\bar{E}_s$ is the value at x_s of the solution of some differential equation. This result will be proved in the next section, and the quoted equation is the error equation (1.10.3).

On using this result eq. (1.9.16) becomes

$$\Delta f_s = g_s E_s + O(h^{2m}). \tag{1.9.17}$$

We can further argue that, since h is supposed to be sufficiently small, the interval $[x_k, x_{k+i}] = [x_k, x_k + ih]$ is also narrow enough to admit that the numbers $g_s \bar{E}_s$ are approximately equal for $s = k, k+1, \ldots$ $\ldots, k+i$. (Rigorously speaking we have $g_s \bar{E}_s = g_k \bar{E}_k + O(h)$ for any $s = k, k+1, \ldots, k+i$.) On using this point eq. (1.9.14) reads

$$a_i E_{k+i} + a_{i-1} E_{k+i-1} + \ldots + a_0 E_k = h^{m+1} g_k \bar{E}_k (b_i + b_{i-1} + \ldots$$

$$\ldots + b_0) + h^{m+1} C_{m+1} y^{(m+1)}(x_k) + O(h^{m+2}). \tag{1.9.18}$$

The same arguments can be repeated for the predictor equation (1.9.12) as well, and the result is

$$a_i E^p_{k+i} + a_{i+1} E_{k+i-1} + \dots + a_0 E_n = h^{m+1} g_k \bar{E}_k (b^p_{i-1} + b^p_{i-2} + \dots$$

$$\dots + b^p_0) + h^{m+1} C^p_{m+1} y^{(m+1)}(x_k) + 0(h^{m+2}) \qquad (1.9.19)$$

where $E^p_{k+i} = y(x_{k+i}) - \bar{y}^p_{k+1}$.
We also note the following identity

$$b_i + b_{i-1} + \dots + b_0 = b^p_{i-1} + b^p_{i-2} + \dots + b^p_0 \qquad (1.9.20)$$

which results at once from the fact that both predictor and corrector are convergent and then necessarily consistent. Thus, conditions $C_1 = 0$ and $C^p_1 = 0$ (see eq. (1.6.9)) are automatically satisfied, which implies that each side of eq. (1.9.20) equals the same sum $a_1 + 2a_2 + \dots + ia_i$.

Now subtract eq. (1.9.18) from eq. (1.9.19). By virtue of identity (1.9.20) the corresponding terms cancel each other and the result is

$$a_i(E^p_{k+1} - E_{k+i}) = h^{m+1}(C^p_{m+1} - C_{m+1})\, y^{(m+1)}(x_k) + 0(h^{m+2}) \quad (1.9.21)$$

On recognizing here that $E^p_{k+i} - E_{k+i} = \bar{y}_{k+i} - \bar{y}^p_{k+i}$ we obtain formula (1.9.11).

1.10. USER–ORIENTED INVESTIGATION OF THE ACCUMULATED ERROR FOR MULTISTEP METHODS

In sections 1.6 and 1.7 we were concerned with the main heuristic features of the multistep algorithms to solve the initial value problem for ordinary equations of the first order, viz.

$$y' = f(x, y), \quad x \in [a, b], \quad y(a) = y_0. \qquad (1.10.1)$$

As a matter of fact, although features as convergence, consistency, order of the method, local truncation error, stability etc., are of major importance to understand the field from the mathematical point of view, a typical user of such methods may find most of these issues as rather academic.

Specifically, a typical (hurried) user probably thinks as follows: "I have a certain problem of the form (1.10.1) and a certain computer to solve it. The mathematicians offer to me a large variety of possible numerical algorithms of the form (1.6.1) (the methods listed in sections 1.4 and 1.5 of this book, for instance). Moreover, the results of the mathematical investigations in sections 1.6 and 1.7 guarantee that each of these algorithms is convergent, which means that I am free to choose any of them. Then let me make a certain choice and use it with some step size h. How accurate will be the solution I get? Do I obtain it with t significant figures, as I wish?"

He may ask himself even more directly: "Which h must I use to get the solution with the desired accuracy?"

Although such questions seem so simple and direct, the answer is not easy at all. The main reason is that the answer depends very much on the solution of the problem and this is not yet known. Thus the answer can be formulated only after a sufficient number of numerical tests have been completed and their results compared.

We give below a possible schedule of such runs, the one which is probably the most often used in practice.

It is based on the following

THEOREM. *If* (i) *algorithm* (1.6.1) *to solve problem* (1.10.1) *is convergent and of order m,*

(ii) *it is used on a real computer, i.e. with round-off errors included,*

(iii) *the value of h is sufficiently small,*

(iv) *the starting values* $\tilde{y}_0, \tilde{y}_1, \ldots, \tilde{y}_{i-1}$ *differ from the exact values* $y(a), y(x_1), \ldots, y(x_{i-1})$ *by terms of order* $m' > m$ *plus round-off errors, viz.*

$$\delta_0 \equiv y(a = x_0) - \tilde{y}_0 = r_0$$

$$\delta_s(h) \equiv y(x_s) - \tilde{y}_s = 0(h^{m'}) + r_s, \quad s = 1, 2, \ldots, i-1$$

then, if $\tilde{y}_s$, $s = i, i+1, \ldots n$, *is the numerical value of the solution obtained from algorithm* (1.6.1) *on the real computer, the total error* $\Delta_s = y(x_s) - \tilde{y}_s$ *is given by the formula*

$$\Delta_s = E_s + r_s, \quad s = i, i+1, \ldots, n \tag{1.10.2}$$

where E_s *and* r_s *are the accumulated truncation error and accumulated round-off error, respectively.*

Further on, E_s *can be written as* $E_s = h^m \bar{E}_s$ *where the number* $\bar{E}_s$ *is the value at* $x_s = a + sh$ *of the solution of the initial value problem*

$$\bar{E}' = g(x, y(x))\, \bar{E} + \bar{C} y^{(m+1)}(x), \quad x \in [a, b], \quad \bar{E}(a) = 0, \tag{1.10.3}$$

called the error equation. The function $g(x, y(x)) = \left.\dfrac{\partial f(x, y)}{\partial y}\right|_{y=y(x)}$ *is related to the equation to be solved and* $\bar{C}$, *defined as*

$$\bar{C} = \frac{C_{m+1}}{b_i + b_{i-1} + \ldots + b_0}, \tag{1.10.4}$$

is related to the algorithm.

The accumulated round-off error r_s, $s = i, i+1, \ldots, n$, *is bounded as follows*

$$|\bar{r}_s| < h^{-1} \bar{R}_s \tag{1.10.5}$$

where $\bar{R}_s$ *depends on the word-length of the computer, on the specific form of the algorithm and of the equation to be solved.*

In short, this theorem shows that the total error can be written as a sum of two terms and further on it indicates eqs. (1.10.3) and (1.10.5) as practical ways to evaluate the size of these two terms.

Its most restrictive assumption is (iii) i.e. that the step size h must be sufficiently small. Then the user is justified to ask himself if this restriction is not too severe, in the sense that he may well reach the required accuracy at values of h which are so large that eq. (1.10.2) is of no use to confirm such evidence. Such a case may actually happen in practice but this takes place merely when the user requires only very poor accuracy for his results. If higher accuracy is required, then naturally smaller values of h are involved, so that the chances that the formula (1.10.2) is a practical tool to evaluate the error are accordingly increased. As a matter of fact, decades of experimental work indicate that in most typical cases, when accuracies of three, four or five significant figures are required, the suitable value of h falls well in the range of validity of this formula, thus giving rise to the considerable popularity for the schedule of runs based upon it.

The schedule theoretically consists of assigning some test value for h, viz. h_{test}, then check whether h_{test} is sufficiently small in the sense discussed above and next compute $\bar{E}_s$ and $\bar{R}_s$.

In practice at least one piece of this schedule is often disregarded via a suitable programming. This is the evaluation of the round-off bound $\bar{R}_s$ which is no longer necessary because the round-off errors can be always made as small as necessary by simply programming the algorithm in multiple precision arithmetic. To check which word-length is suitable for the problem under investigation we have to run the algorithm twice, in single precision and in double precision, respectively, and then compare the results. If the values of $\tilde{y}_s$, $s = 0, 1,2, \ldots, n$ corresponding to these two runs agree in the desired level of accuracy we can conclude that the single precision arithmetic is sufficient. If not, double or more extended precision is needed, a matter which can be settled via a similar test.

Once the suitable arithmetic was found out, all subsequent runs may be well performed with the corresponding version of the program and at the same time we can neglect the last term in eq. (1.10.2).

The second step is to check whether h_{test} is small enough to ensure the validity of the expression $E_s = h^m\bar{E}_s$ for the accumulated truncation error.

Two distinct ways are equally possible.

One of them is the explicit computation of eq. (1.10.3).

In a first run we compute eq. (1.10.3) with the chosen h_{test} and store the values of $\bar{E}_s^1$, $s = i, i + 1, \ldots n$. The upper label 1 stands for that we refer to this *first* run. This computation is suggested to be run in parallel with the one for the original equation (1.10.1), since the latter yields quantities which are of direct use in the right side of error equation, namely $\tilde{y}_s^1$ and $y_s^{1(m+1)}$, $s = i, i + 1, \ldots n$. (A possible procedure to evaluate $y_s^{1(m+1)}$ is that explained in Sec. 1.9).

In a *second* run we repeat the computation of eq. (1.10.3) but with $h = h_{\text{test}}/2$. Finally compare the values of the resultant $\bar{E}_{2s}^2$ with the previously stored $\bar{E}_s^1$, $s = i, i + 1, \ldots, n$. (Note that both refer to the same mesh points viz.: $x_s^1 = x_{2s}^2$.) There are two cases:

(i) $\bar{E}_s^1$ and $\bar{E}_{2s}^2$, $s = i, i + 1, \ldots n$, agree in a decent number of significant figures. This indicates that the chosen h_{test} is indeed sufficiently small to make eq. (1.10.2) valid.

Let t^a be the number of agreeing significant figures in $\bar{E}_s^1$ and $\bar{E}_{2s}^2$. It can then be said that the product $h_{\text{test}}^m\bar{E}_s^1$ represents the error (with t^a exact figures) of the computed $\tilde{y}_s^1$. This also enables us to obtain an extrapolated, more accurate value for the numerical solution. In fact $\tilde{y}_s^{1*} = \tilde{y}_s^1 + h_{\text{test}}^m \bar{E}_s^1$ is a more accurate approximation to $y(x_s^1)$ than the original $\tilde{y}_s^1$.

As an example, suppose that for some problem of the form (1.10.1) we use a second-order algorithm and run the program with $h_{\text{test}} = 1/4$

to get some $\tilde{y}_s^1 = 15.73745$. (We also assume that the round-off error was checked before being negligible.) As for $\bar{E}_s^1$ and $\bar{E}_{2s}^2$ we get $\bar{E}_s^1 = 0.3232$ and $\bar{E}_{2s}^2 = 0.3429$. We see that $\bar{E}_s^1$ and $\bar{E}_s^2$ agree in one significant figure. Then compute $h_{\text{test}}^2 \bar{E}_s^1$ and retain only the first significant figure, to yield 0.02. It can be concluded that only the first three figures in $\tilde{y}_s^1$ are exact, but in the extrapolated value $\tilde{y}_s^{1*} = \tilde{y}_s^1 + 0.02 = 15.75745$ four figures are exact. If we are not yet satisfied with this accuracy we can further use eq. (1.10.2) to predict the suitable h. Suppose for instance that we need that the absolute error at x_s^1 be not higher than $5 \cdot 10^{-4}$. Then the suitable h for the above example must satisfy the equation $5 \cdot 10^{-4} = 0.2\, h^2$ i.e. $h = 1/20$.
(ii) $\bar{E}_s^1$ and $\bar{E}_{2s}^2$ exhibit large differences. This means that h_{test} is too large and we must accordingly repeat the computation with smaller value of h_{test} until we fall into the previous case.

The second way avoids the explicit computation of the error equation (1.10.3). The procedure, known as Richardson's extrapolation to the limit, is as follows.

Take *three* successive partitions $h_1 = h_{\text{test}}$, $h_2 = h_1/2$, $h_3 = h_2/2$ and integrate eq.(1.10.1) with the chosen algorithm. Let $\tilde{y}_s^1$, $\tilde{y}_{2s}^2$, and $\tilde{y}_{4s}^3$ be the numerical values of the solution at $X = x_s^1 = x_{2s}^2 = x_{4s}^3$ corresponding to the three runs.

If h_{test} is sufficiently small (which implies that $\bar{E}_s^1 \simeq \bar{E}_{2s}^2 \simeq \bar{E}_{4s}^3$) we must have

$$\frac{\Delta_{2s}^2}{\Delta_s^1} \equiv \frac{y(X) - \tilde{y}_{2s}^2}{y(X) - \tilde{y}_s^1} \simeq \left(\frac{1}{2}\right)^m, \quad \frac{\Delta_{4s}^3}{\Delta_{2s}^2} \equiv \frac{y(X) - \tilde{y}_{4s}^3}{y(X) - \tilde{y}_{2s}^2} \simeq \left(\frac{1}{2}\right)^m. \tag{1.10.6}$$

To check for the error we take equal sign in eqs (1.10.6) to find out the extrapolated solutions

$$\tilde{y}^*(X) = \frac{\tilde{y}_{2s}^2 - \left(\frac{1}{2}\right)^m \tilde{y}_s^1}{1 - \left(\frac{1}{2}\right)^m}, \quad \tilde{\tilde{y}}^*(X) = \frac{\tilde{y}_{4s}^3 - \left(\frac{1}{2}\right)^m \tilde{y}_{2s}^2}{1 - \left(\frac{1}{2}\right)^m}. \tag{1.10.7}$$

The agreement of $\tilde{y}^*(X)$ and $\tilde{\tilde{y}}^*(X)$ gives a measure of the extent in which formula (1.10.2) is valid. In fact, if $\tilde{y}^*(X)$ and $\tilde{\tilde{y}}^*(X)$ agree in a decent number of figures we can further use $\tilde{\tilde{y}}^*(X)$ (which is expected to be a more accurate approximation of the exact solution $y(X)$ than $\tilde{y}^*(X)$) as the exact solution. Then, as a by-product, eq. (1.10.2) gives an approximation for $\bar{E}_s$, viz.:

$$\bar{E}_s^1 \simeq \frac{\tilde{\tilde{y}}^*(X) - \tilde{y}_s^1}{h_{\text{test}}^m}. \tag{1.10.8}$$

If necessary, this can be further used, as explained above, to estimate the value of h which is appropriate to reach the desired higher accuracy.

A strong disagreement between $\tilde{y}^*(X)$ and $\tilde{\tilde{y}}^*(X)$ indicates that the chosen h_{test} is too large. In this case the entire procedure must be repeated with a new, smaller value of h_{test}.

One final remark must be added here, and it concerns both ways discussed before. It is based upon the fact that when h is decreased the genuine accumulated error (i.e. first term in r.h.s. of eq. (1.10.2)) *decreases* as h^m while the accumulated round-off error r_s *increases* as h^{-1}, cf. eq. (1.10.5). Then the influence of the round-off errors which was checked to be negligible for the run with h_{test}, may become significant for the runs with $h \ll h_{\text{test}}$. However, a separate check for round-off errors at this h is typically unnecessary since eq. (1.10.5) allows to estimate that these errors will be h_{test}/h times larger than the ones observed in the original test. Naturally, if these estimated values are found larger than the desired accuracy, then more extended precision arithmetic must be used for the run with h.

The proof of the stated theorem proceeds in two steps. In the first step we will neglect the round-off error to derive the first term in r.h.s. of eq. (1.10.2) from the associated eq. (1.10.3). In the second step we will include the round-off errors to derive the bound (1.10.5).

For both stages the knowledge of the following *lemma* is necessary.

LEMMA. *Consider the following i-step recurrence relation*

$$a_i z_{k+i} + a_{i-1} z_{k+i-1} + \ldots + a_0 z_k = h(\beta_{i,k} z_{k+i} + \beta_{i-1,k} z_{k+i-1} +$$

$$+ \ldots + \beta_{0,k} z_k) + \lambda_k, \quad k = 0, 1, 2, \ldots, n - i \tag{1.10.9}$$

with given parameters a_s, $\beta_{s,k}$, $s = 0, 1, 2, \ldots i$, and λ_k. We assume that its characteristic polynomial, viz.

$$P(d) = a_i d^i + a_{i-1} d^{i-1} + \ldots + a_0 \tag{1.10.10}$$

admits only roots which are in modulus at most equal to 1 and that the roots of modulus 1 are simple.

The first assertion of the lemma is that if we develop $1/\left(d^i P\left(\frac{1}{d}\right)\right)$ *in powers of d, viz.*

$$\frac{1}{d^i P\left(\frac{1}{d}\right)} = \frac{1}{a_i + a_{i-1}d + \ldots + a_0 d^i} = \gamma_0 + \gamma_1 d + \gamma_2 d^2 + \ldots \tag{1.10.11}$$

then all coefficients $\gamma_0, \gamma_1, \gamma_2, \ldots$ *are finite.*

Let further denote

$$\Gamma = \sup_{j=0,\,1,\,2,\ldots} |\gamma_j| < \infty$$

and also admit that there exist three nonnegative constants B^*, β, *and* Λ *such that*

$$|\beta_{i,k}| + |\beta_{i-1,k}| + \ldots + |\beta_{0,k}| \leqslant B^*,\ |\beta_{i,k}| \leqslant \beta \text{ and } |\lambda_k| \leqslant \Lambda \tag{1.10.12}$$

for any $k = 0, 1, 2, \ldots, n - i$.

We also assume that the starting values in the recurrence relation (1.10.10) *are bounded, viz.*

$$|z_s| \leqslant Z,\ s = 0, 1, 2, \ldots, i - 1. \tag{1.10.13}$$

The second assertion of the lemma is that if $h < |a_i|\beta^{-1}$ *the resultant* z_s $(s \geqslant i)$ *are bounded as follows*

$$|z_s| \leqslant k^* \exp(shL^*),\ s = i, i + 1, \ldots, n, \tag{1.10.14}$$

where

$$\Gamma^* = \frac{\Gamma}{1 - h|a_i|^{-1}\beta},\quad L^* = \Gamma^* B^*,$$
$$A = |a_i| + |a_{i-1}| + \ldots + |a_0|,\ k^* = \Gamma^*(s\Lambda + iAZ). \tag{1.10.15}$$

We omit the proof which is given in detail in Chapter 5 of ref. [1].

First step: genuine accumulated truncation error

As stated, at this step we thoroughly disregard the round-off errors, that is we consider algorithm (1.6.1) with the starting conditions $\bar{y}_0, \bar{y}_1, \ldots, \bar{y}_{i-1}$ such as

$$\bar{y}_0 = y_0, \delta_s = y(x_s) - \bar{y}_s = y(a+hs) - \bar{y}_s = 0(h^{m'}), s = 1, 2, \ldots, i-1 \tag{1.10.16}$$

We make use of the formula (1.6.13) which connects the errors associated to the m-order algorithm (1.6.1), viz.

$$a_i E_{k+i} + a_{i-1} E_{k+i-1} + \ldots + a_0 E_k = h(b_i \Delta f_{k+i} + \ldots + b_0 \Delta f_k)$$

$$+ h^{m+1}(C_{m+1} y_k^{(m+1)} + 0(h)), \; k = 0, 1, 2, \ldots, n-i \tag{1.10.17}$$

where

$$E_s \equiv y(x_s) - \bar{y}_s, \Delta f_s \equiv f(x_s, y(x_s)) - f(x_s, \bar{y}_s), \; s = 0, 1, 2, \ldots, n. \tag{1.10.18}$$

(We see that $E_s = \delta_s$ for $s \leqslant i-1$.)

Our immediate purpose is to convert eq. (1.10.17) into a form which is convenient for the application of the previous lemma. To this aim we introduce the numbers v_s, $s = 0, 1, 2, \ldots, n$, defined as follows

$$v_s = \begin{cases} \dfrac{\Delta f_s}{E_s} & \text{if } E_s \neq 0, \\ 0 & \text{if } E_s = 0, \end{cases} \tag{1.10.19}$$

and also denote

$$D_k(h) = C_{m+1} y_k^{(m+1)} + 0(h). \tag{1.10.20}$$

Eq. (1.10.17) then becomes

$$a_i E_{k+i} + a_{i-1} E_{k+i-1} + \ldots + a_0 E_k = h(b_i v_{k+i} E_{k+i} + \ldots +$$

$$+ b_0 v_k E_k) + h^{m+1} D_k(h) \tag{1.10.21}$$

which is of the form (1.10.9) in which $\beta_{s,k} = b_s v_{k+s}$, $s = 0, 1, 2, \ldots, i$, and $\lambda_k = h^{m+1}D_k(h)$.

Next check whether the conditions of the lemma are satisfied.

First of all recall that algorithm (1.6.1) is assumed to be convergent. Then it is also stable so that the condition imposed on its characteristic polynomial $P(d)$ is automatically satisfied. (See eq. (1.7.26) and the discussion following it.) Secondly, we recall that the function $f(x, y)$ occurring in eq. (1.10.1) has to satisfy the Lipschitz condition, as discussed in "Preliminaries". This implies that all v_s are bounded by the Lipschitz constant L,

$$|v_s| \leqslant L, \quad s = 0,1, 2, \ldots, n. \tag{1.10.22}$$

Thirdly, the starting values are bounded

$$|\delta_s| < h^{m'}E_I(h) \text{ with } E_I(h) = \sup_{s=1,2,\ldots,i-1} \{|\delta_s|h^{-m'}\}. \tag{1.10.23}$$

Thus we are in a position to apply the lemma with the following identifications

$$Z = h^{m'}E_I(h), \; B^* = BL, \; \beta = |b_i|\, L, \; A = h^{m+1}D(h) \tag{1.10.24}$$

where

$$D(h) = \sup_{k=0,1,\ldots,n-i} |D_k(h)|, \; B = |b_i| + |b_{i-1}| + \ldots + |b_0|. \tag{1.10.25}$$

With these identifications eq. (1.10.14) reads

$$|E_s| \leqslant \Gamma^*(h^m(x_s - a)D(h) + Ah^{m'}E_I(h)i)\exp[(x_s - a)\Gamma^*BL] \tag{1.10.26}$$

where use was made of the fact that $sh = x_s - a$.

Most of the parameters entering eq. (1.10.26) are independent of h. In fact A, B, m and i depend only on the algorithm constants while L on the equation to be solved. The h dependent parameters are (except for h itself) Γ^*, $D(h)$ and $E_I(h)$. When h is decreased Γ^* tends to Γ while $D(h)$ and $E_I(h)$ tend to some nonvanishing constants denoted as $\bar{D}$ and $\bar{E}_I$, respectively. Therefore when h is so small that the deviation of the three h dependent parameters from their limits becomes numerically negligible, eq. (1.10.26) becomes

$$|E_s| \leqslant h^m B_A(x_s) + h^{m'}B_I(x_s), \; s = i, i+1, \ldots, n \tag{1.10.27}$$

where

$$B_A(x_s) = (x_s - a)\, \Gamma \bar{D} \exp[(x_s - a)\, \Gamma BL], \tag{1.10.28 a}$$

$$B_I(x_s) = \Gamma A\, \bar{E}_I i \exp[(x_s - a)\, \Gamma BL]. \tag{1.10.28 b}$$

The two terms in r.h.s. of eq. (1.10.27) can be interpreted as representing the influence of the inaccuracy of the very algorithm (notice the factor $\bar{D}$ in eq. (1.10.28 a)) and of the inaccuracy in starting values, respectively. As a matter of fact, the latter data are furnished by some extra numerical method and therefore we can always arrange things in such a way that the accuracy of the starting values is sufficiently high, specifically such that $m' > m$. If so, the last term can be safely neglected, i.e.

$$|E_s| \leqslant h^m B_A(x_s). \tag{1.10.29}$$

This equation can be rewritten in a form which is more transparent to investigate the convergence properties. In fact, following the procedure associated to the definition of convergence (see it in section 1.6) we take some arbitrary fixed X and various finer and finer equidistant partitions to cover $[a, X]$, viz.: $x_0 = a$, $x_1 = a + h, \ldots, x_n = a + nh = X$. Eq. (1.10.29) becomes

$$|E(X; h)| \leqslant h^m B_A(X) \tag{1.10.30}$$

and it not only that confirms that $\lim_{h \to 0} |E(X; h)| = 0$ as it must be for a convergent algorithm, but shows something more, namely that the decrease is at least as fast as h^m. The reader might ask: Why "at least as fast as h^m" and not simply "as fast as h^m"? The reason is that eq. (1.10.30) represents only a bound and a behaviour proportional to $h^{\bar{m}}$, $\bar{m} > m$, is still possible since it does not contradict the bound.

Because of this and also because B_A is also suspected to strongly overestimate the real size of the accumulated error we are tempted to search for finer bounds. Fortunately, it is possible to reach even more, namely just an exact representation for the error. Such an investigation, due to Henrici [1], proceeds as follows: Consider again eq. (1.10.17) and preserve the assumption that $m' > m$. Eqs (1.10.16) and (1.10.27) can be equivalently written as $E_s = 0(h^{m'})$ for $s < i$ and

$E_s = O(h^{m'})$ for $s \geqslant i$, respectively. Since $m' > m$ these two results can be collectively written as

$$E_s = O(h^m) \text{ for any } s = 0, 1, 2, \ldots, n. \tag{1.10.31}$$

Now we develop $f(x, y)$ in Taylor series against y

$$f(x, y + \Delta y) = f(x, y) + \Delta y \frac{\partial f(x, y)}{\partial y} + \frac{1}{2} \Delta y^2 \frac{\partial^2 f(x, y)}{\partial y^2}\bigg|_{y=y^*} \tag{1.10.32}$$

where $y^* \in (y, y + \Delta y)$. On identifying $y = \bar{y}_s$, $\Delta y = E_s$ and $g = \left.\frac{\partial f(x, y)}{\partial y}\right|_{y=\bar{y}_s}$, and using eq. (1.10.31) we get

$$\Delta f_s = g_s E_s + \frac{1}{2} E_s^2 \frac{\partial^2 f(x, y)}{\partial y^2}\bigg|_{y=y^*} = g_s E_s + O(h^{2m}), \; s = 0, 1, 2, \ldots, n. \tag{1.10.33}$$

This is further introduced in eq. (1.10.17). The result is

$$a_i E_{k+i} + a_{i-1} E_{k+i-1} + \ldots + a_0 E_k = h(b_i g_{k+i} E_{k+i} + \ldots + b_0 g_k E_k +$$
$$+ O(h^{2m})) + C_{m+1} h^{m+1} y_k^{(m+1)} + O(h^{m+2}) \quad k = 0, 1, 2, \ldots, n - i. \tag{1.10.34}$$

The term $O(h^{2m})$ in the parentheses will be disregarded since the order of $h \cdot O(h^{2m})$ is at least as high as the order of the retained term $O(h^{m+2})$. (Remember that $m \geqslant 1$.)

The next step of the procedure is to "dissolve" the term $C_{m+1} h^{m+1} y_k^{(m+1)}$ into suitable pieces contributing in the parenthesis multiplying h.

To this aim we use the Taylor series for $y^{(m+1)}(x)$

$$y^{(m+1)}(x + \Delta x) = y^{(m+1)}(x) + \Delta x \cdot y^{(m+2)}(x^*)$$

with $x^* \in (x, x + \Delta x)$, in which identify $x = x_k$ and $\Delta x = jh$, $j = 1, 2, \ldots, i$. We have

$$y^{(m+1)}(x_k + jh) \equiv y_{k+j}^{(m+1)} = y_k^{(m+1)} + jh y^{(m+2)}(x^*) = y_k^{(m+1)} + O(h) \tag{1.10.35}$$

We also use the obvious identity

$$C_{m+1}h^{m+1}y_k^{(m+1)} = \frac{C_{m+1}}{b_0 + b_1 + \ldots + b_i} h^{m+1}(b_0 y_k^{(m+1)} +$$
$$+ b_1 y_k^{(m+1)} + \ldots + b_i y_k^{(m+1)}) \tag{1.10.36}$$

in which we introduce the representation (1.10.35) to get the desired "dissolved" representation

$$C_{m+1}h^{m+1}y_k^{(m+1)} = b_0\bar{C}h^{m+1}y_k^{(m+1)} + b_1\bar{C}h^{m+1}y_{k+1}^{(m+1)} + \ldots +$$
$$+ b_i\bar{C}h^{m+1}y_{k+i}^{(m+1)} + 0(h^{m+2}) \tag{1.10.37}$$

with $\bar{C}$ defined in eq. (1.10.4). On introducing it in eq. (1.10.34) we obtain

$$a_iE_{k+i} + a_{i-1}E_{k+i-1} + \ldots + a_0E_k = h(b_iW_{k+i} + b_{i-1}W_{k+i-1} +$$
$$+ \ldots + b_0W_k) + 0(h^{m+2}) \quad k = 0, 1, 2, \ldots, n - i \tag{1.10.38 a}$$

where

$$W_s = g_sE_s + \bar{C}h^m y_s^{(m+1)}. \tag{1.10.38 b}$$

This is finally written in terms of the new numbers $\bar{E}_s = h^{-m}E_s$, $s = 0, 1, \ldots, n$

$$a_i\bar{E}_{k+i} + a_{i-1}\bar{E}_{k+i-1} + \ldots + a_0\bar{E}_k =$$
$$= h(b_i\bar{W}_{k+i} + b_{i-1}\bar{W}_{k+i-1} + \ldots + b_0\bar{W}_k) + 0(h^2) \tag{1.10.39 a}$$

where

$$\bar{W}_s = g_s\bar{E}_s + \bar{C}y_s^{(m+1)} \tag{1.10.39 b}$$

As for the starting values in eq. (1.10.39 a), these are $\bar{E}_0 = h^{-m}E_0 = 0$ and $\bar{E}_s = h^{-m}E_s = h^{-m}0(h^{m'}) = 0(h^{m'-m})$ for $s = 1, 2, \ldots, i - 1$. Notice also that, because we have assumed that $m' > m$, we have

$$\lim_{h\to 0} \bar{E}_s = 0 \; (= \bar{E}_0) \tag{1.10.40}$$

for any $s = 1, 2, \ldots, i - 1$.

Now think of h as being so small that the term $O(h^2)$ in the recurrence relation (1.10.39 a) is numerically so small that it can be neglected. If so, we recognize at once that eqs (1.10.39 a) and (1.10.39 b) with the starting values which satisfy eq. (1.10.40) represent just the i-step algorithm (1.6.1) for the solution of the initial value problem

$$\bar{E}' = g(x, y(x))\bar{E} + \bar{C}y^{(m+1)}(x), \ x \in [x_0 = a, x_n = b], \ \bar{E}(a) = 0. \tag{1.10.41}$$

This is nothing else than the error equation (1.10.3). Thus the proof is just completed for the fact that when the round-off errors are absent, then the (genuine) accumulated error is of the form

$$E_s = h^m \bar{E}_s, \ s = i, i + 1, \ldots, n, \tag{1.10.42}$$

where $\bar{E}_s$ is the value at x_s of the solution of the error equation (1.10.3).

Before going on to investigate the influence of the round-off errors we pause a bit to compare the results obtained so far from bound arguments to ones from the fine investigation. The first remark is that the fine investigation has shown that the genuine accumulated error decreases with h just as fast as h^m; *not* "at least as fast as h^m" as it resulted from the bound arguments. It is however important to realize that the bound investigations should have not been omitted whereas only due to them we were justified to write eq. (1.10.31) which, in turn, was a necessary ingredient in the finer investigation.

The second remark consists of illustrating how large actually is the bound weight $B_A(x_s)$, eq. (1.10.29 a) in comparison with the real asymptotic weight $\bar{E}(x_s)$.

Consider the problem

$$y' = \lambda y, \quad x \in [a = 0, b], \quad y(0) = y_0 = 1, \quad \lambda < 0$$

with the exact solution $y(x) = \exp(\lambda x)$. The same test problem was also considered in section 1.7. As there, we consider the two step Adams-Bashforth algorithm i.e.: $i = 2$, $a_2 = 1$, $a_1 = -1$, $a_0 = 0$, $b_2 = 0$, $b_1 = 3/2$, and $b_0 = -1/2$ and also assume that the starting values coincide with the exact solution at the corresponding mesh points.

To evaluate $B_A(x_s)$ we first calculate the constants Γ, $\bar{D}$, B and L. The characteristic polynomial of this algorithm is $P(d) = d(d - 1)$ so that $1/\left(d^2 P\left(\frac{1}{d}\right)\right) = (1 - d)^{-1}$.

On using the identity

$$(1 - d)^{-1} = 1 + d + d^2 + d^3 + \dots$$

it results that $\Gamma = 1$.

The quoted algorithm is of second order, $m = 2$ and, as shown in eq. (1.7.4) we have $C_{m+1} = \frac{5}{12}$. Thus, eq. (1.10.20) gives

$$D_k(h) = \frac{5}{12} \frac{d^3[\exp(\lambda x)]}{dx^3}\bigg|_{x=x_k} + O(h) = \frac{5}{12} \lambda^3 \exp(\lambda x_k) + O(h).$$

Since λ is negative we have

$$D(h) = \sup_{k=0,1,2,\dots n-i} |D_k(h)| = |D_0(h)| = \frac{5}{12} |\lambda|^3 + O(h),$$

whose asymptotic value is $\bar{D} = \frac{5}{12} |\lambda|^3$.

As for B we have in order $B = |b_2| + |b_1| + |b_0| = 2$, while the Lipschitz constant is $L = |\lambda|$.

On using the above values we find

$$B_A(x_s) = \frac{5}{12} |\lambda|^3 x_s \exp(2|\lambda| x_s).$$

To evaluate $\bar{E}(x_s)$ we first construct the error equation. Since $\bar{C} = C_{m+1}/(b_2 + b_1 + b_0) = C_{m+1} = \frac{5}{12}$, $y'''(x) = \lambda^3 \exp(\lambda x)$ and $g = \frac{\partial(\lambda y)}{\partial y} = \lambda$, eq. (1.10.3) reads

$$\bar{E}' = \lambda \bar{E} + \frac{5}{12} \lambda^3 \exp(\lambda x), \quad x \in [0, b], \ \bar{E}(0) = 0,$$

with the solution

$$\bar{E}(x) = \frac{5}{12}\lambda^3 x \exp(\lambda x).$$

The ratio

$$|B_A(x_s)/\bar{E}(x_s)| = \exp(3|\lambda|x_s)$$

confirms that for large x_s the discrepancy becomes very large indeed.

Second step: accumulated round-off errors

Consider an arbitrary fixed value of k and assume that all computations with algorithm (1.6.1) up to this k were affected by round-off errors. Then the entrance values into the algorithm are some numbers $\tilde{y}_k$, $\tilde{y}_{k+1}, \ldots, \tilde{y}_{k+i-1}$ which differ from $\bar{y}_k, \bar{y}_{k+1}, \ldots, \bar{y}_{k+i-1}$ which should have resulted from an absolutely accurate computation. We thus can write

$$\tilde{y}_s = \bar{y}_s - r_s \quad s = k, k+1, \ldots, k+i-1, \tag{1.10.43}$$

where r_s represents the accumulated round-off error.

The objective of the kth iteration with algorithm (1.6.1) is to compute the value of the solution at x_{k+i} and to do it the algorithm has to perform a set of operations such as to compute the values of $f(x_s, \tilde{y}_s)$, to multiply them with the weights b_j, to add the partial results, next to multiply the sum by h, and so on. Of course, on a real computer each of these operations is affected by round-off errors. The final result of this step will be some $\tilde{y}_{k+i}$ which is displaced from $\bar{y}_{k+i}$ by the accumulated round off error r_{k+i} which has two sources: one is the inaccuracy in the entrance data and the other is the round-off error affected computation of the formula (1.6.1) at the step k.

Let us now reintroduce this $\tilde{y}_{k+i}$ in algorithm (1.6.1) and compute separately the right side and the left side of it,

$$LHS = a_i\tilde{y}_{k+i} + a_{i-1}\tilde{y}_{k+i-1} + \ldots + a_0\tilde{y}_k,$$

$$RHS = h(b_i\tilde{f}_{k+i} + b_{i-1}\tilde{f}_{k+i-1} + \ldots + b_0\tilde{f}_k), \tag{1.10.44}$$

$$\tilde{f}_s = f(x_s, \tilde{y}_s), \; s = k, k+1, \ldots, k+i,$$

on *an ideal computer*, i.e. with no round-off error. The values of *LHS* and *RHS* will be different, of course. Then, to restore the equality we have to add to one of them (to *LHS*, say) a compensating quantity R_k, i.e.

$$LHS + R_k = RHS.$$

That is to say that the value of $\tilde{y}_{k+i}$ resulting by computing eq. (1.6.1) on a *real* computer is identical with the one that results from an *ideal* computer but with a *modified recurrence relation*, viz.

$$a_i\tilde{y}_{k+i} + \ldots + a_0\tilde{y}_k + R_k = h(b_i\tilde{f}_{k+i} + b_{i-1}\tilde{f}_{k+i-1} + \ldots + b_0\tilde{f}_k), \tag{1.10.45}$$

and such an assertion is true for any k. Thus to investigate the accumulated round-off error we simply subtract eq. (1.10.45) from eq. (1.6.1) and allow k to run from zero to $n - i$

$$a_ir_{k+i} + a_{i-1}r_{k+i-1} + \ldots + a_0r_k = h(b_i\Delta\tilde{f}_{k+i} + \ldots + b_0\Delta\tilde{f}_k) + R_k,$$
$$k = 0, 1, \ldots, n - i \tag{1.10.46}$$

with

$$r_s = \bar{y}_s - \tilde{y}_s, \Delta\tilde{f}_s = f(x_s, \bar{y}_s) - f(x_s, \tilde{y}_s), \; s = 0, 1, 2, \ldots, n. \tag{1.10.47}$$

We next convert eq. (1.10.46) into a form that is suited for the previous lemma. To this aim we denote

$$v_s = \begin{cases} r_s^{-1}\Delta\tilde{f}_s & \text{if} \quad r_s \neq 0 \\ 0 & \text{if} \quad r_s = 0. \end{cases}$$

Eq. (1.10.46) then reads

$$a_ir_{k+i} + a_{i-1}r_{k+i-1} + \ldots + a_0r_k = h(b_iv_{k+i}r_{k+i} + \ldots$$
$$\ldots + b_0v_kr_k) + R_k \tag{1.10.48}$$

The conditions of the lemma are satisfied. Indeed, the arguments developed in connection with eq (1.10.21) can be repeated word by word to conclude that: Γ exists and is finite, $|v_s| \leqslant L$ for any $s = 0, 1, 2, \ldots, n$, $B^* = BL$, and $\beta = |b_i|L$. As for Z and Λ, if we denote

$$r_I = \sup \{|r_0|, |r_1|, \ldots, |r_{i-1}|\}, \tag{1.10.49 a}$$

and

$$R_A = \sup \{|R_0|, |R_1|, \ldots, |R_{n-i}|\}, \tag{1.10.49 b}$$

then $Z = r_I$ and $\Lambda = R_A$.

Eq. (1.10.14) then gives

$$|r_s| \leqslant \Gamma^*[(x_s - a)\, h^{-1} R_A + A r_I i] \exp[(x_s - a)\, LB\Gamma^*] \tag{1.10.50}$$

$$s = i, i + 1, \ldots, n.$$

The bracket in the member on the right of eq. (1.10.50) consists of two terms. The first contains R_A and thus it represents the contribution of the round-off errors associated with the very algorithm. The presence of the factor r_I in the second term suggests that this term accounts for the starting value errors. What is important to note here is that when s increases but x_s is kept fixed ($x_s = X$, say) (in such a procedure h is accordingly decreased since $hs = X - a$) then the first component *increases* as h^{-1} while the second remains bounded. Thus when h assumes sufficiently small values only the first component is numerically relevant and, at the same time, approximation $\Gamma^* \simeq \Gamma$ makes sense. Under such conditions we may write

$$|r_s| \leqslant h^{-1} \bar{R}_s, \quad s = i, i + 1, \ldots, n \tag{1.10.51}$$

with

$$\bar{R}_s = (x_s - a) \Gamma R_A \exp[(x_s - a) LB\Gamma]. \tag{1.10.52}$$

Eq. (1.10.51) represents a bound for the accumulated round-off error and, as any bound, it may be well conservative from a numerical point of view. A finer description of the round-off errors is also possible in principle and the procedure will be similar to the one for the genuine accumulated truncation error. We have to notice however that such an investigation will be more difficult since it will need also some extra assumptions concerning the size and distribution of the local round-off errors R_k, $k = 0, 1, \ldots, n - i$. However, such an investigation does not seem to be particularly relevant for a typical user since in practice, as discussed, the influence of the round-off error can be kept under control by using suitable precision arithmetic. For this reason we omit it and send the interested reader to ref. [1].

The final step of the proof consists of bringing together the results of the two previous steps. We have in order

$$\Delta_s = y(x_s) - \tilde{y}_s = (y(x_s) - \bar{y}_s) + (\bar{y}_s - \tilde{y}_s) = E_s + r_s,$$

and, as we have seen, if h is sufficiently small, then E_s is determined by eqs. (1.10.42) and (1.10.41) while r_s is bounded as in eq. (1.10.51), Q.E.D.

1.11. MULTISTEP METHODS FOR SECOND–ORDER DIFFERENTIAL EQUATIONS OF SPECIAL FORM

As discussed in section 1.8, the typical procedure to solve numerically the initial value problem for second-order differential equations of general form $y'' = f(x, y, y')$ is to write such an equation as a system of two equations of the first-order to which we apply any of the algorithms of form (1.6.1).

The case of second-order differential equations is of large practical interest in applied science and in engineering and it often happens there that the differential equation to be solved is of special form, namely the function f depends on x and y, but *not* of y'.

This circumstance led the mathematicians to devise specific algorithms to solve second-order differential equations of this special form, with the hope that such specific algorithms will be simpler and more efficient than the typical procedure used for the general form.

We actually consider the initial value problem

$$y'' = f(x, y), \; x \in [a, b] \tag{1.11.1}$$

with the initial conditions

$$y(a) = y_0, \; y'(a) = y_0' \tag{1.11.2}$$

and devise multistep algorithms to solve it.

We follow the same strategy as under (i) in section 1.1. Thus we find out an equivalent of integral form for eq. (1.11.1), and next approximate the function f occurring in the integrand by an interpolation polynomial. We start with the evaluation of the integral

$$\mathrm{I} = \int_X^{X+\Delta X} (X + \Delta X - t) f(t, y(t))\, dt \tag{1.11.3}$$

for any X and $X + \Delta X$ inside $[a, b]$.
We write I as a sum of two integrals and integrate by parts for the second, viz.

$$I = (X + \Delta X) \int_X^{X+\Delta X} y''(t)\, dt - \int_X^{X+\Delta X} t y''(t) dt = (X + \Delta X) [y'(X+\Delta X) -$$

$$- y'(X)] - \left[t y'(t) \Big|_X^{X+\Delta X} - \int_X^{X+\Delta X} y'(t)\, dt \right]$$

$$= (X + \Delta X) [y'(X + \Delta X) - y'(X)] - [(X + \Delta X) y'(X + \Delta X) -$$

$$- X y'(X) - y(X + \Delta X) + y(X)] = y(X + \Delta X) - y(X) - \Delta X y'(X). \tag{1.11.4}$$

We thus have obtained that

$$y(X + \Delta X) - y(X) = \Delta X y'(X) + \int_X^{X+\Delta X} (X + \Delta X - t) f(t, y(t))\, dt. \tag{1.11.5 a}$$

On replacing ΔX by $- \Delta X$ this becomes

$$y(X - \Delta X) - y(X) = - \Delta X y'(X) + \int_X^{X-\Delta X} (X - \Delta X - t) f(t, y(t))\, dt, \tag{1.11.5 b}$$

so that on adding eqs (1.11.5 a) and (1.11.5 b) we get a formula in which the first derivative of y disappears,

$$y(X + \Delta X) - 2y(X) + y(X-\Delta X) = \int_X^{X+\Delta X} (X + \Delta X - t) f(t, y(t))\, dt$$

$$+ \int_X^{X-\Delta X} (X-\Delta X - t) f(t, y(t))\, dt. \tag{1.11.6}$$

This is the integral form of eq. (1.11.1), as was eq. (1.1.1) for eq. $y' = f(x, y)$.

Eq. (1.11.6) is used as a basis to obtain the numerical algorithms. As announced, the procedure consists of replacing the function f which occurs in the two integrals by an interpolating polynomial. To put it into practice we take equidistant mesh points on $[a, b]$, i.e., $x_0 = a$, $x_1, x_2, \ldots, x_i = x_0 + ih, \ldots, x_n = b$, and consider a current mesh point x_k to use the interpolation polynomial of the function f relative to the $r + 1$ knots $x_k, x_{k-1}, \ldots, x_{k-r}$. We also take the integration limits as some mesh points, namely $X = x_{k-q}$ and $X+\Delta X = x_{k+p}$ (q and p nonnegative integers). Now evaluate the integrals by using the formula (1.2.12) of the interpolating polynomial. For the first integral we have in order

$$I_1 = \int_{x_{k-q}}^{x_{k+p}} (x_{k+p} - x) f(x, y(x))\, dx =$$

$$= \sum_{j=0}^{r} (-1)^j \nabla^j f_k \int_{x_{k-q}}^{x_{k+p}} (x_{k+p} - x) \binom{-s}{j} dx$$

$$= h^2 \sum_{j=0}^{r} (-1)^j \nabla^j f_k \int_{-q}^{p} (p - s) \binom{-s}{j} ds. \tag{1.11.7}$$

(To get the last member of this chain we changed the variable of integration from x to $s = (x - x_k)/h$.)

For the second integral we have in order

$$I_2 = \int_{x_{k-q}}^{x_{k-2q-p}} (x_{k-2q-p} - x) f(x, y(x))\, dx =$$

$$= \sum_{j=0}^{r} (-1)^j \nabla^j f_k \int_{x_{k-q}}^{x_{k-2q-p}} (x_{k-2q-p} - x) \binom{-s}{j} dx =$$

$$= h^2 \sum_{j=0}^{r} (-1)^j \nabla^j f_k \int_{-q}^{-2q-p} (-2q - p - x) \binom{-s}{j} ds$$

$$= h^2 \sum_{j=0}^{r} (-1)^j \nabla^j f_k \int_{-q}^{p} (p - s') \binom{s' + 2q}{j} ds'. \tag{1.11.8}$$

(The last member results from the previous one by changing the integration variable s to the new $s' = -2q - s$.)

On introducing these results into eq. (1.11.6) we obtain the general algorithm searched for. It is

$$\bar{y}_{k+p} - 2\bar{y}_{k-q} + \bar{y}_{k-2q-p} = h^2 \sum_{j=0}^{r} \alpha_j \nabla^j f_k \tag{1.11.9}$$

with

$$\alpha_j = (-1)^j \int_{-q}^{p} (p - s) \left[\binom{-s}{j} + \binom{s + 2q}{j} \right] ds. \tag{1.11.10}$$

This can be equally expressed in terms of the value of f at the knots, viz.

$$\bar{y}_{k+p} - 2\bar{y}_{k-q} + \bar{y}_{k-2q-p} = h^2 \sum_{j=0}^{r} \beta_j f_{k-j} \tag{1.11.11}$$

where β_j are connected with $\alpha_0, \alpha_1, \ldots, \alpha_r$ via eq. (1.3.5).

On assigning various particular values for the nonnegative integers p and q we obtain various particular formulas. We actually assign just the same particular values as in sections 1.4 and 1.5 to distinguish open (or explicit) and closed (or implicit) formulas.

Open formulas

We first assign $p = 1$ and $q = 0$, as for the Adams-Bashforth algorithm. The resultant algorithm,

$$\bar{y}_{k+1} - 2\bar{y}_k + \bar{y}_{k-1} = h^2 \sum_{j=0}^{r} \alpha_j \nabla^j f_k \tag{1.11.12}$$

is known as *Störmer* (or *Adams-Störmer) method* [8]. The weights α_j up to $j = 6$ are listed in Table 7.

For $r = 0$ and $r = 1$ this formula reduces to the simple rule

$$\bar{y}_{k+1} - 2\bar{y}_k + \bar{y}_{k-1} = h^2 f_k. \tag{1.11.13}$$

Next assign $p = 1$ and $q = 1$, as for the Nyström method. The algorithm reads

$$\bar{y}_{k+1} - 2\bar{y}_{k-1} + \bar{y}_{k-3} = h^2 \sum_{j=0}^{r} \alpha_j \nabla^j f_k, \tag{1.11.14}$$

and its weights are also listed in table 7. This algorithm, proposed by Henrici [1, p. 293], is recommended only for $r \geqslant 2$.

Closed formulas

We now take $p = 0$ and $q = 1$ to obtain an analog for the Adams-Moulton method. Eq. (1.11.9) becomes

$$\bar{y}_k - 2\bar{y}_{k-1} + \bar{y}_{k-2} = h^2 \sum_{j=0}^{r} \alpha_j \nabla^j f_k; \tag{1.11.15}$$

(see table 7 for the weights α_j, $j \leqslant 6$). Algorithm (1.11.15) is known as the *method of Cowell* [9]. It is not used for $r = 0$, while for $r = 1$

it reduces to eq. (1.11.13). The most popular version is that for $r=2$ or $r = 3$, viz.

$$\bar{y}_k - 2\bar{y}_{k-1} + \bar{y}_{k-2} = \frac{1}{12} h^2(f_k + 10f_{k-1} + f_{k-2}). \tag{1.11.16}$$

Finally, when $p = 0$ and $q = 2$, an analog of the Milne-Simpson method results, namely

$$\bar{y}_k - 2\bar{y}_{k-2} + \bar{y}_{k-4} = h^2 \sum_{j=0}^{r} \alpha_j \nabla^j f_k, \tag{1.11.17}$$

see table 7 for its weights. This algorithm is proposed by *Henrici* [1, p. 294] who recommends only the versions with $r \geqslant 3$.

Table 7

Coefficients α_j, $j = 0, 1, \ldots, 6$ for the Adams-Störmer, Henrici (eq. 1.11.14)), Cowell, and Henrici (eq. (1.11.17)) methods

j	0	1	2	3	4	5	6
Adams-Störmer	1	0	$\frac{1}{12}$	$\frac{1}{12}$	$\frac{19}{240}$	$\frac{3}{40}$	$\frac{863}{12096}$
Henrici, eq. (1.11.14)	4	-4	$\frac{4}{3}$	0	$\frac{1}{15}$	$\frac{1}{15}$	$\frac{61}{945}$
Cowell	1	-1	$\frac{1}{12}$	0	$-\frac{1}{240}$	$-\frac{1}{240}$	$-\frac{221}{60480}$
Henrici, eq. (1.11.17)	4	-8	$\frac{16}{3}$	$-\frac{4}{3}$	$\frac{1}{15}$	0	$-\frac{2}{945}$

The practical application of these methods is very similar with the one associated to the multistep algorithms for the first-order equations. For instance the starting values must be calculated by some extra procedure, and again the method of truncated Taylor series is a serious candidate to do it. As a matter of fact, this is the only problem for open versions, while an additional problem occurs in the case of closed methods. In fact, the latter are not self-acting and thus they must be assisted by some open method to furnish them an initial guess for the value of the solution at each new mesh point. This is completely analogous with what happened with closed methods for first-order equations and the procedure to be used is the same predictor-corrector procedure, as discussed in section 1.9. For instance, a good predictor for the Cowell method is its open counterpart, that is the Adams-Störmer method.

The general theory of multistep methods for special second-order equations is very similar with the one of the multistep methods to solve first-order equations. In the following we review it briefly to insist upon differences.

The multistep algorithms to solve eqs. (1.11.1)—(1.11.2) are collectively written in the form:

$$a_i\bar{y}_{k+i} + a_{i-1}\bar{y}_{k+i-1} + \ldots + a_0\bar{y}_k = h^2(b_if_{k+i} + b_{i-1}f_{k+i-1} + \\ + \ldots + b_0f_k) \qquad k = 0, 1, 2,\ldots, n - i. \tag{1.11.18}$$

where a_s, b_s, $s = 0, 1, 2,\ldots, i$ are specified numbers, $|a_0| + |b_0| \neq 0$ and $a_i \neq 0$. Notice that the only but very important difference as compared to algorithm (1.6.1) for first-order equations is that now a factor h^2 occurs instead of h there. The definition of the convergence of algorithm (1.11.18) is a straightforward extension of the definition for algorithm (1.6.1) given in section 1.6. (The difference originates in the fact that the initial value problem for eq. (1.11.1) consists of *two* initial conditions, eq. (1.11.2) not only *one*, as it was in the case of first-order equation.)

Thus, to define the convergence of algorithm (1.11.18) we choose some fixed, arbitrary $X \in (a, b]$ and consider equidistant partitions with n elementary subintervals on $[a, X]$. (This construction implies that h and n are connected through $nh = X - a$.) Finally make n arbitrarily large, or, conversely, h arbitrarily small, and suppose that the starting values $y_0, \bar{y}_1, \ldots, \bar{y}_{i-1}$ are such that

$$\lim_{h\to 0} \bar{y}_s = y_0, \quad s = 1, 2,\ldots, i - 1 \tag{1.11.19 a}$$

and

$$\lim_{h\to 0} \frac{\bar{y}_s - y_0}{sh} = y_0', \quad s = 1, 2, \ldots, i - 1. \tag{1.11.19 b}$$

Under such conditions, and if the solution $y(x)$ of the problem (1.11.1)—(1.11.2) exists, algorithm (1.11.18) is called convergent if $\lim_{h\to 0} \bar{y}_n = y(X)$.

The theorem according to which a multistep algorithm is convergent if and only if it is both consistent and stable remains valid for algorithm (1.11.18). However, the definitions of consistency and stability must be slightly modified.

Consistency

We associate to eq. (1.11.18) the following functional

$$L(z(x);h) = (a_i z(x+ih) + a_{i-1}z(x+(i-1)h) + \ldots + a_0 z(x)) - h^2(b_i z''(x+ih) + b_{i-1}z''(x+(i-1)h) + \ldots + b_0 z''(x)). \quad (1.11.20)$$

This is a direct generalization of eq. (1.6.7) with h replaced by h^2 and the first derivatives by second derivatives.

On following the same procedure as for eq. (1.6.7) i.e. use Taylor series expansion for z and appropriately organize the right side of eq. (1.11.20) we get

$$L(z(x);h) = C_0 z(x) + C_1 z'(x)h + C_2 z''(x)h^2 + \ldots + C_p z^p(x)h^p + \ldots \quad (1.11.21)$$

where

$$C_0 = a_0 + a_1 + \ldots + a_i$$

$$C_1 = a_0 + 2a_1 + \ldots + ia_i \quad (1.11.22)$$

$$\vdots$$

$$C_p = \frac{1}{p!}(a_1 + 2^p a_2 + \ldots + i^p a_i) - \frac{1}{(p-2)!}(0^{p-2}b_0 + + 1^{p-2}b_1 + \ldots + i^{p-2}b_i), \quad p = 2,3,\ldots$$

Algorithm (1.11.18) is called of order m if the first nonvanishing weight in expansion (1.11.21) is C_{m+2}. Note again a difference: for first-order equations the algorithm (1.6.1) was of order m when the first nonvanishing coefficient was C_{m+1}; notice that the expressions of the coefficients also differ. (Compare with eq. (1.6.9).)

The method (1.11.18) is called *consistent* when it is of the order 1, at least. Thus the necessary conditions of consistence are $C_0 = C_1 = C_2 = 0$. (Recall that in the previous case only the first *two* coefficients had to be zero.)

As for the local truncation error, it is again defined as the difference

$$e_{k+i} = y(x_{k+i}) - \bar{y}_{k+i} \tag{1.11.23}$$

in which $\bar{y}_{k+i}$ is the value resulted from eq. (1.11.18) provided that all the other quantities occurring in that equation coincide with the exact values. For small h, e_{k+i} can be expressed with sufficient numerical accuracy in terms of the first nonvanishing term in eq. (1.11.21). Specifically, if m is the order of the method then

$$e_{k+i} \simeq \frac{C_{m+2}}{a_i} y_k^{(m+2)} h^{m+2} \tag{1.11.24}$$

The quantity $C = C_{m+2}/a_i$ is called the error constant. In table 8 we list the number of steps, orders and error constants for the above methods in terms of r.

Table 8

Number of steps, order and error constant in terms of r, for several special methods to solve second-order equations of special form

Method	number of steps i	order m	error constant C
Adams-Störmer $r = 0$, $r = 1$	2	2	$\frac{1}{12}$
$r > 1$	$r + 1$	$r+1$	α_{r+1}
Henrici, eq. (1.11.14) $r = 2$	4	4	$\frac{1}{15}$
$r > 2$	$r + 1$	$r+1$	α_{r+1}
Cowell $r = 2$	2	4	$-\frac{1}{240}$
$r > 2$	r	$r+1$	α_{r+1}
Henrici, eq. (1.11.17) $r = 4$	4	6	$-\frac{2}{945}$
$r > 4$	r	$r+1$	α_{r+1}

It can be seen again that the closed versions are more accurate than their open counterparts. (Compare for instance Adams-Störmer (AS) and Cowell (C) methods. For one and the same i, AS is of order i while C is of order $i + 1$, for $i = 2$ even higher, namely of the fourth order. For one and the same r, they are of the same order $r + 1$; yet C should be preferred since $|\alpha_{r+1}^{C}| < |\alpha_{r+1}^{AS}|$.)

Stability

The stability properties of the i-step algorithm (1.11.18) are governed by the same characteristic polynomial

$$P(d) = a_i d^i + a_{i-1} d^{i-1} + \ldots + a_0, \tag{1.11.25}$$

as it was for algorithm (1.6.1).

However, the condition of stability is slightly different: Algorithm (1.11.18) is stable provided the modulus of no root of $P(d)$ exceeds 1 and that the multiplicity of the roots of modulus 1 *is at most 2*. (The difference actually consists of the last part of the statement; remember that for algorithm (1.6.1) the roots of modulus 1 had to be simple.)

The reader can easily check that all the versions included in table 8 are consistent and stable, so that they are convergent.

Accumulated error

The main problem for a typical user of algorithm (1.11.18) is similar to the one for the user of algorithm (1.6.1), and it is to find a measure for the accuracy of the results which he actually obtains from the computer.

To acquire it a suitable schedule of test runs is necessary, and this is similar with the schedule discussed in detail in section 1.10 in connection with first-order equations.

At the basis of this schedule is a theorem which is very similar to the one proved in section 1.10.

We give it below without proof.

THEOREM. *If:* (i) *algorithm* (1.11.18) *to solve eqs* (1.11.1)—(1.11.2), *is convergent and of order m*,

(ii) *computations are made with round-off errors*,

(iii) *h is sufficiently small,*
(iv) *starting values* $\tilde{y}_0, \tilde{y}_1, \ldots, \tilde{y}_{i-1}$ *are displaced from the exact values as follows:*

$$\delta_0 = y(a) - \tilde{y}_0 = r_0,$$

$$\delta_s(h) = y(x_s) - \tilde{y}_s = 0(h^{m'}) + r_s, \quad s = 1, 2, \ldots, i-1,$$

with $m' > m + 1$,
then, if $\tilde{y}_s$ *is the numerical value at* x_s, $s \geqslant i$, *of the solution resulted from algorithm* (1.11.18) *on the real computer, the error* $\Delta_s = y(x) - \tilde{y}_s$ *is given by formula*

$$\Delta_s = E_s + r_s, \quad s = i, i+1, \ldots, n \tag{1.11.26 a}$$

where $E_s = h^m \bar{E}_s$ *and* $\bar{E}_s$ *is the value at* $x_s = a + sh$ *of the solution of the initial value problem*

$$\bar{E}'' = g(x, y(x))\, \bar{E} + \bar{C}^{(m+2)}(x), \; x \in [a, b], \; \bar{E}(a) = 0, \; \bar{E}'(a) = 0 \tag{1.11.26 b}$$

called the error equation. The function g is defined as $g(x, y) = \dfrac{\partial f(x, y)}{\partial y}$ *and the constant* $\bar{C}$ *is given by*

$$\bar{C} = \frac{C_{m+2}}{b_0 + b_1 + \ldots + b_i} \tag{1.11.27}$$

The accumulated round-off error r_s *is bounded as follows*

$$|r_s| < h^{-2} \bar{R}_s \tag{1.11.28}$$

where $\bar{R}_s$ *depends on word-length of the computer, specified form of algorithm and equation to be solved.*

Notice that the basic differences are: $m' > m + 1$ (before we had $m' > m$); the error equation is a second-order equation; $\bar{C}$ contains C_{m+2} instead of C_{m+1} before; bound (1.11.28) for $|r_s|$ increases as h^{-2}, not as h^{-1} before.

The fact that the accumulated round-off error for algorithm (1.11.18) increases as h^{-2} when h is decreased is rather discouraging. A possibility however exists to restore the growth of the accumulated round-off errors at the decent law h^{-1} (as it was in the case of the algorithm (1.6.1)). This is based on the fact that the unpleasant law h^{-2} is entirely due to the presence of the factor h^2 in front of the second member of algorithm (1.11.18). It is actually suggested to try expressing eq. (1.11.18) as a suitable set of two coupled recurrent relations whose right member is *linear* in h, as it was for the recurrent relation (1.6.1).

Such a procedure, first investigated by Henrici [1, p. 327], leads to the so-called *summed form* of algorithm (1.11.18).

To derive the summed form we write down the original algorithm eq. (1.11.18) for $k = 0, 1, 2, \ldots, N$ and call S_N the sum of the resulting equations. For reason of clarity, we work out the two members separately. Thus we first consider the sum on the left

$$S_N = a_i(\bar{y}_i + \bar{y}_{i+1} + \ldots + \bar{y}_{N+i}) + a_{i-1}(\bar{y}_{i-1} + \bar{y}_i + \ldots + \bar{y}_{N+i-1}) + \ldots + a_0(\bar{y}_0 + \bar{y}_1 + \ldots + \bar{y}_i)$$

and rearrange it to get

$$\begin{aligned} S_N = {} & a_0\bar{y}_0 + (a_1 + a_0)\,\bar{y}_1 + \ldots + (a_{i-1} + a_{i-2} + \ldots + a_0)\,\bar{y}_{i-1} \\ & + (a_i + a_{i-1} + \ldots + a_0)(\bar{y}_i + \bar{y}_{i+1} + \ldots + \bar{y}_N) \\ & + (a_i + a_{i-1} + \ldots + a_1)\bar{y}_{N+1} + (a_i + a_{i-1} + \ldots + a_2)\bar{y}_{N+2} + \\ & + \ldots + a_i\bar{y}_{N+i}. \end{aligned} \tag{1.11.29}$$

Now use the fact that algorithm (1.11.18) is assumed consistent, hence

$$a_0 + a_1 + \ldots + a_i = 0$$

which is equation $C_0 = 0$ with C_0 defined in eq. (1.11.22). This allows us to define the following numbers

$$\begin{aligned} a'_{i-1} & \equiv a_i = - a_{i-1} - a_{i-2} - \ldots - a_0 \\ a'_{i-2} & \equiv a_i + a_{i-1} = - a_{i-2} - a_{i-3} - \ldots - a_0 \\ & \vdots \\ a'_0 & \equiv a_i + a_{i-1} + \ldots + a_1 = - a_0 \end{aligned} \tag{1.11.30}$$

With this notation eq. (1.11.29) becomes

$$S_N = a'_{i-1}\bar{y}_{N+i} + a'_{i-2}\bar{y}_{N+i-1} + \dots + a'_0\bar{y}_{N+1}$$
$$- a'_{i-1}\bar{y}_{i-1} - a'_{i-2}\bar{y}_{i-2} - \dots - a'_0\bar{y}_0. \tag{1.11.31}$$

Note the interesting structure of this equation: the first line of its right member connects the values of the solution at $x_{N+i}, x_{N+i-1}, \dots, x_{N+1}$ while the second line connects the initial values $\bar{y}_{i-1}, \bar{y}_{i-2} \dots, \bar{y}_0$.

Now sum the terms on the right of eq. (1.11.18) to get

$$S_N = h^2\{b_i(f_i + f_{i+1} + \dots + f_{N+i}) + b_{i-1}(f_{i-1} + f_{i-2} + \dots$$
$$\dots + f_{N+i-1}) + \dots + b_0(f_0 + f_1 + \dots + f_N)\} \tag{1.11.32}$$

in which we want to organize the terms occurring on the right in a similar manner, that is to formally separate the data at initial points $x_{i-1}, x_{i-2}, \dots, x_0$.

To this aim we consider some (for the moment arbitrary) numbers $\alpha \neq 0$ and H, define F_m as

$$F_m \equiv H + \frac{h}{\alpha}\sum_{j=0}^{m} f_j, \quad m = 0, 1, 2, \dots, N + i, \tag{1.11.33}$$

and express eq. (1.11.32) in terms of these F_m. We have in order

$$S_N = \alpha h[b_i(F_{N+i} - F_{i-1}) + b_{i-1}(F_{N+i-1} - F_{i-2}) + \dots + b_0(F_N - H)]$$
$$= \alpha h(b_iF_{N+i} + b_{i-1}F_{N+i-1} + \dots + b_0F_N) - \tag{1.11.34}$$
$$- \alpha h(b_iF_{i-1} + b_{i-1}F_{i-2} + \dots + b_0H)$$

and we see that the last member of this chain is formally of the very desired structure.

The next step of the derivation is to impose a meaningful condition aimed to fix the values of α and H. In fact, as our very purpose here is to obtain a recurrent relation involving the current mesh points, the natural condition is that the blocks of terms relating initial values in eqs (1.11.31) and (1.11.34) are equal, viz.

$$a'_{i-1}\bar{y}_{i-1} + a'_{i-2}\bar{y}_{i-2} + \ldots + a'_0\bar{y}_0 = \alpha h(b_i F_{i-1} + b_{i-1}F_{i-2} + \ldots$$

$$\ldots + b_0 H). \tag{1.11.35}$$

On reexpressing $F_{i-1}, F_{i-2}, \ldots$ in terms of the original $f_{i-1}, f_{i-2}, \ldots,$ the right member of eq. (1.11.35) becomes

$$\alpha h(b_i F_{i-1} + b_{i-1}F_{i-2} + \ldots + b_0 H) = h^2(b'_{i-1}f_{i-1} + \ldots + b'_0 f_0) +$$

$$+ \alpha h(b_0 + b_1 + \ldots + b_i)\, H \tag{1.11.36}$$

where

$$\begin{aligned} b'_{i-1} &= b_i \\ b'_{i-2} &= b_i + b_{i-1} \\ &\vdots \\ b'_0 &= b_i + b_{i-1} + \ldots + b_1, \end{aligned} \tag{1.11.37}$$

so that eq. (1.11.35) can be solved for the product αH to yield

$$\alpha H = \frac{1}{h(b_0 + b_1 + \ldots + b_i)} \{a'_{i-1}\bar{y}_{i-1} + a'_{i-2}\bar{y}_{i-2} + \ldots$$

$$\ldots + a'_0\bar{y}_0 - h^2(b'_{i-1}f_{i-1} + \ldots + b'_0 f_0)\}, \tag{1.11.38}$$

an equation which may be computed without difficulty since the expression on the right is a known function of the starting values. Note also for completeness that $b_0 + b_1 + \ldots + b_i \neq 0$ for any convergent algorithm of form (1.11.18).

Since eq. (1.11.38) gives only the product αH we choose to regard α as a scale factor which may be adjusted at will. Once $\alpha \neq 0$ fixed the value of H results uniquely.

Finally (i) equalize eqs (1.11.31) and (1.11.34) in which we take into account the existence of condition (1.11.35), and (ii) write eq. (1.11.33) in a convenient recurrent form. These two stages lead in order to the following set of two coupled recurrent relations

$$a'_{i-1}\bar{y}_{N+i} + a'_{i-2}\bar{y}_{N+i-1} + \ldots + a'_0\bar{y}_{N+1} =$$
$$= \alpha h(b_i F_{N+i} + b_{i-1}F_{N+i-1} + \ldots + b_0 F_N) \qquad (1.11.39\text{ a})$$

$$F_{-1} = H,\ F_{N+i} - F_{N+i-1} = \frac{h}{\alpha} f_{N+i},\ N = 0, 1, 2, \ldots, n - i \qquad (1.11.39\text{ b})$$

which defines the *summed form* of algorithm (1.11.18).

Note that the expression on the right in each of eqs (1.11.39 a, b) is linear in h, while it is quadratic in the original algorithm (1.11.18). For this reason, although the two algorithms are mathematically equivalent, the *summed form* is the method of choice on a real computer since, as expected, it is less sensitive to the propagation of the round-off errors *.

The practical application is as follows: we calculate $f_0, f_1, \ldots, f_{i-1}$ from the starting values, fix some value for α and determine H from eq. (1.11.38). Then $F_0, F_1, \ldots, F_{i-1}$ are calculated from eq. (1.11.39 b). The preliminary stage completed, we begin the very evaluation through formulas (1.11.39 a, b). First apply eq. (1.11.39 a) at $N = 0$ to compute $\bar{y}_i$. (If algorithm (1.11.39 a) is open (that is $b_i = 0$) the computation of $\bar{y}_i$ is direct, while if it is closed $(b_i \neq 0)$ a predictor-corrector procedure is needed. In the latter case the predictor-corrector procedure involves both recurrent relations. Specifically, we take some guess value y_i^* compute $f_i^* \equiv f(x_i, y_i^*)$ and also $F_i^* \equiv F_{i-1} + \dfrac{h}{\alpha} f_i^*$. This is further used as entrance data for F_i in eq. (1.11.39 a) to obtain some new, better approximation y_i^{**}. The process is repeated as many times as necessary to secure the convergence.) This is further used to compute f_i and then to compute F_i from eq. (1.11.39 b) at the same $N = 0$. The resultant F_i is next introduced in eq. (1.11.39 a) at $N = 1$ to compute $\bar{y}_{i+1}$ and so on.

As an illustration we derive below the summed form of the Cowell method with $r = 2$, eq. (1.11.16), which we conveniently write here again as

$$\bar{y}_{k+2} - \frac{1}{12} h^2 f_{k+2} = 2\bar{y}_{k+1} + \frac{5}{6} h^2 f_{k+1} - \left(\bar{y}_k - \frac{1}{12} h^2 f_k\right) \qquad (1.11.40)$$

* A rigorous proof of this fact is made in [1].

Upon denoting

$$z = y - \frac{1}{12} h^2 f(x, y) \tag{1.11.41}$$

eq. (1.11.40) becomes

$$\bar{z}_{k+2} - 2\bar{z}_{k+1} + \bar{z}_k = h^2 f_{k+1}. \tag{1.11.42}$$

Now replace the argument y in function f by the new defined z, to obtain a function g which will also depend parametrically on h. For instance, if $f(x, y) = b(x)y$ then $y = z/\left(1 - \frac{1}{12} h^2 b(x)\right)$ so that $f(x, y) = b(x)z/\left(1 - \frac{1}{12} h^2 b(x)\right) \equiv g(x, z; h)$. With this notation eq. (1.11.42) reads

$$\bar{z}_{k+2} = 2\bar{z}_{k+1} - \bar{z}_k + h^2 g_{k+1}, \; g_{k+1} \equiv g(x_{k+1}, \bar{z}_{k+1}; h). \tag{1.11.43}$$

[Note in passing that this formula can be viewed as algorithm (1.11.13) to solve the following second-order differential equation which depends parametrically on h

$$z'' = g(x, z; h), \quad x \in [a, b],$$

with suitable initial conditions $z(a) = y_0 - \frac{1}{12} h^2 f(a, y_0)$, $z'(a) = y_0' - \frac{1}{12} h^2 f'(x, y)_{x=a,\, y=y_0}$.]

Now we write the summed form of algorithm (1.11.43) and to this aim we first calculate H. Eq. (1.11.38) gives

$$H = \frac{1}{\alpha h} (\bar{z}_1 - \bar{z}_0 - h^2 g_0) \tag{1.11.44}$$

so that $F_0 = H + \frac{h}{\alpha} g_0 = \frac{1}{\alpha h} (\bar{z}_1 - \bar{z}_0)$. Then, the two equations (1.11.39 a) and (1.11.39 b) are in order

$$\bar{z}_{N+1} - \bar{z}_N = \alpha h F_N \tag{1.11.45 a}$$

$$F_{N+1} - F_N = \frac{h}{\alpha} g_{N+1}, \; N = 0, 1, \ldots, n-1 \tag{1.11.45 b}$$

with given initial values $\bar{z}_0$ and F_0.

Finally denote $S_j = \alpha h F_j$, to get the popular summed form

$$\bar{z}_{N+1} = \bar{z}_N + S_N \tag{1.11.46 a}$$

$$S_{N+1} = S_N + h^2 g_{N+1},\ N = 0, 1, \ldots, n-1 \tag{1.11.46 b}$$

with given initial values $\bar{z}_0$ and $S_0(= \bar{z}_1 - \bar{z}_0)$.

It is interesting to note here that the summed form is open while the original algorithm, eq. (1.11.40) was of closed type. This fact is however of no practical help because the very link between the solutions y and z of the two algorithms is given by eq. (1.11.41) which is obviously implicit. Thus, after computing $\bar{z}_2, \bar{z}_3, \bar{z}_4, \ldots$ via the summed form, the values of $\bar{y}_2, \bar{y}_3, \bar{y}_4, \ldots$, which we actually desire, result by solving the *implicit* equation (1.11.41). This can be solved numerically as shown in section 1.9 and a reasonable initial guess is simply $y = z$.

1.12. THE DE VOGELAERE METHOD

This is a numerical method for the solution of the initial value problem for second-order equations of the same special form as in section 1.11, that is in which the first derivative is absent. Thus the problem to be solved by de Vogelaere's method is the same (1.11.1) with the initial condition (1.11.2) which we rewrite here again

$$y'' = f(x, y),\ x \in [a, b],\ y(a) = y_0,\ y'(a) = y'_0. \tag{1.12.1}$$

The method differs from all the numerical methods described in section 1.11 in that, although it is also a multistep method, its algorithm is not of the form (1.11.18). Due to that, it exhibits certain advantages as, for instance, a specific simple procedure to start the integration and also flexibility in changing the step size during the integration, if necessary. For these reasons de Vogelaere's method [10] enjoys large popularity, particularly among physicists who are involved in the solution of the Schrödinger equation, see for instance [11, 12, 13].

Since the method does not belong to the class of standard multistep algorithms, its mathematical theory is naturally different from the theory associated with the general multistep algorithm (1.11.18). Specifically, we will avoid using concepts as consistency or stability, since these were defined namely to cope with algorithms of the form (1.11.18). Instead, to characterize the convergence properties of this method we will derive and use formulas for local and accumulated error.

To proceed with, we first present a derivation of de Vogelaere's algorithm. For the moment consider *equidistant* points $x_0 = a, x_1, x_2, \ldots$ $\ldots, x_{2n} = b$ with the step size h. As a matter of fact, it is important to note from the very beginning that for this method the odd-labelled points and even-labelled points play distinct roles, and that the last point for the computation must be even-labelled.

At each iteration k ($k=0,1,2,\ldots,n-1$) the de Vogelaere algorithm consists of *three* formulas which compute in order $\bar{y}_{2k+1}$, $\bar{y}_{2k+2}$, and $\bar{y}'_{2k+2}$ in terms of given $\bar{y}_{2k-1}$, $\bar{y}_{2k}$, and $\bar{y}'_{2k}$. (It should be remarked that the derivatives at the odd-labelled mesh points are never involved in this algorithm. Also, as it will be seen later on, the values of the resultant solution at even-labelled points are more accurate than the ones at odd-labelled points.)

To derive the algorithm we will frequently use the Taylor series expansion of $y(x)$ and of $f(x) \equiv f(x, y(x))$ around x_{2k}, viz.

$$y(x_{2k} + \delta) = y(x_{2k}) + \delta y'(x_{2k}) + \frac{\delta^2}{2!} y''(x_{2k}) + \ldots + \frac{\delta^p}{p!} y^{(p)}(x_{2k}) + \ldots \tag{1.12.2}$$

and

$$f(x_{2k} + \delta) = f(x_{2k}) + \delta f'(x_{2k}) + \frac{\delta^2}{2!} f''(x_{2k}) + \ldots + \frac{\delta^p}{p!} f^{(p)}(x_{2k}) + \ldots \tag{1.12.3}$$

We will also have in mind that y and f are connected by the very equation, viz. $y''(x) = f(x)$, $y'''(x) = f'(x)$, $y^{(4)}(x) = f''(x)$, and so on.

Specifically, all these elements are used below to convert three expressions involving the value of f at points x_{2k-1}, x_{2k}, x_{2k+1}, and x_{2k+2}, to forms involving the solution itself.

The first expression is

$$I_1 = \frac{h^2}{6}(4f(x_{2k}) - f(x_{2k-1})) \tag{1.12.4}$$

which we want to convert into a suitable form, with accuracy of order h^5. On using eq. (1.12.3) with $\delta = -h$, we have in order

$$\frac{h^2}{6}(4f(x_{2k}) - f(x_{2k-1})) =$$

$$= \frac{h^2}{6}\left[4f(x_{2k}) - \left(f(x_{2k}) - hf'(x_{2k}) + \frac{h^2}{2}f''(x_{2k}) + O(h^3)\right)\right]$$

$$= \frac{h^2}{6}\left[3f(x_{2k}) + hf'(x_{2k}) - \frac{h^2}{2}f''(x_{2k})\right] + O(h^5)$$

$$= \frac{h^2}{2!}f(x_{2k}) + \frac{h^3}{3!}f'(x_{2k}) + \frac{h^4}{4!}f''(x_{2k}) - \left(\frac{1}{12} + \frac{1}{4!}\right)h^4f''(x_{2k}) + O(h^5)$$

$$= \frac{h^2}{2!}y''(x_{2k}) + \frac{h^3}{3!}y'''(x_{2k}) + \frac{h^4}{4!}y^{(4)}(x_{2k}) - \frac{1}{8}h^4f''(x_{2k}) + O(h^5)$$

$$= y(x_{2k+1}) - y(x_{2k}) - hy'(x_{2k}) - \frac{1}{8}h^4f''(x_{2k}) + O(h^5). \tag{1.12.5}$$

The first and the last members of this chain thus give

$$y(x_{2k+1}) = y(x_{2k}) + hy'(x_{2k}) + \frac{h^2}{6}(4f(x_{2k}) - f(x_{2k-1})) +$$

$$+ \frac{1}{8}h^4f''(x_{2k}) + O(h^5). \tag{1.12.6}$$

The second expression is

$$I_2 = \frac{h^2}{3}\,(4f(x_{2k+1}) + 2f(x_{2k})) \tag{1.12.7}$$

and we work it out by a similar procedure (but with $\delta = h$ in eq. (1.12.3)) in which we retain terms up to order h^6. We have in order

$$\frac{h^2}{3}\,(4f(x_{2k+1}) + 2f(x_{2k})) =$$

$$= \frac{h^2}{3}\left[4\left(f(x_{2k}) + hf'(x_{2k}) + \frac{h^2}{2}f''(x_{2k}) + \frac{h^3}{6}f'''(x_{2k}) + O(h^4)\right) + 2f(x_{2k})\right]$$

$$= \frac{h^2}{3}\left[6f(x_{2k}) + 4hf'(x_{2k}) + 2h^2f''(x_{2k}) + \frac{2h^3}{3}f'''(x_{2k})\right] + O(h^6)$$

$$= \frac{(2h)^2}{2!}f(x_{2k}) + \frac{(2h)^3}{3!}f'(x_{2k}) + \frac{(2h)^4}{4!}f''(x_{2k}) + \frac{(2h)^5}{5!}f'''(x_{2k})$$

$$+ \left(\frac{2}{9} - \frac{2^5}{5!}\right) h^5 f'''(x_{2k}) + O(h^6) = y(x_{2k+2}) - y(x_{2k}) - 2hy'(x_{2k}) -$$

$$- \frac{2}{45}\,h^5 f'''(x_{2k}) + O(h^6), \tag{1.12.8}$$

that is

$$y(x_{2k+2}) = y(x_{2k}) + 2hy'(x_{2k}) + \frac{h^2}{3}\,(4f(x_{2k+1}) + 2f(x_{2k})) +$$

$$+ \frac{2}{45}\,h^5 f'''(x_{2k}) + O(h^6). \tag{1.12.9}$$

The third expression is

$$I_3 = \frac{h}{3}\,(f(x_{2k}) + 4f(x_{2k+1}) + f(x_{2k+2})) \tag{1.12.10}$$

and we evaluate it up to terms of $O(h^6)$,

$$\frac{h}{3}(f(x_{2k}) + 4f(x_{2k+1}) + f(x_{2k+2})) = \frac{h}{3}\Big\{f(x_{2k}) + 4\Big[f(x_{2k}) +$$

$$+ hf'(x_{2k}) + \frac{h^2}{2}f''(x_{2k}) + \frac{h^3}{6}f'''(x_{2k}) + \frac{h^4}{24}f^{(4)}(x_{2k}) + O(h^5)\Big]$$

$$+ \Big[f(x_{2k}) + (2h)f'(x_{2k}) + \frac{(2h)^2}{2}f''(x_{2k}) + \frac{(2h)^3}{6}f'''(x_{2k})$$

$$+ \frac{(2h)^4}{24}f^{(4)}(x_{2k}) + O(h^5)\Big]\Big\} = (2h)f(x_{2k}) + \frac{(2h)^2}{2!}f'(x_{2k})$$

$$+ \frac{(2h)^3}{3!}f''(x_{2k}) + \frac{(2h)^4}{4!}f'''(x_{2k}) + \frac{(2h)^5}{5!}f^{(4)}(x_{2k})$$

$$+ \left(\frac{5}{18} - \frac{2^5}{5!}\right)h^5 f^{(4)}(x_{2k}) + O(h^6) = (2h)\,y''(x_{2k}) + \frac{(2h)^2}{2!}y'''(x_{2k})$$

$$+ \frac{(2h)^3}{3!}y^{(4)}(x_{2k}) + \frac{(2h)^4}{4!}y^{(5)}(x_{2k}) + \frac{(2h)^5}{5!}y^{(6)}(x_{2k}) + \frac{1}{90}h^5 f^{(4)}(x_{2k})$$

$$+ O(h^6) = y'(x_{2k+2}) - y'(x_{2k}) + \frac{1}{90}h^5 f(x_{2k}) + O(h^6), \tag{1.12.11}$$

so that

$$y'(x_{2k+2}) = y'(x_{2k}) + \frac{h}{3}(f(x_{2k}) + 4f(x_{2k+1}) + f(x_{2k+2})) -$$

$$- \frac{1}{90}h^5 f^{(4)}(x_{2k}) + O(h^6). \tag{1.12.12}$$

The de Vogelaere algorithm is the set of three formulas (1.12.6), (1.12.9) and (1.12.12) in which we neglect the terms of order h^4, h^5, and h^5, respectively. Specifically, the algorithm is as follows

Given $\bar{y}_{2k-1}$, $\bar{y}_{2k}$ and $\bar{y}'_{2k}$, first compute $f_{2k-1} = f(x_{2k-1}, \bar{y}_{2k-1})$ and $_{2k} = f(x_{2k}, \bar{y}_{2k})$ and calculate $\bar{y}_{2k+1}$ by the formula

$$\bar{y}_{2k+1} = \bar{y}_{2k} + h\bar{y}'_{2k} + \frac{h^2}{6}(4f_{2k} - f_{2k-1}). \tag{1.12.13 a}$$

On using it, we next compute $f_{2k+1} = f(x_{2k+1}, \bar{y}_{2k+1})$ to obtain $\bar{y}_{2k+2}$ by the formula

$$\bar{y}_{2k+2} = \bar{y}_{2k} + 2h\bar{y}'_{2k} + \frac{h^2}{3}(4f_{2k+1} + 2f_{2k}). \tag{1.12.13 b}$$

Finally compute $f_{2k+2} = f(x_{2k+2}, \bar{y}_{2k+2})$ and calculate $\bar{y}'_{2k+2}$ by means of the formula

$$\bar{y}'_{2k+2} = \bar{y}'_{2k} + \frac{h}{3}(f_{2k} + 4f_{2k+1} + f_{2k+2}). \tag{1.12.13 c}$$

In the following we undertake the error analysis for this algorithm. Our analysis will cover the local truncation, accumulated and round-off errors. Some of these issues are also investigated in detail in [14].

Local truncation errors

We preserve the same definition for the local truncation error at the k step as before, namely as being the deviation of the output values from the exact ones when all the input values are supposed to be exact and the computation is supposed free of round-off errors. In our case the local truncation error consists of a set of three deviations, viz.

$$e_{2k+1} = y(x_{2k+1}) - \bar{y}_{2k+1}, \tag{1.12.14 a}$$

$$e_{2k+2} = y(x_{2k+2}) - \bar{y}_{2k+2}, \tag{1.12.14 b}$$

$$e'_{2k+2} = y'(x_{2k+2}) - \bar{y}'_{2k+2}, \tag{1.12.14 c}$$

calculated under the assumption that $\bar{y}_{2k-1} = y(x_{2k-1})$, $\bar{y}_{2k} = y(x_{2k})$ and $\bar{y}'_{2k} = y'(x_{2k})$. To evaluate e_{2k+1} we use the pair of eqs. (1.12.6) and (1.12.13 a) and remark that, since the input values are assumed to be exact, the quantities f_{2k} and f_{2k-1} which occur in eq. (1.12.13 a) just coincide with $f(x_{2k})$ and $f(x_{2k-1})$ in eq. (1.12.6). Then, on subtracting eq. (1.12.13 a) from eq. (1.12.6) all terms in r.h.s. cancel each other except the last two terms in eq. (1.12.6), viz.

$$e_{2k+1} \equiv y(x_{2k+1}) - \bar{y}_{2k+1} = \frac{1}{8} h^4 f''(x_{2k}) + O(h^5). \tag{1.12.15 a}$$

In the following, particularly to evaluate e_{2k+2} and e'_{2k+2}, we will repeatedly use the familiar formula

$$f(x, y_2) - f(x, y_1) = \left.\frac{\partial f(x, y)}{\partial y}\right|_{y=y_1} (y_2 - y_1) + O((y_2 - y_1)^2) \tag{1.12.16}$$

which is nothing else but the truncated Taylor expansion of $f(x, y)$ around y_1.

Now evaluate e_{2k+2} and to this aim we examine formula (1.12.13 b) to remark that all quantities in its r.h.s. coincide with the exact ones except for f_{2k+1} which differs from $f(x_{2k+1})$. The deviation is measured by eq. (1.12.16) in which $x = x_{2k+1}$, $y_2 = y(x_{2k+1})$ and $y_1 = \bar{y}_{2k+1}$

$$f(x_{2k+1}) - f_{2k+1} = \left.\frac{\partial f(x_{2k+1}, y)}{\partial y}\right|_{y=\bar{y}_{2k+1}} \cdot e_{2k+1} + O(e_{2k+1}^2),$$

and since $e_{2k+1} \sim O(h^4)$ (see eq. (1.12.15 a)) we have $f(x_{2k+1}) = f_{2k+1} + O(h^4)$. Then, on subtracting eq. (1.12.13 b) from eq. (1.12.9), we have in order

$$e_{2k+2} \equiv y(x_{2k+2}) - \bar{y}_{2k+2} = \frac{4}{3} h^2 (f(x_{2k+1}) - f_{2k+1}) +$$

$$+ \frac{2}{45} h^5 f'''(x_{2k}) + O(h^6) = \frac{2}{45} h^5 f'''(x_{2k}) + O(h^6). \tag{1.12.15 b}$$

To evaluate e'_{2k+2} we appropriately use eq. (1.12.16) to find out that $f(x_{2k+2}) = f_{2k+2} + O(h^5)$. If we further subtract eq. (1.12.13 c) from eq. (1.12.12) we get

$$e'_{2k+2} \equiv y'(x_{2k+2}) - \bar{y}'_{2k+2} = \frac{4h}{3}(f(x_{2k+1}) - f_{2k+1}) +$$

$$+ \frac{h}{3}(f(x_{2k+2}) - f_{2k+2}) - \frac{1}{90} h^5 f'''(x_k) + O(h^6) =$$

$$= \frac{h^5}{6} \left.\frac{\partial f(x_{2k+1}, y)}{\partial y}\right|_{y=\bar{y}_{2k+1}} \cdot f''(x_{2k}) - \frac{1}{90} h^5 f^{(4)}(x_{2k}) + O(h^6). \tag{1.12.15 c}$$

To summarize, we have found out that the local truncation errors in $\bar{y}_{2k+2}, \bar{y}'_{2k+2}$, and $\bar{y}_{2k+1}$ are of order h^5, h^5, and h^4, respectively.

It is important to remark now that the above formulas of the local truncation errors remain the same (in the sense that the leading term of each of them remains unchanged) also when the input value $\bar{y}_{2k-1}$ slightly differs from $y(x_{2k-1})$, more exactly when $y(x_{2k-1}) = \bar{y}_{2k-1} + O(h^q)$, $q \geqslant 3$. To see this, let us denote $\bar{y}^*_{2k+1}$ as being the value obtained from eq. (1.12.13 a) when $\bar{y}_{2k}$ and $\bar{y}'_{2k}$ are exact but $\bar{y}_{2k-1}$ is displaced as just said. On subtracting eq. (1.12.13 a) from eq. (1.12.6) we get

$$y(x_{2k+1}) - \bar{y}^*_{2k+1} = -\frac{h^2}{6}(f(x_{2k-1}) - f_{2k-1}) + \frac{1}{8} h^4 f'''(x_{2k}) + O(h^5) \tag{1.12.17}$$

and since, as resulting from eq. (1.12.16), $f(x_{2k-1}) - f_{2k-1} = O(h^q)$, $q \geqslant 3$, eq. (1.12.17) becomes

$$e^*_{2k+1} \equiv y(x_{2k+1}) - \bar{y}^*_{2k+1} = \frac{1}{8} h^4 f'''(x_{2k}) + O(h^5),$$

i.e. e^*_{2k+1} has the same leading term as e_{2k+1}.

Similar examination shows that the leading term is also identical in pairs (e^*_{2k+2}, e_{2k+2}) and $(e^{*\prime}_{2k+2}, e'_{2k+2})$.

This remark has two practical consequences. The first consequence is that it allows a simple way to start the integration. Indeed, the first iteration with the de Vogelaere algorithm needs $\bar{y}_0, \bar{y}'_0$ which are just the known initial conditions, and $\bar{y}_{-1}$ which is the numerical representation of $y(x_0 - h)$. The above remark is equivalent to saying that the accuracy of the algorithm is not damaged at all if $\bar{y}_{-1}$ is taken as the Taylor series representation of $y(x_0 - h)$ around x_0, with only *three* terms retained,

$$\bar{y}_{-1} = \bar{y}_0 - h\bar{y}_0' + \frac{h^2}{2}\bar{y}_0'' = \bar{y}_0 - h\bar{y}_0' + \frac{h^2}{2}f(x_0, \bar{y}_0). \tag{1.12.18}$$

The second consequence is that the de Vogelaere algorithm is very flexible with respect to changing step sizes.

Suppose, for instance, that in the $(k - 1)$th iteration we used it at a step size h_{k-1}. So, at the moment, we know $\bar{y}_{2k}$, $\bar{y}_{2k}'$, and $\bar{y}_{2k-1}$ which are approximations to $y(x_{2k})$, $y'(x_{2k})$ and $y(x_{2k}-h_{k-1})$, respectively. At the kth iteration, we want to use a new step size $h_k \neq h_{k-1}$. To go on, the algorithm requires the knowledge of $\bar{y}_{2k}$, $\bar{y}_{2k}'$ and of $\bar{y}_{2k-1}$ but note that the latter means now a representation of $y(x_{2k} - h_k)$ and therefore the value of $\bar{y}_{2k-1}$ resulted from the previous iteration is useless. Then the question is to find a way to compute the new value of $\bar{y}_{2k-1}$.

To answer it we notice that, at each k, the de Vogelaere algorithm can be viewed as an approximate procedure to solve the given equation $y'' = f(x, y)$ on the local interval $[x_{2k}, x_{2k+2}]$ with known initial conditions $\bar{y}_{2k}$ and $\bar{y}_{2k}'$. Expressed in other words, it can be viewed as being *the first* iteration to solve the same differential equation for $x \geqslant x_{2k}$, with the initial conditions $\bar{y}_{2k}$ and $\bar{y}_{2k}'$. We are thus in the conditions of the previous consequence so that, as explained before, a sufficiently accurate representation of $\bar{y}_{2k-1}$ is

$$\bar{y}_{2k-1} = \bar{y}_{2k} - h_k\bar{y}_{2k}' + \frac{h_k^2}{2}f(x_{2k}, \bar{y}_{2k}), \tag{1.12.19}$$

a formula which can be computed so easily, indeed.

Accumulated errors

As seen before, the order of local error of $\bar{y}_{2k}$ and $\bar{y}_{2k}'$ is higher than that of $\bar{y}_{2k+1}$ and due to this fact the user will be the best served if he cares to assign *even labels* to the mesh points which are of primary importance for him. For the same reason we examine here only the accumulated error at even-labelled mesh points. Our procedure is based on the following general

LEMMA. *Consider the real interval* $[a, b]$ *and a certain partition of it*, $x_0 = a, x_1, x_2, \ldots, x_n = b$, *which is in general not equidistant. Let* $g(x)$ *be some function on* $[a, b]$ *and consider the following three initial value problems*

$$\varepsilon_s'' = g(x)\varepsilon_s, \quad x \in [x_s, x_{s+1}], \quad \varepsilon_s(x_s) = E_s, \quad \varepsilon_s'(x_s) = E_s', \tag{1.12.20}$$

where E_s *and* E_s', $s = 0, 1, 2, \ldots, n-1$, *are given numbers*,

$$F'' = g(x)F, \quad x \in [a, b], \quad F(b) = 0, \quad F'(b) = -1, \tag{1.12.21}$$

and

$$D'' = g(x)D, \quad x \in [a, b], \quad D(b) = 1, \quad D'(b) = 0. \tag{1.12.22}$$

Then the following identities exist

$$\varepsilon_{n-1}(x_n) = E_0'F(x_0) - E_0F'(x_0) + \sum_{s=0}^{n-2} [F'(x_{s+1})\, d_{s+1} - F(x_{s+1})d_{s+1}'], \tag{1.12.23}$$

$$\varepsilon_{n-1}'(x_n) = E_0'D(x_0) - E_0D'(x_0) + \sum_{s=0}^{n-2} [D'(x_{s+1})\, d_{s+1} - D(x_{s+1})\, d_{s+1}'], \tag{1.12.24}$$

where

$$d_{s+1} = \varepsilon_s(x_{s+1}) - E_{s+1} \tag{1.22.25 a}$$

and

$$d_{s+1}' = \varepsilon_s'(x_{s+1}) - E_{s+1}'. \tag{1.12.25 b}$$

Proof. Let $G(x)$ satisfy the differential equation

$$G'' = g(x)G, \quad x \in [a, b]. \tag{1.12.26}$$

(F and D are the particular cases of G when the initial conditions are appropriately imposed).

Take some current s, multiply eq. (1.12.20) by $G(x)$ and eq. (1.12.26) by $\varepsilon_s(x)$ and then subtract. The result is

$$\varepsilon_s''(x)G(x) - \varepsilon_s(x)G''(x) = 0, \quad x \in [x_s, x_{s+1}].$$

Next integrate the latter from x_s to some $x \in [x_s, x_{s+1}]$,

$$\int_{x_s}^{x} [\varepsilon_s''(x')\, G(x') - \varepsilon_s(x')\, G''(x')]\, dx' =$$

$$\int_{x_s}^{x} \frac{d}{dx'} [\varepsilon_s'(x')\, G(x') - \varepsilon_s(x')\, G'(x')]\, dx'$$

$$= W_s(x) - W_s(x_s) = \text{const.}, \tag{1.12.27}$$

where

$$W_s(x) = \varepsilon_s'(x)G(x) - \varepsilon_s(x)G'(x). \tag{1.12.28}$$

Now, since eq. (1.12.27) is valid for any x, in particular for $x = x_s$, the constant is simply zero. We thus have obtained the result that $W_s(x)$ is constant for all $x \in [x_s, x_{s+1}]$, and this is a very familiar result indeed, namely the so-called Wronskian theorem. On further taking $x = x_{s+1}$ we have

$$W_s(x_{s+1}) = W_s(x_s). \tag{1.12.29}$$

This identity is now summed over all values of s,

$$W_0(x_1) + W_1(x_2) + W_2(x_3) + \dots + W_{n-2}(x_{n-1}) + W_{n-1}(x_n) =$$

$$= W_0(x_0) + W_1(x_1) + W_2(x_2) + \dots + W_{n-2}(x_{n-2}) + W_{n-1}(x_{n-1})$$

from which $W_{n-1}(x_n)$ is explicited

$$W_{n-1}(x_n) = W_0(x_0) + \sum_{s=0}^{n-2} (W_{s+1}(x_{s+1}) - W_s(x_{s+1})). \tag{1.12.30}$$

We next work out the terms in the sum to account for the initial conditions. We have in order

$$W_{s+1}(x_{s+1}) - W_s(x_{s+1}) = \varepsilon'_{s+1}(x_{s+1})\, G(x_{s+1}) - \varepsilon_{s+1}(x_{s+1})\, G'(x_{s+1})$$

$$- (\varepsilon'_s(x_{s+1})\, G(x_{s+1}) - \varepsilon_s(x_{s+1})\, G'(x_{s+1}))$$

$$= (\varepsilon'_{s+1}(x_{s+1}) - \varepsilon'_s(x_{s+1}))\, G(x_{s+1}) - (\varepsilon_{s+1}(x_{s+1}) - \varepsilon_s(x_{s+1}))\, G'(x_{s+1})$$

$$= (E'_{s+1} - \varepsilon'_s(x_{s+1}))\, G(x_{s+1}) - (E_{s+1} - \varepsilon_s(x_{s+1}))\, G'(x_{s+1})$$

$$= -\, d'_{s+1} G(x_{s+1}) + d_{s+1} G'(x_{s+1}) \tag{1.12.31}$$

Eq. (1.12.30) hence becomes

$$\varepsilon'_{n-1}(x_n)\, G(x_n) - \varepsilon_{n-1}(x_n)\, G'(x_n) = E'_0 G(x_0) - E_0 G'(x_0)$$

$$+ \sum_{s=0}^{n-2} (d_{s+1} G'(x_{s+1}) - d'_{s+1} G(x_{s+1})). \tag{1.12.32}$$

Identities (1.12.23) and (1:12.24) are particular cases of this equation. In fact, in the particular case when $G(x_n) = 0$ and $G'(x_n) = -1$, that is $G \equiv F$, the left side of this equation is simply $\varepsilon_{n-1}(x_n)$, while when $G(x_n) = 1$ and $G'(x_n) = 0$, which means $G \equiv D$, it is $\varepsilon'_{n-1}(x_n)$. The proof is thus completed.

To bring the problem of evaluating the accumulated errors into a form tractable via eq. (1.12.23) and eq. (1.12.24) we need some additional ingredients.

First of all we tackle the following question: how can we measure the difference between the solutions of one and the same second-order differential equation defined on a sufficiently narrow interval when the initial conditions are different, but sufficiently close to each other?

Specifically, consider the following two initial value problems

$$y'' = f(x, y),\ x \in [\alpha, \beta],\ y(\alpha) = Y,\ y'(\alpha) = Y', \tag{1.12.33 a}$$

and

$$z'' = f(x, z),\ x \in [\alpha, \beta],\ z(\alpha) = Z,\ z'(\alpha) = Z'. \tag{1.12.33 b}$$

(Note that f is the same in both equations.)

Consider the Taylor expansion of $f(x, z)$ with respect to the second variable,

$$f(x, z) = f(x, y) + \frac{\partial f(x, y)}{\partial y} \cdot (z - y) + \frac{\partial^2 f(x, y)}{\partial f^2} (z - y)^2 + \ldots \tag{1.12.34}$$

and actually assume that $|Y - Z|$ and $|Y' - Z'|$ are so small and that the domain $[\alpha, \beta]$ is so narrow that $|y(x) - z(x)|$ remains small enough for any $x \in [\alpha, \beta]$ to secure that only two terms in the Taylor series (1.12.34) are numerically relevant.
Under such conditions eq. (1.12.34) reads

$$f(x, y) - f(x, z) = g(x)(y - z), \text{ with } g(x) = \left.\frac{\partial f(x, y)}{\partial y}\right|_{y=y(x)} . \tag{1.12.35}$$

On using this formula we subtract eq. (1.12.33 b) from eq. (1.12.33 a) to get the following initial value problem for the deviation $\varepsilon(x) \equiv y(x) - z(x)$

$$\varepsilon'' = g(x)\varepsilon, \; x \in [\alpha, \beta], \; \varepsilon(\alpha) = Y - Z, \; \varepsilon'(\alpha) = Y' - Z', \tag{1.12.36}$$

that is to measure the deviation we have to solve this problem.

Now particularize this result for the case of eq. (1.12.1) in the elementary interval $[x_{2k}, x_{2k+2}]$.

The exact solution of this equation automatically satisfies the following local problem

$$y'' = f(x, y), \; x \in [x_{2k}, x_{2k+2}], \; Y = y(x_{2k}), \; Y' = y'(x_{2k}) \tag{1.12.37 a}$$

while what the de Vogelaere scheme tentatively solves at the kth iteration is the problem

$$z'' = f(x, z), \; x \in [x_{2k}, x_{2k+2}], \; Z = \bar{y}_{2k}, \; Z' = \bar{y}'_{2k}, \tag{1.12.37 b}$$

where $\bar{y}_{2k}$ and $\bar{y}'_{2k}$ are numbers resulted as output of the previous iteration with the quoted algorithm. Then the deviation $\varepsilon_{2k}(x)$ is the solution of the associated problem

$$\varepsilon''_{2k} = g(x)\varepsilon_{2k}, \; x \in [x_{2k}, x_{2k+2}], \; \varepsilon_{2k}(x_{2k}) = E_{2k} \equiv y(x_{2k}) - \bar{y}_{2k},$$
$$\varepsilon'_{2k}(x_{2k}) = E'_{2k} \equiv y'(x_{2k}) - \bar{y}'_{2k}. \tag{1.12.38}$$

Clearly, E_{2k} and E'_{2k} represent just the accumulated errors at x_{2k}. As for eq. (1.12.37 b) this is solved only approximately by the quoted algorithm. Then the resultant values $\bar{y}_{2k+2}$ and $\bar{y}'_{2k+2}$ do not coincide with the exact values $z(x_{2k+2})$ and $z'(x_{2k+2})$. The corresponding errors

$$\Delta z_{2k+2} \equiv z(x_{2k+2}) - \bar{y}_{2k+2}, \ \Delta z'_{2k+2} \equiv z'(x_{2k+2}) - \bar{y}'_{2k+2} \tag{1.12.39}$$

are obviously given by eqs (1.12.15 b) and (1.12.15 c) in which y is replaced by z, for instance

$$\Delta z_{2k+2} = \frac{2}{45} h^5 f'''(x_{2k}, z(x_{2k})) + O(h^6). \tag{1.12.40}$$

The accumulated errors at any even-labelled mesh point x_{2p} thus read

$$E_{2p} \equiv y(x_{2p}) - \bar{y}_{2p} = \varepsilon_{2(p-1)}(x_{2p}) + \Delta z_{2p} \tag{1.12.41 a}$$

and

$$E'_{2p} \equiv y'(x_{2p}) - \bar{y}'_{2p} = \varepsilon'_{2(p-1)}(x_{2p}) + \Delta z'_{2p} \tag{1.12.41 b}$$

$$p = 1, 2, 3, \ldots, k, \ldots .$$

To get explicit formulas for the accumulated errors at x_{2k} we use identities (1.12.23) and (1.12.24) in which we take the end point x_n in the lemma as the current point x_{2k} and identify $x_s = x_{2p}$. The initial conditions in the set of the local equations (1.12.20) are $E_0 = E'_0 = 0$ (since the initial conditions are embedded exactly in the de Vogelaere scheme) and E_{2p} and E'_{2p} defined in eqs (1.12.41 a) and (1.12.41 b) for $p = 1, 2, \ldots, k-1$.

Identity (1.12.23) yields the expression of E_{2k}. We have in order

$$E_{2k} \equiv y(x_{2k}) - \bar{y}_{2k} = \varepsilon_{2(k-1)}(x_{2k}) + \Delta z_{2k} = \Delta z_{2k} + \sum_{p=1}^{k-1} (F(x_{2p}) \Delta z'_{2p}$$

$$- F'(x_{2p}) \Delta z_{2p}) = \sum_{p=1}^{k} (F(x_{2p}) \Delta z'_{2p} - F'(x_{2p}) \Delta z_{2p}) \tag{1.12.42 a}$$

(To write down the last member in this chain we made use of the specified conditions for F at x_{2k}.)

Similarly, eq. (1.12.24) furnishes the following formula for E'_{2k}

$$E'_{2k} \equiv y'(x_{2k}) - y'_{2k} = \sum_{p=1}^{k} (D(x_{2p})\, \Delta z'_{2p} - D'(x_{2p})\, \Delta z_{2p}). \qquad (1.12.42\text{ b})$$

The final aim is to express E_{2k} and E'_{2k} in terms of the genuine local truncation errors. We actually show below that if the step sizes are small enough then Δz_{2p} and $\Delta z'_{2p}$ which occur in eqs. (1.12.42 a) and (1.12.42 b) can be safely replaced through the genuine local errors e_{2p} and e'_{2p}, respectively.

In a first step we show that the order of the de Vogelaere method is $m = 4$. For simplicity, we assume that the partition is equidistant, $x_{2p} = x_0 + 2ph$. Eq. (1.12.40) and a similar formula for $\Delta z'_{2k+2}$ ensure that two constants $Z_{\max}$ and $Z'_{\max}$ exist such that

$$|\Delta z_{2p}| \leqslant h^5 Z_{\max}, \quad |\Delta z'_{2p}| \leqslant h^5 Z'_{\max}$$

for any $p = 1, 2, \ldots, k$. In turn it may also be said that two constants $E_{\max}$ and $E'_{\max}$ equally exist such that

$$|F(x_{2p})\, \Delta z'_{2p} - F'(x_{2p})\, \Delta z_{2p}| \leqslant h^5 E_{\max}$$

and

$$|D(x_{2p})\, \Delta z'_{2p} - D'(x_{2p})\, \Delta z_{2p}| \leqslant h^5 E'_{\max}$$

for any $p = 1, 2, \ldots, k$. Thus E_{2k} and E'_{2k} are bounded as follows

$$|E_{2k}| \leqslant kh^5 E_{\max} = \frac{x_{2k} - a}{2h} h^5 E_{\max} = \frac{1}{2} h^4 (x_{2k} - a)\, E_{\max}, \qquad (1.12.43\text{ a})$$

$$|E'_{2k}| \leqslant \frac{1}{2} h^4 (x_{2k} - a)\, E'_{\max}, \qquad (1.12.43\text{ b})$$

and hence the method is of order $m = 4$.

The second step consists of explicitly writing the genuine local truncation errors in eqs. (1.12.42 a) and (1.12. 42 b). We write for instance eq. (1.12.42 b) as follows

$$E_{2k} = \sum_{p=1}^{k} (F(x_{2p})\, e'_{2p} - F'(x_{2p})\, e_{2p}) + \Delta E_{2k} \qquad (1.12.44)$$

with

$$\Delta E_{2k} \equiv \sum_{p=0}^{k-1} (F(x_{2p+2})(\Delta z'_{2p+2} - e'_{2p+2}) - F'(x_{2p+2})(\Delta z_{2p+2} - e_{2p+2}))$$

and show that the deviation ΔE_{2k} is of higher order than the very order $m = 4$ of E_{2k}.

We actually consider the expression

$$\Delta z_{2p+2} - e_{2p+2} = \frac{2}{45} h^5[f'''(x_{2p}, y(x_{2p})) - f'''(x_{2p}, z(x_{2p}))] + O(h^6)$$

and formally think of $f'''(x, y(x))$ as being the particular case $y = y(x)$, $y' = y'(x)$ of some function A of three variables x, y and y'.

Example. Consider the second-order differential equation $y'' = x + y^2$, that is $f(x, y) = x + y^2$, and construct the total derivatives of f with respect to x up to the third order: $f' = 1 + 2yy'$, $f'' = (f')' = 2y'^2 + 2yy'' = 2y'^2 + 2y(x + y^2) = 2y'^3 + 2xy + 2y^3$; $f''' = (f'')' = 4y'y'' + 2y + 2xy' + 6y^2y' = 2y + 6xy' + 10y^2y'$. We thus obtain that $f'''(x, y(x))$ is the value of $A(x, y, y') = 2y + 6xy' + 10y^2y'$ when y and y' are the values at x of the solution of the given differential equation and of the first derivative of the solution.

On using such representation and the appropriately truncated Taylor series for A we have:

$$\begin{aligned} f'''(x_{2p}, y(x_{2p})) - f'''(x_{2p}, z(x_{2p})) &= A(x_{2p}, y(x_{2p}), y'(x_{2p})) \\ - A(x_{2p}, z(x_{2p}), z'(x_{2p})) &\simeq A_y(x_{2p}, y(x_{2p}), y'(x_{2p}))\ E_{2p} \\ &+ A_{y'}(x_{2p}, y(x_{2p}), y'(x_{2p}))\ E'_{2p}, \end{aligned}$$

so that on using eqs (1.12.43 a) and (1.12.43 b) we get at once

$$\Delta z_{2p+2} - e_{2p+2} = O(h^9). \tag{1.12.45}$$

A similar examination also shows that $\Delta z'_{2p+2} - e'_{2p+2} = O(h^9)$. These evaluations are finally introduced in expression of ΔE_{2k} which consists of a sum of $k = (x_{2k} - a)/2h$ such terms. The result is clearly $\Delta E_{2k} = O(h^{9-1}) = O(h^8)$. The same procedure can be repeated word by word for E'_{2k} and the result is that the error introduced there if Δz_{2p+2} and $\Delta z'_{2p+2}$ are replaced by e_{2p+2} and e'_{2p+2} is also of order eight, thus negligible for small h. With these points in mind we further regard the sum in (1.12.44) as a quadrature sum. We thus have

$$E_{2k} = \frac{1}{2h} \int_{x_0}^{x_{2k}} [F(x)\, e'(x;h) - F'(x)\, e(x;h)]\, dx \tag{1.12.46 a}$$

and

$$E'_{2k} = \frac{1}{2h} \int_{x_0}^{x_{2k}} [D(x)\, e'(x;h) - D'(x)\, e(x;h)]\, dx \tag{1.12.46 b}$$

for any $h < \bar{h}$ where $\bar{h}$ is some threshold value which guarantees that the quadrature sums and the integrals are numerically equivalent. Naturally, in this case $e(x;h)$ and $e'(x;h)$ are given by the leading terms in eqs (1.12.15 b) and (1.12.15 c), viz.

$$e(x;h) = \frac{2}{45} h^5 y^{(5)}(x), \; e'(x;h) = -\frac{1}{90} h^5 y^{(6)}(x) + \frac{1}{6} h^5 \frac{\partial f}{\partial y} y^{(4)}(x) \tag{1.12.47}$$

so that

$$E_{2k} = h^4 \bar{E}^F(x_{2k}), \; E'_{2k} = h^4 \bar{E}^D(x_{2k}) \tag{1.12.48}$$

where

$$\bar{E}^G(x_{2k}) = \int_{x_0}^{x_{2k}} \left[\left(-\frac{1}{180} y^{(6)}(x) + \frac{1}{12} \frac{\partial f}{\partial y} y^{(4)}(x) \right) G(x) - \right.$$

$$\left. - \frac{1}{45} y^{(5)}(x)\, G'(x) \right] dx \tag{1.12.49}$$

for $G \equiv F$ and $G \equiv D$, respectively.

At this point we make a short digression to recall that the definition of convergence given in section 1.11 referred to the multistep algorithm

(1.11.18) and that the order of the method was there defined as the value of m such that C_{m+2} is the first nonvanishing term in series (1.11.21), a series actually constructed to examine the *local* truncation error and the consistency for that algorithm. On the other hand, in the case of first-order differential equations the order was defined as the value of m such that C_{m+1} is the first nonvanishing term in expansion (1.6.8). For both cases the real common relevance of m is unveiled only in connection with the *accumulated* error. In fact the structure of E_s which occurs in eqs (1.10.2) and (1.11.26) (viz.: $E_s = h^m \bar{E}_s$) has shown that in the absence of round-off errors the accumulated error is proportional to h^m in both cases.

Note that this chain, that is to first define the order m from local arguments and then to show its major relevance as governing the behaviour of accumulated errors, is mainly due to the constructivist nature of the very mathematical theory developed there. An *a posteriori* look over the whole matter suggests that it is more natural to define the order m just as the power of h at which the accumulated error at some arbitrary $X \in (a, b]$ tends to zero when h tends to zero so that $nh = X - a$, provided, of course, that the starting values are sufficiently accurate. This new definition does not contradict at all the two above definitions and has the advantage that it can be extended to refer to any other type of numerical methods as for instance de Vogelaere method or the one-step methods to be discussed in Chapter 2.

With this new definition for m, any stable numerical method * will be convergent if $m > 0$, and since m typically assumes integer values, we must have $m \geqslant 1$.

In particular, eq. (1.12.48) shows that the de Vogelaere scheme with the initial conditions $\bar{y}_0 = y_0$, $\bar{y}_0' = y_0'$, and $\bar{y}_{-1}$ given by eq. (1.12.18), is convergent and of order $m = 4$.

Round-off errors

Let us now assume that the computation by means of the de Vogelaere scheme is affected by round-off errors and let us denote these round-off affected representations of the input values $\bar{y}_{2k-1}$, $\bar{y}_{2k}$ and $\bar{y}_{2k}'$ by $\tilde{y}_{2k-1}$, $\tilde{y}_{2k}$ and $\tilde{y}_{2k}'$, respectively. Our first purpose is to find out the

---

* The condition of stability deserves special consideration only in connection with the multistep methods of sections 1.6 and 1.11 and for the special one-step methods of section 2.6; the genuine one-step methods are always stable.

corresponding modification in the algorithm when its formulas are computed on a real computer. Consider for instance eq. (1.12.13 a). On computing its r.h.s. on a real computer we obtain the rounded representation $\tilde{y}_{2k+1}$ of $\bar{y}_{2k+1}$. If we hypothetically compute the same r.h.s. but with absolute accuracy we obtain some different value, $\tilde{y}^*_{2k+1}$, say. Then, on denoting $R_{2k+1} = \tilde{y}^*_{2k+1} - \tilde{y}_{2k+1}$, the computation of formula (1.12.13 a) on a real computer is equivalent to the mathematically exact computation by means of a modified formula, viz.

$$\tilde{y}_{2k+1} = \tilde{y}_{2k} + h\tilde{y}'_{2k} + \frac{h^2}{6}(4f(x_{2k}, \tilde{y}_{2k}) - f(x_{2k-1}, \tilde{y}_{2k-1})) - R_{2k+1}, \tag{1.12.50}$$

where R_{2k+1} manifestly depends on the accuracy of the used computer.

On subtracting eq. (1.12.50) from eq. (1.12.13 a) we get the following formula relating the round-off errors $r_s = \bar{y}_s - \tilde{y}_s$ and $r'_s = \bar{y}'_s - \tilde{y}'_s$

$$r_{2k+1} = r_{2k} + hr'_{2k} + R_{2k+1}. \tag{1.12.51}$$

(The terms of order h^2 have been disregarded.)
Similarly, from eqs. (1.12.13 b) and (1.12.13 c) we get

$$r_{2k+2} = r_{2k} + 2hr'_{2k} + R_{2k+2} \tag{1.12.52}$$

and

$$r'_{2k+2} = r'_{2k} + \frac{h}{3}(g_{2k}r_{2k} + 4g_{2k+1}r_{2k+1} + g_{2k+2}r_{2k+2}) + R'_{2k+2}, \tag{1.12.53}$$

where

$$g_s = \begin{cases} r_s^{-1}(f(x_s, \bar{y}_s) - f(x_s, \tilde{y}_s)) & \text{if } r_s \neq 0 \\ 0 & \text{if } r_s = 0 \end{cases} \tag{1.12.54}$$

so that in any case $|g_s| \leqslant L$, where L is the Lipschitz constant of $f(x, y)$, for $x \in [a, b]$.

Eq. (1.12.53) is further worked out by using eqs (1.12.51) and (1.12.52) to yield

$$r'_{2k+2} = r'_{2k} + \frac{h}{3}[(g_{2k} + 4g_{2k+1} + g_{2k+2})\, r_{2k} + 4g_{2k+1} R_{2k+1} + \\ + g_{2k+2}R_{2k+2}] + R'_{2k+2}. \tag{1.12.55}$$

On taking the absolute values in eqs (1.12.52) and (1.12.55), these lead to the following inequalities

$$|r_{2k+2}| \leqslant |r_{2k}| + 2h\,|r'_{2k}| + |R_{2k+2}|, \tag{1.12.56 a}$$

$$|r'_{2k+2}| \leqslant |r'_{2k}| + 2Lh|r_{2k}| + \frac{h}{3} L(4|R_{2k+1}| + |R_{2k+2}|) + |R'_{2k+2}|. \tag{1.12.56 b}$$

(To derive the latter we also used the inequality $|g_s| \leqslant L$.)
Next add the last two inequalities,

$$|r_{2k+2}| + |r'_{2k+2}| \leqslant |r_{2k}|(1 + 2Lh) + |r'_{2k}|(1 + 2h) + \lambda_{2k+2}, \tag{1.12.57}$$

where

$$\lambda_{2k+2} = |R_{2k+2}| + \frac{hL}{3}(4|R_{2k+1}| + |R_{2k+2}|) + |R'_{2k+2}| \tag{1.12.58}$$

is a quantity which collects the round-off errors associated to the computation of the three formulas in the kth step.
Now we write

$$\mathscr{L} = \begin{cases} L & \text{if } \; L \geqslant 1 \\ 1 & \text{if } \; L < 1. \end{cases} \tag{1.12.59}$$

Inequality (1.12.57) then becomes

$$|r_{2k+2}| + |r'_{2k+2}| \leqslant (1 + 2\mathscr{L}h)\,(|r_{2k}| + |r'_{2k}|) + \lambda_{2k+2}, \quad k = 0, 1, 2, \ldots, n-1, \tag{1.12.60}$$

and upon writing $\Lambda = \sup\limits_{k=0,1,\ldots,n-1} \lambda_{2k+2}$, it is of the form

$$z_{k+1} \leqslant (1 + \alpha)\; z_k + b, \quad k = 0, 1, 2, \ldots, n-1 \tag{1.12.61}$$

where $z_k = |r_{2k}| + |r'_{2k}|$, $\alpha = 2\mathscr{L}h$ and $b = \Lambda$.
The recurrent inequality (1.12.61) is consistent with the bound

$$z_k \leqslant z_0 \exp(k\alpha) + \frac{b}{\alpha}\,(\exp(k\alpha) - 1). \tag{1.12.62}$$

To see this * we consider the following auxiliary difference equation

$$w_{k+1} = (1 + \alpha)w_k + b, \quad k = 0, 1, 2, \ldots, n - 1. \tag{1.12.63}$$

We see by induction that $z_k \leqslant w_k$ for all k.
The solution of the homogeneous equation

$$w_{k+1} = (1 + \alpha)w_k$$

is

$$w_k = c(1 + \alpha)^k$$

where c is arbitrary. A particular solution of the original equation (1.12.63) is

$$w_k = -\frac{b}{\alpha}$$

and thus the general solution of the original equation is

$$w_k = c(1 + \alpha)^k - \frac{b}{\alpha}\cdot \tag{1.12.64}$$

As for the starting value, we require $w_0 = z_0$ so that

$$c = z_0 + \frac{b}{\alpha}\cdot$$

Then

$$z_k \leqslant w_k = z_0(1 + \alpha)^k + \frac{b}{a}((1 + \alpha)^k - 1) \leqslant z_0\exp(\alpha k) +$$

$$+ \frac{b}{\alpha}(\exp(\alpha k) - 1), \tag{1.12.65}$$

and this is just eq. (1.12.62).

---

* This proof is taken from [7].

Coming back to the recurrent inequality (1.12.60) we have

$$|r_{2k}| + |r'_{2k}| \leqslant (|r_0| + |r'_0|)\exp(2\mathscr{L}hk) + \frac{\Lambda}{2h\mathscr{L}}(\exp(2\mathscr{L}hk) - 1)$$

$$= (|r_0| + |r'_0|)\exp[\mathscr{L}(x_{2k} - a)] + \frac{\Lambda}{2h\mathscr{L}}(\exp[\mathscr{L}(x_{2k} - a)] - 1),$$

$$k = 0, 1, 2, \ldots, n. \tag{1.12.66}$$

Note that for fixed $X = x_{2k}$, $|r_{2k}| + |r'_{2k}|$ increases as h^{-1} when h decreases thus resembling the behaviour of the round-off errors for multistep algorithms written in summed form.

On collecting eqs (1.12.48), (1.12.49) and eq. (1.12.65) we establish the following theorem which is an analog of the theorem given in section 1.11 in connection with algorithm (1.11.18):

THEOREM. *If the de Vogelaere scheme, eqs* (1.12.13 a, b, c), *is applied to solve problem* (1.12.1) *on a real computer and*
(i) *the starting values* $\tilde{y}_0$, $\tilde{y}_0$ *and* $\tilde{y}_{-1}$ *are the numerical, round-off affected representations of* $\bar{y}_0 = y_0$, $\bar{y}'_0 = y'_0$ *and of* $\bar{y}_{-1}$ *as given by eq.* (1.13.18);
(ii) *the partition is equidistant and* h *is sufficiently small,*
then the global accumulated errors at the even-labelled points x_{2k}, $k = 1, 2, \ldots, n$ *are*

$$y(x_{2k}) - \tilde{y}_{2k} = h^4\bar{E}^F(x_{2k}) + r_{2k} \tag{1.12.67 a}$$

$$y'(x_{2k}) - \tilde{y}'_{2k} = h^4\bar{E}^D(x_{2k}) + r'_{2k} \tag{1.12.67 b}$$

in which the magnified errors $\bar{E}^F(x_{2k})$ *and* $\bar{E}^D(x_{2k})$ *are given by the integral*

$$\bar{E}^G(x_{2k}) = \int_{x_0}^{x_{2k}} \left[\left(-\frac{1}{180}y^{(6)}(x) + \frac{1}{12}g(x)y^{(4)}(x)\right)G(x) - \right.$$

$$\left. -\frac{1}{45}y^{(5)}(x)G'(x)\right]dx \tag{1.12.68}$$

where G *satisfies the differential equation*

$$G'' = g(x)G,\ x \in [x_0, x_{2k}],\ g(x) = \left.\frac{\partial f(x, y)}{\partial y}\right|_{y=y(x)}$$

and $G \equiv F$ *when the initial conditions are* $G(x_{2k}) = 0$ *and* $G'(x_{2k}) = -1$, *while* $G \equiv D$ *when the initial conditions are* $G(x_{2k}) = 1$ *and* $G'(x_{2k}) = 0$. *As for the round-off errors* r_{2k} *and* r'_{2k} *these are collectively bounded as follows*

$$|r_{2k}| + |r'_{2k}| \leqslant (|r_0| + |r'_0|) \exp[(x_{2k} - a)\mathscr{L}] +$$

$$+ \frac{\Lambda}{2h\mathscr{L}}(\exp[(x_{2k} - a)\mathscr{L}] - 1).$$

For completeness we briefly consider the modification of the previous error analysis when the de Vogelaere method is used with variable step size.

We assume for convenience that the variable step width is a function of x, viz. $h_V(x; h)$ factorized as follows

$$h_V(x; h) = S(x)\, h \tag{1.12.69}$$

where the parameter h is the largest envisaged step width and $S(x)$ is some weighting function $0 < S(x) \leqslant 1$. In practice eq. (1.12.69) must be understood as defining recurrently both the local step size h_k and the corresponding positions of the mesh points

$$h_k = S(x_{2k})h, \ x_{2k+1} = x_{2k} + h_k, \ x_{2k+2} = x_{2k} + 2h_k, \ k = 0, 1, 2, \ldots \tag{1.12.70}$$

The local truncation errors remain of the same form as in eqs (1.12.15 a, b, c) except for h which must be now replaced by h_k. Eq. (1.12.15 c), for instance, becomes

$$e_{2k+2} = \frac{2}{45} h^5 S^5(x_{2k}) f'''(x_{2k}) + 0(h^6) \tag{1.12.71}$$

As for the accumulated error we focus on eq. (1.12.44) (in which ΔE_{2k} is neglected) and which we appropriately transform into a quadrature sum. To this aim we simply multiply and divide each member of the sum by the associated increment $\Delta x_{2p+2} \equiv x_{2p+2} - x_{2p} = 2S(x_{2p})h$

$$E_{2p} = \frac{1}{2h} \sum_{p=0}^{k-1} \frac{1}{S(x_{2p})} (F(x_{2p+2})\, e'_{2p+2} - F'(x_{2p+2})\, e_{2p+2})\, \Delta x_{2p+2} \tag{1.12.72}$$

which, when the maximal h is sufficiently small, reads

$$E_{2k} = \frac{1}{2h} \int_{x_0}^{x_{2k}} \frac{1}{S(x)} \left[F(x)\, e'(x; h) - F'(x)\, e(x; h)\right] dx. \tag{1.12.73 a}$$

Similarly we get for E'_{2k} the formula

$$E'_{2k} = \frac{1}{2h} \int_{x_0}^{x_{2k}} \frac{1}{S(x)} \left[D(x)\, e'(x; h) - D'(x)\, e(x; h)\right] dx. \tag{1.12.73 b}$$

Finally, since the two equations under (1.12.47) are now

$$e(x; h) = \frac{2}{45} h^5 S^5(x)\, y^{(5)}(x),\ e'(x; h) = h^5 S^5(x) \left[-\frac{1}{90} y^{(5)}(x) + \right.$$

$$\left. + \frac{1}{6} \frac{\partial f}{\partial y} y^{(4)}(x) \right] \tag{1.12.74}$$

we find out that when variable step size is used the two accumulated truncation errors remain of the same form $E_{2k} = h^4 \bar{E}^F(x_{2k})$, $E'_{2k} = h^4 \bar{E}^D(x_{2k})$ except that the integrand in eq. (1.12.68) must be multiplied by $S^4(x)$, that is

$$\bar{E}^G(x_{2k}) = \int_{x_0}^{x_{2k}} S^4(x) \left[\left(-\frac{1}{180} y^{(6)}(x) + \frac{1}{12} g(x)\, y^{(4)}(x) \right) G(x) - \right.$$

$$\left. - \frac{1}{45} y^{(5)}(x)\, G'(x) \right] dx. \tag{1.12.75}$$

The task of investigating the modification of the analysis of the propagated round-off errors is left to the reader. He will find out essentially the same behaviour as in the case of a constant step size, that is $|r_{2k}| + |r'_{2k}|$ increases as h^{-1} when h is decreased.

1.13. METHODS TO SOLVE LINEAR SECOND-ORDER DIFFERENTIAL EQUATIONS

As discussed in "Preliminaries" and in section 1.8, any separable differential equation of higher order can be equivalently written as a system of first-order equations, so that any of the methods developed in sections 1.4 and 1.5 is theoretically suitable to solve such an equation.

In section 1.11 we have seen that for differential equations of second order and of special form (that is where the first derivative is absent), it is also possible to devise direct suitable algorithms. Although such algorithms can be applied at once also on the further simplified case when the equation is linear the latter case is met in practice so often that we find useful to consider it here in greater detail.

Before doing so we still consider the *general* linear second-order equation, that is with first-order derivative included, to devise an ad hoc algorithm, commonly associated with the names of Fox and Goodwin [15].

To be specific, our purpose is now to find out an algorithm for the following scalar initial value problem

$$y'' + a(x)y' + b(x)y + c(x) = 0, \ x \in [a, b], \ y(a) = y_0, \ y'(a) = y'_0 \tag{1.13.1}$$

where $a(x)$, $b(x)$ and $c(x)$ are given function on the quoted interval $[a, b]$.

To find out the corresponding algorithm we first need some elements of the theory of central differences. (Such elements are treated in any standard book on numerical analysis, see for instance [7].)

Let $x_0, x_1, x_2, \ldots, x_i, x_{i+1}, \ldots$ be equidistant points on the real axis with the step size h, that is $x_i = x_0 + ih$. By definition, the first central difference at x_i of a function $g(x)$ is given by the expression

$$\delta g(x_i) \equiv \delta g_i \equiv g(x_i + h/2) - g(x_i - h/2) = g_{i+1/2} - g_{i-1/2} \tag{1.13.2}$$

The second-order central difference is the first central difference of the previous one,

$$\delta^2 g_i = \delta(\delta g_i) = \delta(g_{i+1/2} - g_{i-1/2}) = \delta g_{i+1/2} - \delta g_{i-1/2}$$

$$= g_{i+1/2+1/2} - g_{i+1/2-1/2} - (g_{i-1/2+1/2} - g_{i-1/2-1/2}) = g_{i+1} -$$

$$- 2g_i + g_{i-1} \tag{1.13.3}$$

and so on. In general the nth central difference is defined as

$$\delta^n g_i = \delta(\delta^{n-1} g_i). \tag{1.13.4}$$

The operator δ is called a central difference operator.
We also introduce the operator μ defined as follows

$$\mu g_i \equiv \frac{1}{2}(g_{i+1/2} + g_{i-1/2}), \tag{1.13.5}$$

and a direct calculation shows that

$$\mu\delta g_i = \frac{1}{2}(g_{i+1} - g_{i-1}). \tag{1.13.6}$$

Operators μ and δ are also connected through a simple operator relation. To find it we apply μ^2 on g_i

$$\mu^2 g_i = \mu(\mu g_i) = \mu\left[\frac{1}{2}(g_{i+1/2} + g_{i-1/2})\right] = \frac{1}{4}[(g_{i+1/2+1/2} + g_{i-1/2+1/2})$$

$$+ (g_{i-1/2+1/2} + g_{i-1/2-1/2})] = \frac{1}{4}(g_{i+1} + 2g_i + g_{i-1}).$$

The same result is also obtained if $\left(1 + \frac{\delta^2}{4}\right)$ is applied on g_i, hence

$$\mu^2 = 1 + \frac{\delta^2}{4} \tag{1.13.7}$$

which is the announced relation.

We now search for a connection between operators μ and δ, on one hand, and the differential operator $D \equiv \frac{d}{dx}$, on the other.

We assume that $g(x)$ is differentiable indefinitely many times and use its Taylor series expansion,

$$g(x_i + \Delta) = g(x_i) + \frac{1}{1!}\Delta g'(x_i) + \frac{1}{2!}\Delta^2 g''(x_i) + \ldots$$

$$\ldots + \frac{1}{j!}\Delta^j g^{(j)}(x_i) + \ldots \tag{1.13.8}$$

On recognizing that the right member of this relation is formally the Taylor series expansion of $\exp(\Delta D)g(x_i)$, eq. (1.13.9) alternatively reads

$$g(x_i + \Delta) = \exp(\Delta D)g(x_i). \tag{1.13.9}$$

Accordingly, eq. (1.13.6) becomes

$$\mu\delta g_i = \frac{1}{2}(\exp(hD) - \exp(-hD))g_i = \sinh(hD)g_i, \tag{1.13.10}$$

that is $\mu\delta = \sinh(hD)$, and hence $hD = \sinh^{-1}(\mu\delta)$. Now, on taking into account the power series representation of $\sinh^{-1}(x)$,

$$\sinh^{-1}(x) = x - \frac{x^3}{6} + \frac{3x^5}{40} - \ldots \tag{1.13.11}$$

we get

$$hD = \mu\delta - \frac{\mu^3\delta^3}{6} + \frac{3\mu^5\delta^5}{40} - \ldots$$

in which, on using eq. (1.13.7), we eliminate the powers of μ higher than 2, to get

$$hD = \mu\left(\delta - \frac{\delta^3}{6} + \frac{\delta^5}{30} - \ldots\right), \tag{1.13.12}$$

a formula which yields D in terms of μ and δ.

To express the second-order derivative in terms of the same μ and δ, we take the square of eq. (1.13.12). The result is

$$h^2D^2 = \delta^2 - \frac{\delta^4}{12} + \frac{\delta^6}{90} - \ldots, \tag{1.13.13}$$

and the same procedure can be repeated to obtain the expressions of the derivatives of any higher order.

On applying operator identities (1.13.12) and (1.13.13) on g_i we obtain

$$hDg_i \equiv h\frac{dg}{dx}\bigg|_{x=x_i} = \frac{1}{2}(g_{i+1} - g_{i-1}) + \mu\left(-\frac{\delta^3}{6} + \frac{\delta^5}{30} - \ldots\right)g_i, \tag{1.13.14 a}$$

and

$$h^2D^2g_i \equiv h^2 \frac{d^2g}{dx^2}\bigg|_{x=x_i} = g_{i+1} - 2g_i + g_{i-1} + \left(-\frac{\delta^2}{12} + \frac{\delta^6}{90} - \dots\right) g_i, \tag{1.13.14 b}$$

respectively.
Note also the following relations

$$\delta^2 = h^2D^2 + \frac{h^4D^4}{12} + \frac{h^6D^6}{360} + \dots \tag{1.13.15}$$

$$\delta^3 = h^3D^3 + \frac{h^5D^5}{4} + \frac{h^7D^7}{10} + \dots \tag{1.13.16}$$

$$\delta^4 = h^4D^4 + \frac{h^6D^6}{6} + \frac{h^8D^8}{80} + \dots \tag{1.13.17}$$

which can be derived on using the series expansion for $\mu\delta = \sinh(hD)$. These relations are useful in that they show that the last terms in right sides of eqs (1.13.14 a) and (1.13.14 b) are of order h^3 and h^4, respectively.

To derive the Fox-Goodwin algorithm for problem (1.13.1) we define the equidistant partition $x_0 = a, x_1, x_2, \dots, x_n = b$, multiply eq. (1.13.1) at some current mesh point x_{k+1} by h^2,

$$h^2y''_{k+1} + ha_{k+1} \cdot hy'_{k+1} + h^2b_{k+1}y_{k+1} + h^2c_{k+1} = 0,$$

and replace here $h^2y''_{k+1}$ and hy'_{k+1} by formulas (1.13.14 b) and (1.13.14 a). The result is

$$y_{k+2} - 2y_{k+1} + y_k + \frac{h}{2} a_{k+1}(y_{k+2} - y_k) + h^2b_{k+1}y_{k+1} + h^2c_{k+1} +$$
$$+ \left(-\frac{\delta^4}{12} + \frac{\delta^6}{90} - \dots\right) y_{k+1} + ha_{k+1}\mu\left(-\frac{\delta^3}{6} + \frac{\delta^5}{30} - \dots\right) y_{k+1} = 0 \tag{1.13.18}$$

which, upon solving for y_{k+2}, becomes

$$y_{k+2} = \left(1 + \frac{h}{2} a_{k+1}\right)^{-1} \left[-\left(1 - \frac{h}{2} a_{k+1}\right) y_k + \right.$$

$$\left. + (2 - h^2 b_{k+1}) y_{k+1} - h^2 c_{k+1} \right] + e_{k+2} \tag{1.13.19}$$

with

$$e_{k+2} = \left(1 + \frac{h}{2} a_{k+1}\right)^{-1} \left[\left(\frac{\delta^4}{12} - \frac{\delta^6}{90} + \dots\right) + \right.$$

$$\left. + h a_{k+1} \mu \left(\frac{\delta^3}{6} - \frac{\delta^5}{30} + \dots\right) \right] y_{k+1}. \tag{1.13.20}$$

The Fox-Goodwin method is the recurrent relation (1.13.19) in which e_{k+2} is neglected, hence

$$\bar{y}_{k+2} = \left(1 + \frac{h}{2} a_{k+1}\right)^{-1} \left[-\left(1 - \frac{h}{2} a_{k+1}\right) \bar{y}_k + \right.$$

$$\left. + (2 - h^2 b_{k+1}) \bar{y}_{k+1} - h^2 c_{k+1} \right] \tag{1.13.21}$$

$$k = 0, 1, 2, \dots, n-2.$$

The quantity e_{k+2} represents just the local truncation error of the method. (To see this remember that the local truncation error is defined as the deviation $y_{k+2} - \bar{y}_{k+2}$ when all entrance data in the algorithm at the step k are exact. In our case eq. (1.13.19) is an exact representation of the original differential equation while eq. (1.13.21) is the associated algorithm. Then, on subtracting the latter from the former with the assumption that $y_k = \bar{y}_k$ and $y_{k+1} = \bar{y}_{k+1}$, we get at once $y_{k+2} - \bar{y}_{k+2} = e_{k+2}$, and this proves the assertion.)

On using eqs (1.13.16) and (1.13.17), eq. (1.13.20) can be equivalently re-written as

$$e_{k+2} = h^4\left(\frac{1}{12} D^4 + \frac{1}{6} a_{k+1}\mu D^3\right) y_{k+1} + 0(h^5) =$$

$$= \frac{1}{6} h^4 \left(\frac{1}{2} y_{k+1}^{(4)} + a_{k+1} y_{k+1}'''\right) + 0(h^5), \tag{1.13.22}$$

which means that the local truncation error is of order h^4.

As for the accumulated error we notice without proof that it is of order h^2, that is the quoted algorithm is of second order. Although the order is so small, formula (1.1ɔ.21) is simple enough to enjoy real popularity in practical applications. Note also in passing that the above derivation of the algorithm can be repeated word by word for equations of more general form, viz. $y'' + a(x)y' + f(x, y) = 0$ that is in which only y' dependence is linear. In particular, the local truncation error remains exactly of the form (1.13.22).

We next consider some special cases of eq. (1.13.1) and to start with we first examine the case when $a(x) = 0$. The resultant equation is of the form (1.11.1) with

$$f(x, y) = - b(x)y - c(x) \tag{1.13.23}$$

so that all methods discussed in section 1.11 can be applied directly. As a matter of fact, the Fox-Goodwin algorithm becomes identical with the Adams-Stormer formula (1.11.13).

An important general remark which is so useful in practical applications is that when $f(x, y)$ is linear in y the closed methods become open so that they can be applied directly (that is without any additional predictor-corrector procedure). As illustration take closed eq. (1.11.16) which we re-write here again

$$\bar{y}_{k+1} - 2\bar{y}_k + \bar{y}_{k-1} = \frac{1}{12} h^2(f_{k+1} + 10f_k + f_{k-1}). \tag{1.13.24}$$

On using the form (1.13.23), this equations becomes

$$\bar{y}_{k+1} - 2\bar{y}_k + \bar{y}_{k-1} = -\frac{1}{12} h^2(b_{k+1}\bar{y}_{k+1} + 10b_k\bar{y}_k + b_{k-1}\bar{y}_{k-1} +$$

$$+ c_{k+1} + 10c_k + c_{k-1}), \tag{1.13.25}$$

$$k = 1, 2, \ldots, n-1$$

from which the desired $\bar{y}_{k+1}$ is explicited at once,

$$\bar{y}_{k+1} = \left(1 + \frac{1}{12} h^2 b_{k+1}\right)^{-1} \left[\left(2 - \frac{5}{6} h^2 b_k\right) \bar{y}_k - \right.$$

$$\left. - \left(1 + \frac{1}{12} h^2 b_{k-1}\right) \bar{y}_{k-1} - \frac{h^2}{12} c_{k+1} - \frac{5}{6} h^2 c_k - \frac{h^2}{12} c_{k-1}\right]$$

$$k = 1, 2, \ldots, n - 1. \qquad (1.13.26)$$

Algorithm (1.13.25) (or (1.13.26)) is known as the *Numerov* method, [16] and this is perhaps the most popular particularization of algorithm (1.11.18) to integrate such special second-order linear differential equations, see for instance [17]. The method can be also generalized to solve systems of differential equations of higher order in which only derivatives of even order occur, see [18]. There exist also versions which take into account more than three successive mesh points, see [19].

Another popular algorithm to solve special second-order equations is the de Vogelaere algorithm discussed in section 1.12. Remember that this is also a multistep method but not of the form (1.11.18). It is then instructive to compare the two algorithms for efficiency.

To this aim we consider the following simple scalar problem

$$y'' = -y, \; x \geqslant 0, \; y(0) = 0, \; y'(0) = 1 \qquad (1.13.27)$$

with the exact solution $y(x) = \sin x$.

To put both methods on equal footing we take the same step size h, and the partition $x_0 = 0$, $x_1 = h$, $x_2 = 2h, \ldots, x_{2k} = 2kh$, $x_{2k+1} = (2k+1)h, \ldots$ to finally compare the accumulated errors only at even-labelled mesh points x_{2k}, $k = 1, 2, \ldots$. We also assume that the round-off errors are negligible.

Numerov method. The accumulated error is given by formula (1.11.26 a) where $\bar{E}(x)$ is the solution of problem (1.11.26 b) with the following identifications: $m = 4$, $\bar{C} = -1/240$ (remember that the Numerov method is identical with the Cowell method with $r = 2$, and also use that for this method $a_2 = 1$ and $b_0 + b_1 + b_2 = 1$ then consult Table 8), $g(x) = -1$, $y^{(m+2)} = y^{(6)} = -\sin x$. Hence the magnified error function is

$$\bar{E}(x) = \frac{1}{480}(\sin x - x\cos x) \tag{1.13.28}$$

so that the accumulated error at x_{2k} is

$$E_{2k}^N \equiv y(x_{2k}) - \bar{y}_{2k}^N = \frac{h^4}{480}(\sin x_{2k} - x_{2k}\cos x_{2k}). \tag{1.13.29}$$

(An alternative derivation of this formula is given in [20].)

de Vogelaere method. Again $g(x) = -1$ in eqs (1.12.36), (1.12.21), (1.12.22) and (1.12.26). Then if $b = x_{2k}$ the solution of eq. (1.12.21) is $F(x) = \sin(x_{2k} - x)$. Note also that $y^{(4)}(x) = \sin x$ and $y^{(5)} = \cos x$. Thus, for $G = F$, eq. (1.12.68) gives

$$\bar{E}^F(x_{2k}) = \int_{x_0}^{x_{2k}} \left\{\left[\frac{1}{180}\sin x - \frac{1}{12}\sin x\right]\sin(x_{2k} - x) + \right.$$

$$\left. + \frac{1}{45}\cos x \cdot \cos(x_{2k} - x)\right\} dx$$

$$= \frac{1}{90}\int_{x_0}^{x_{2k}} [2\cos x\cos(x_{2k} - x) - 7\sin x\sin(x_{2k} - x)]\, dx$$

$$= \frac{1}{180}(9x_{2k}\cos x_{2k} - 5\sin x_{2k}) \tag{1.13.30}$$

so that

$$E_{2k}^V \equiv y(x_{2k}) - \bar{y}_{2k}^V = h^4\bar{E}^F(x_{2k}) = \frac{h^4}{180}(9x_{2k}\cos x_{2k} - 5\sin x_{2k}). \tag{1.13.31}$$

On comparing formulas (1.13.29) and (1.13.31) we conclude that for problem (1.13.27) the Numerov method is more accurate than the other method, and the conclusion remains unchanged also for the general scalar case with f given by eq. (1.13.23).

Another important factor in the comparison of the two methods for efficiency is the computation time consumed to advance the solution up to x_{2k}.

For single equations both methods require approximately equal efforts. In fact, the most time consuming portion of each run is the evaluation of the functions $b(x)$ and $c(x)$. Now, on using an optimal programming for both algorithms, which means that the values of $b(x)$ and $c(x)$ at each mesh point are computed once and stored for further re-use, we see that both methods require the same number of $2k+1$ evaluations of $b(x)$ and $c(x)$, viz.: $b_0, c_0, b_1, c_1, \ldots, b_{2k}, c_{2k}$. Also, the number of remaining arithmetic operations is approximately the same in the two algorithms (see Chapter 5).

The picture changes in case of systems of equations. To see this we consider the vector problem

$$\mathbf{y}'' + \mathbf{b}(x)\,\mathbf{y} + \mathbf{c}(x) = \mathbf{0},\ x \in [a, b],\ \mathbf{y}(a) = \mathbf{y}_0,\ \mathbf{y}'(a) = \mathbf{y}_0' \tag{1.13.32}$$

where $\mathbf{b}(x)$ is an N by N matrix of functions, $\mathbf{c}(x)$, $\mathbf{y}$, $\mathbf{y}_0$, and $\mathbf{y}_0'$ are N dimensional vectors. Algorithm (1.13.26) now is:

$$\begin{aligned}\bar{\mathbf{y}}_{k+1} = \left(\mathbf{I} + \frac{1}{12} h^2 \mathbf{b}_{k+1}\right)^{-1} \Bigg[\left(2\mathbf{I} - \frac{5}{6} h^2 \mathbf{b}_k\right) \bar{\mathbf{y}}_k - \\ - \left(\mathbf{I} + \frac{1}{12} h^2 \mathbf{b}_{k-1}\right) \bar{\mathbf{y}}_{k-1} - \frac{h^2}{12}\mathbf{c}_{k+1} - \frac{5h^2}{6}\mathbf{c}_k - \frac{h^2}{12}\mathbf{c}_{k-1}\Bigg],\end{aligned} \tag{1.13.33}$$

where $\mathbf{I}$ is the N by N unit matrix. The point is that the computation now involves some *matrix inversion*, an operation which is time consuming itself *. To save the computing time, special matrix inversion procedures, based upon the fact that the matrix to be inverted is dominantly diagonal, are recommended; see for instance [21]. At the same time note that the de Vogelaere method never involves such an operation; the only matrix operations involved here are multiplication and addition.

Note also that the de Vogelaere algorithm yields the values of both y and y' simultaneously while the Numerov algorithm yields only the values of y.

---

* This feature is more general in the sense that any closed algorithm of the form (1.11.18),when specialized for vector problem (1.13.32), leads to a formula which needs some matrix inversion.

Such a feature, together with the remarkable flexibility of the de Vogelaere method with respect to changing step sizes, are arguments which may make the user to give preference to the de Vogelaere algorithm even in the scalar case.

The further simplified linear second order differential equation is the one in which $c(x) = 0$. The corresponding scalar problem

$$y'' + b(x)y = 0,\ x \in [a, b],\ y(a) = y_0,\ y'(a) = y'_0, \tag{1.13.34}$$

as well as its straightforward extension to the vector case, typically occurs in quantum mechanics contexts. Its numerical solution encounters significant difficulties when $|b(x)|$ is large. Specifically, we shall see in chapter 3, section 3,1, that the larger $|b(x)|$ the less efficient the methods developed in this chapter. In that chapter we shall develop a perturbation formalism, to finally use its results in order to significantly restore the accuracy of the multistep algorithms, in particular of the Numerov method, see section 3.8.

REFERENCES

1. P. Henrici, *Discrete Variable Methods in Ordinary Differential Equations*, John Wiley and Sons, New York, 1964.
2. F. Bashforth and J. C. Adams, *Theories of Capillary Action*. Cambridge University Press, 1883.
3. E. J. Nystrom, *Über die numerische Integration von Differentialgleichungen*, Acta Soc. Sci. Fenn., **50**, No. 13, 1 (1925).
4. F. R. Moulton, *New Methods in Exterior Ballistics*, Univ. Chicago Press, 1926.
5. W. E. Milne, *Numerical Integration of Ordinary Differential Equations*, Amer. Math. Monthly, **33**, 455 (1826), see also his book *Numerical Solution of Differential Equations*, John Wiley and Sons, New York, 1953.
6. G. Dalquist, *Convergence and stability in the numerical integration of ordinary differential equations*. Math. Scand. **4**, 33 (1956); *Stability and error bounds in the nnmerical integration of ordinary differential equations*, K. Tekn. Hogsk, Handl., Nr. 130 (1959); Proceedings of Symposium of Appl. Math., American. Math. Soc., Providence, R.I., 1963.
7. G. M. Phillips and P. J. Taylor, *Theory and Applications of Numerical Analysis*, Academic Press, London–New York–San Francisco, 1973.

8. C. Störmer, *Sur les trajectoires des corpuscules électrisés*, Arch. sci. phys. nat., Genève, **24,** 5; **24,** 113; **24,** 221(1907); also *Méthodes d'intégration numérique des équations différentielles ordinaires*, Comptes Rendus, congr. intern. math., Strasbourg, 1921.
9. P. H. Cowell and C. A. D. Crommelin, *Investigation of the motion of Halley's comet from 1759 to 1910*, Appendix to Greenwich Observations for 1909, Edinburgh, p. 84 (1910).
10. R. de Vogelaere, *A method for the numerical integration of differential equations of second order without explicit first derivatives*, J. Res. Nat. Bur. Standards, **54,** 119 (1955).
11. W. A. Lester, Jr., *De Vogelaere's method for the numerical integration of second order differential equations without explicit first derivatives: application to coupled equations arising from the Schrödinger equation*, J. Computational Physics, **3,**, 322 (1968).
12. J.-M. Launay, *Body-fixed formulation of rotational excitation: exact and centrifugal decoupling results for CO-He*, J. Phys. B: Atom. Molec. Phys., **9,** 1823(1976).
13. L. Verlet, *Computer "experiments" on classical fluids. I. Thermodynamical properties of Lennard-Jones molecules*, Phys. Rev., **159,** 98 (1967).
14. J. P. Coleman and Julie Mohamed, *On de Vogelare's method for* $y'' = f(x, y)$ Math. of Computation, **32,** 751 (1978).
15. L. Fox and E. T. Goodwin, *Some new methods for the numerical integration of ordinary differential equations*, Proc. Camb. Phyl. Soc., **45,** 373 (1949).
16. B. V. Numerov, *A method of extrapolation of perturbations*, Monthly Notices Roy. Ast. Soc. **84,** 592(1924); see also Astron. Nachricht., **230,** 359 (1927), and Bull. Acad. Sci. USSR, **1,** 1 (1932).
17. M. A. Melkanoff, T. Sawada, J. Raynal, *Nuclear optical model calculations*, Methods in Computational Physics, **6,** 1 (1966).
18. M. G. Salvadori and M. L. Baron, *Numerical methods in engineering*, Prentice Hall International, Englewood Cliffs, London, 1961.
19. F. Y. Hajj, H. Kobeisse, N. R. Nassif, *On the numerical solution of Schroedinger's radial equation*, J. Computational Physics, **16,** 150 (1974).
20. I. H. Sloan, *Errors in the Numerov and Runge-Kutta methods*, J. Computational Physics, **2,** 414 (1968).
21. A. C. Allison, *The numerical solution of coupled differential equations arising from the Schrödinger equation*, J. Computational Physics, **6,** 378 (1970).

Chapter 2

One-step methods

The one-step methods compute the solution at x_{k+1} on using information only from the very preceding mesh point x_k, thus being in contrast with the multistep methods which require information from *several* previous mesh points.

The major advantage of the one-step methods consists in the direct possibility of using nonequal step sizes. As for disadvantages, these come from the increased complexity of the calculations to derive one-step algorithms as well as from the fact that these algorithms are longer and less "aesthetic" than the multistep algorithms. It is also important to mention that the computational effort/step is generally higher in the case of one-step methods than for the multistep methods of the same accuracy.

In this chapter we discuss several usual one-step methods and two classes will be actually delimited. The criterion of such a division is the amount of information required to compute the solution at x_{k+1}. The methods which require the knowledge of only $x_k, \bar{y}_k$ and $h_k \equiv \equiv x_{k+1} - x_k$ will be called genuine one-step methods, while the methods which require some additional data, as for instance the values of the derivatives of y at x_k, will be called special one-step methods.

The main body of the chapter is devoted to the genuine one-step methods. The special one-step methods will be investigated only in short, in the last section.

2.1. GENERALITIES

To solve the initial value problem, eqs. (1) and (3) in "Preliminaries", viz.

$$y' = f(x, y), \ x \in [a, b], \ y(a) = y_0 \tag{2.1.1}$$

a partition of $[a, b]$ is introduced, $x_0 = a, x_1, x_2, \ldots, x_n = b$. The partition is in general non-equidistant and its current step size is denoted by h_k, viz. $h_k \equiv x_{k+1} - x_k$.

Any algorithm in which the approximation $\bar{y}_{k+1}$ to the solution at the point x_{k+1} can be calculated if only $x_k, \bar{y}_k$ and h_k are known is called a *genuine* one-step (or single-step) method. It is convenient to write the functional dependence of $\bar{y}_{k+1}$ against $x_k, \bar{y}_k$ and h_k in the form

$$\bar{y}_{k+1} = \bar{y}_k + h_k \Phi(x_k, \bar{y}_k; h_k). \tag{2.1.2}$$

The function Φ is called the *algorithm* increment function.

We can equally write the *exact* solution of the problem (2.1.1) in a similar form, viz.

$$y(x_{k+1}) = y(x_k) + h_k \Delta(x_k, y(x_k); h_k) \tag{2.1.3}$$

where the function Δ, called the *exact* increment function, is defined as follows

$$\Delta(x_k, y(x_k); h_k) = \begin{cases} \dfrac{y(x_{k+1}) - y(x_k)}{h_k} & \text{if } h_k > 0 \\ \\ f(x_k, y(x_k)) & \text{if } h_k = 0. \end{cases} \tag{2.1.4}$$

The local truncation error e_{k+1} for the algorithm (2.1.2) is defined as the deviation $y(x_{k+1}) - \bar{y}_{k+1}$ determined in the case when the input value $\bar{y}_k$ in eq. (2.1.2) is just the exact value $y(x_k)$ and the whole computation of eq. (2.1.2) is made with absolute accuracy *.

On subtracting eq. (2.1.2) from eq. (2.1.3) we get the following expression for e_{k+1}

$$e_{k+1} = h_k(\Delta(x_k, y(x_k); h_k) - \Phi(x_k, y(x_k); h_k)). \tag{2.1.5}$$

To derive genuine one-step algorithms means essentially to search for expressions of Φ which approximate Δ as closely as possible. The typical procedure consists of just assigning some expression depending on several parameters to finally find out the values of the parameters such that the chosen expression $\Phi(x_k, y(x_k); h_k)$ is numerically as close

---

* Notice that this definition is fully consistent with the definition adopted before in the case of multistep methods.

as possible to the exact $\Delta(x_k, y(x_k); h_k)$. Specifically, both Φ and Δ are developed in powers of h_k and the parameters which occur in Φ are fixed such that the coefficients of $h_k^0, h_k^1, h_k^2, \ldots$ agree with the similar coefficients in Δ up to the highest possible power of h_k. If m is the first power for which the two coefficients disagree we may write

$$\Delta(x_k, y(x_k); h_k) - \Phi(x_k, y(x_k); h_k) = h_k^m \varphi(x_k, y(x_k)) + 0(h_k^{m+1}) \quad (2.1.6)$$

where $\varphi(x, y)$ is some nonidentically vanishing function. Then the local truncation error is

$$e_{k+1} = h_k^{m+1} \varphi(x_k, y(x_k)) + 0(h_k^{m+2}) \quad (2.1.7)$$

and the method is called *of order m.*

At this point is should be recalled that, as also discussed in section 1.12, the order of a method is a quantity which is more closely related to the accumulated errors than to the local errors. Yet, the definition given above for the order is fully consistent with this feature. Specifically, it will be shown later on in this section that, if the leading term in the *local* truncation error is proportional to h_k^{m+1}, the leading term in the *accumulated* error is proportional to h^m, where h is the largest stepsize of the partition.

Before investigating this matter in detail we pause a bit to clarify two general elements associated to the genuine one-step algorithms, namely to define the convergence and to specify the condition to be satisfied by Φ such that the algorithm is convergent.

To start with we recall that the definition of convergence for the multistep algorithm (1.6.1) was given in section 1.6 and it essentially says that this algorithm is convergent if for any $X \in (a, b]$ one has

$$\lim_{\substack{h \to 0 \\ (n \to \infty)}} \bar{y}_n = y(x)$$

(note that n and h are there connected by the formula $nh = X - a$) provided the starting values approach the initial condition as indicated in eq. (1.6.5).

That definition explicitly accounted for the two basic peculiarities of the multistep algorithms, namely that they use *equal* step sizes and require *several* starting values.

To define the convergence of a one-step method we must modify the above definition by appropriately leaving out the two restrictions.

We thus take again some arbitrary fixed $X \in (a, b]$ and consider the problem (2.1.1) on $[a, X]$ only. We further assume that the starting value in the algorithm (2.1.2) is just the given initial condition

$$\bar{y}_0 = y_0 \tag{2.1.8}$$

and introduce (generally nonequidistant) partitions of $[a, X]$:

$$a = x_0,\ x_1 = x_0 + h_0,\ x_2 = x_1 + h_1,\ \ldots,\ X \equiv x_n = x_{n-1} + h_{n-1}.$$

Let us denote

$$h = \max\{h_0, h_1, \ldots, h_{n-1}\} \tag{2.1.9}$$

and run the algorithm on finer and finer partitions, that is with decreasing h. $\left(\text{This also implies increasing } n \text{ in the sense that } h \to 0 \text{ implies } n \to \infty. \text{ Also notice that } X - a = \sum_{k=0}^{n-1} h_k.\right)$

The algorithm (2.1.2) is called *convergent* if

$$\lim_{\substack{h \to 0 \\ (n \to \infty)}} \bar{y}_n = y(X). \tag{2.1.10}$$

Again, it is tacitly assumed here that all runs are free of round-off errors.

As for the condition under which a numerical method is convergent, the equivalent of the theorem regarding the multistep methods (we recall it here: the algorithm (1.6.1) is convergent if and only if it is both consistent and stable) is formulated as follows for the algorithm (2.1.2)

THEOREM. *If the increment function $\Phi(x, y; h)$ is continuous (jointly as a function of its three arguments) in the region $x \in [a, b]$, $y \in (-\infty, +\infty)$, $0 \leq h \leq \bar{h}$ where $\bar{h} > 0$, and satisfies the Lipschitz condition in y,* viz.

$$|\Phi(x, y; h) - \Phi(x, y^*; h)| \leq L|y - y^*| \tag{2.1.11}$$

for all $(x, y; h)$ and $(x, y^*; h)$ in the region just defined, then algorithm (2.1.2) to solve problem (2.1.1) is convergent if and only if

$$\Phi(x, y; 0) = f(x, y). \tag{2.1.12}$$

This condition is usually called the *condition of consistency* of a genuine one-step algorithm *. We omit the proof of this theorem and refer the reader to [1, p. 71].

The two elements considered above provide a basis for the error analysis which follows.

The very purpose of the error analysis is to produce some more quantitative information on the convergence than it is embedded in the heuristic condition (2.1.10). We are actually interested to find a *measure* of the total accumulated error

$$\Delta_k = y(x_k) - \tilde{y}_k, \quad k = 1, 2, \ldots$$

in the practical case when (i) the values of the local step sizes $h_0, h_1, h_2, \ldots$ are given small numbers and (ii) the computation *is* affected by round-off errors.

We consider from the very beginning the case when a variable step size rule $h_V(x; h)$ is used and for convenience this is further assumed to be of the form $h_V(x; h) = S(x)h$ where h is the maximal step size of the run and the step size weight function $S(x)$ is a given function of x such that $0 < S(x) \leqslant 1$ for any $x \in [a, b]$. Clearly, given $S(x)$ and h the partition is generated uniquely, viz.

$$x_0 = a;\ h_k = h_V(x_k; h) = S(x_k)h,\ x_{k+1} = x_k + h_k,\ k = 0, 1, 2, \ldots \tag{2.1.13}$$

The error analysis is divided into two steps. In a first step the round-off errors are disregarded to find out a measure for the genuine accumulated truncation errors. The second step consists of finding a bound for the round-off errors. The results of the two steps are finally collected in the theorem given at the end of this section.

Genuine accumulated truncation errors

As announced, our aim at this stage is to find out a *measure* for the genuine accumulated truncation error

$$E_k \equiv y(x_k) - \bar{y}_k,\ k = 0, 1, 2, \ldots. \tag{2.1.14}$$

but to start with we search for a *bound* of E_k only.

---

* The theorem formulated in the main text essentially says that the consistency is the only condition for the convergence of a genuine one-step method. For comparison it is recalled that the convergence of multistep methods required an extra condition, that of stability. On corroborating the two results it can be concluded that the genuine one-step methods are automatically stable.

We assume that the chosen one-step algorithm is of order $m \geqslant 1$ and make use of the formula (2.1.7) for the *local* truncation error e_{k+1}, which we write here again as

$$e_{k+1} = h^{m+1}\{\varphi(x_k, y(x_k))\,[S(x_k)]^{m+1} + O(h)\}. \tag{2.1.15}$$

Let N_{k+1} be a bound for the expression multiplying h^{m+1},

$$N_{k+1} \geqslant |\varphi(x_k, y(x_k))[S(x_k)]^{m+1} + O(h)|, \tag{2.1.16}$$

$k = 0, 1, 2, \ldots, n-1$. These local bounds are next compared to yield the global bound N, defined as follows

$$N \equiv \max\{N_1, N_2, \ldots, N_n\}. \tag{2.1.17}$$

This construction guarantees that the inequality

$$|e_{k+1}| \leqslant Nh^{m+1} \tag{2.1.18}$$

holds for all $k = 0, 1, 2, \ldots, n-1$.
Now eq. (2.1.2) is subtracted from eq. (2.1.3) to get

$$\begin{aligned} E_{k+1} &= E_k + h_k(\Delta(x_k, y(x_k); h_k) - \Phi(x_k, \bar{y}_k; h_k)) = \\ &= E_k + h_k[\Delta(x_k, y(x_k); h_k) - \Phi(x_k, y(x_k); h_k)] + h_k[\Phi(x_k, y(x_k); h_k) - \\ &- \Phi(x_k, \bar{y}_k; h_k)] = E_k + e_{k+1} + h_k[\Phi(x_k, y(x_k); h_k) - \Phi(x_k, \bar{y}_k; h_k)]. \end{aligned} \tag{2.1.19}$$

By virtue of eq. (2.1.11) the last term is bounded as follows

$$\begin{aligned} h_k|[\Phi(x_k, y(x_k); h_k) - \Phi(x_k, \bar{y}_k; h_k)]| &\leqslant h_k L|y(x_k) - \bar{y}_k| \\ = h_k L|E_k| = hS(x_k)L|E_k| &\leqslant hL|E_k|, \end{aligned} \tag{2.1.20}$$

so that, upon taking the absolute values of the terms in eq. (2.1.19) we get the following recurrence inequality

$$|E_{k+1}| \leqslant (1 + hL)|E_k| + Nh^{m+1},\ E_0 = y(a) - \bar{y}_0 = 0,\ k = 0, 1, 2, \ldots \tag{2.1.21}$$

This is exactly of the form (1.12.61) and hence the bound (1.12.62) is valid, i.e.

$$|E_k| \leqslant h^m N \frac{\exp(Lkh) - 1}{L}. \tag{2.1.22}$$

Since the argument of the exponential function does not contain explicitly the mesh point x_k at which E_k refers directly, we search for a way to bring the product kh to a form in which x_k is explicited. We have in order:

$$kh = \sum_{j=0}^{k-1} h = \sum_{j=0}^{k-1} \frac{h_j}{S(x_j)} = \sum_{j=0}^{k-1} \frac{1}{S(x_j)} (x_{j+1} - x_j) \tag{2.1.23}$$

and since each h_j is supposed to be sufficiently small we can accept that the direct approximation

$$\frac{1}{S(x_j)} (x_{j+1} - x_j) \simeq \int_{x_j}^{x_{j+1}} \frac{1}{S(x)} dx \tag{2.1.24}$$

is sufficiently accurate. We thus have

$$kh \simeq \int_a^{x_k} \frac{1}{S(x)} dx \equiv F(x_k, a; S). \tag{2.1.25}$$

(Note in passing that in the case of equal step sizes we have $S(x) = 1$ and then $F(x_k, a; 1) = x_k - a$.)
On using this result we return to eq. (2.1.22) to conclude that, if h is sufficiently small, the genuine accumulated truncation error at the point x_k is bounded as follows

$$|E_k| \leqslant h^m N G(x_k, a; S) \tag{2.1.26}$$

where

$$G(x_k, a; S) = \frac{\exp(LF(x_k, a; S)) - 1}{L}. \tag{2.1.27}$$

The second sequence of this stage consists of further refining the error estimate by searching for *an approximation* to E_k in lieu of the *bound* (2.1.26). As a matter of fact, the knowledge of the bound (2.1.26) should have not been omitted since it is essential for deriving such an approximation. In fact, it is eq. (2.1.26) that guarantees that some constants C and D exist such that the inequalities

$$|E_k| \leqslant Ch^m \text{ or } |E_k| \leqslant D h_k^m \tag{2.1.28}$$

are valid for small h and any $k = 0, 1, 2, \ldots$.
The starting point of this sequence is eq. (2.1.19) whose last term is conveniently worked out below. Using $y(x_k) = \bar{y}_k + E_k$ and expanding in powers of E_k we get

$$\Phi(x_k, y(x_k); h_k) - \Phi(x_k, \bar{y}_k; h_k) = \Phi_y(x_k, y(x_k); h_k)\, E_k$$

$$- \frac{1}{2}\Phi_{yy}(x_k, y^*; h_k)\, E_k^2 = [\Phi_y(x_k, y(x_k); 0) + h_k \Phi_{yh_k}(x_k, y(x_k); h^*)]E_k$$

$$- \frac{1}{2}\Phi_{yy}(x_k, y^*; h_k)E_k^2 \tag{2.1.29}$$

where y^* is a value between $y(x_k)$ and $\bar{y}_k$, and $0 < h^* < h_k$.

Since $m \geqslant 1$, it follows from eqs. (2.1.6) and (2.1.4) that $\Phi(x, y; 0) = f(x, y)$. (Note in passing that this indicates that any algorithm (2.1.2) of order $m \geqslant 1$ is automatically convergent, see eq. (2.1.12).) We shall consistently write

$$g(x) = f_y(x, y(x)), \tag{2.1.30}$$

hence

$$\Phi_y(x_k, y(x_k); 0) = g(x_k). \tag{2.1.31}$$

Moreover, since eq. (2.1.28) is valid and the second derivatives of Φ are supposed to be bounded, we may write eq. (2.1.29) as follows

$$\Phi(x_k, y(x_k); h_k) - \Phi(x_k, \bar{y}_k; h_k) = g(x_k)E_k + O(h_k^{m+1}). \tag{2.1.32}$$

On collecting the expressions contained in eqs (2.1.7) and (2.1.32) into eq. (2.1.19) we get

$$E_{k+1} = E_k + h_k^{m+1}\varphi(x_k, y(x_k)) + h_k g(x_k)E_k + 0(h_k^{m+2}). \tag{2.1.33}$$

We now introduce new numbers $\bar{E}_s$, to be called magnified errors, by the relation

$$E_s = h^m \bar{E}_s. \tag{2.1.34}$$

With these, eq. (2.1.33) assumes the form

$$\bar{E}_{k+1} = \bar{E}_k + h_k\{g(x_k)\bar{E}_k + [S(x_k)]^m \varphi(x_k, y(x_k))\} + 0(h_k^2), \tag{2.1.35}$$

$$k = 0, 1, 2, \ldots,$$

with the initial condition $\bar{E}_0 = (y(a) - \bar{y}_0)\, h^{-m} = 0$.

When h is so small that the term $0(h_k^2)$ is numerically negligible eq. (2.1.35) is simply the Euler method for the problem

$$\bar{E}' = g(x)\bar{E} + [S(x)]^m \varphi(x, y(x)),\ \bar{E}(a) = 0,\ x \in [a, b]. \tag{2.1.36}$$

The conclusion then is that in the limit of small h the genuine accumulated error at each mesh point x_s is given by eq. (2.1.34) where $\bar{E}_s$ is the value at x_s of the solution of the problem (2.1.36).

Accumulated round-off errors

We consider again the algorithm (2.1.2) and assume that the input $\bar{y}_k$ is now some $\tilde{y}_k$ which is the numerical value of $\bar{y}_k$ as resulted from previous propagation of the solution from x_0 up to x_k when the quoted algorithm is used on a *real* computer. The deviation

$$r_k = \bar{y}_k - \tilde{y}_k \tag{2.1.37}$$

then represents the value at x_k of the accumulated round-off error associated to algorithm (2.1.2).

Our purpose is here to find a bound for r_k. To this end we investigate the computation of the right member of eq. (2.1.2). We first assume that this computation is carried out on the same real computer as all the previous computations up to the current step k. Thus the computer first evaluates $\Phi(x, \tilde{y}_k; h_k)$ to produce some round-off affected value $\tilde{\Phi}$. This is further multiplied by h_k and the result is added to $\tilde{y}_k$ (these operations imply some round-off errors, too) to get $\tilde{y}_{k+1}$.

We secondly assume that the quoted computation in the current step k is free of round-off errors, that is the computer evaluates exactly both $\Phi(x_k, \tilde{y}_k; h_k)$ and the expression $y^*_{k+1} = h_k\Phi(x_k, \tilde{y}_k; h_k) + \tilde{y}_k$.

Briefly, $\tilde{y}_{k+1}$ is affected by both the round-off errors from the previous steps and from the current step k, while in y^*_{k+1} the second source of errors is absent. The deviation

$$R_{k+1} = y^*_{k+1} - \tilde{y}_{k+1} \tag{2.1.38}$$

is called *local* round-off error.

The above discussion is equivalent to saying that the algorithm (2.1.2) carried out on a *real* computer gives the same results as the modified algorithm

$$\tilde{y}_{k+1} = \tilde{y}_k\, \Phi(x_k, \tilde{y}_k; h_k) - R_{k+1} \tag{2.1.39}$$

when the latter is used on an *ideal* computer. Hence, to find an estimate for the accumulated round-off errors we must simply compare the original and the modified algorithms.

We subtract eq. (2.1.39) from eq. (2.1.2) to get

$$r_{k+1} = r_k + h_k(\Phi(x_k, \bar{y}_k; h_k) - \Phi(x_k, \tilde{y}_k; h_k)) + R_{k+1}. \tag{2.1.40}$$

On using eq. (2.1.11) this becomes

$$|r_{k+1}| \leqslant (1 + Lh_k)|r_k| + |R_{k+1}|. \tag{2.1.41}$$

It is next assumed that the local round-off errors are bounded

$$|R_{k+1}| \leqslant R \tag{2.1.42}$$

for any $k = 0, 1, 2, \ldots$ and also replace h_k by the maximal step size h. Eq. (2.1.41) now becomes

$$|r_{k+1}| \leqslant (1 + Lh)|r_k| + R, \quad k = 0, 1, ,2, \ldots \tag{2.1.43}$$

with the starting value $|r_0|$. Since this recurrence relation is of the form (1.12.61) the solution $|r_k|$ is bounded as indicated in eq. (1.12.62), viz.:

$$|r_k| \leqslant |r_0| \exp(Lkh) + R \frac{\exp(Lkh) - 1}{Lh} \tag{2.1.44}$$

or, with the identification (2.1.25),

$$|r_k| \leqslant |r_0| \exp(LF(x_k, a; S)) + \frac{1}{h} R \frac{\exp(LF(x_k, a; S)) - 1}{L} \tag{2.1.45}$$

It is seen that when h is decreased $|r_k|$ increases as h^{-1}.

The results (2.1.34), (2.1.36) and (2.1.45) allow us to formulate the following theorem which is the analog of the basic theorem given on p. 47 for multistep methods

THEOREM. *If* (i) *algorithm* (2.1.2) *to solve the problem* (2.1.1) *is of order* $m \geqslant 1$,

(ii) *it is used on a real computer, i.e. with round-off errors included,*

(iii) *the partition is of the form* (2.1.13) *and its maximal step size* h *is sufficiently small,*

(iv) *the initial condition is some* $\tilde{y}_0$ *affected by round-off error, viz.:* $y(a) - \tilde{y}_0 = r_0$
then the deviation

$$\Delta_k \equiv y(x_k) - \tilde{y}_k, \quad k = 1, 2, 3, \ldots \tag{2.1.46}$$

can be written as a sum,

$$\Delta_k = E_k + r_k, \tag{2.1.47}$$

where (a) *the genuine accumulated truncation error* E_k *is of the form* $E_k = h^m \bar{E}_k$ *and* $\bar{E}_k$ *is the solution at* x_k *of the initial value problem*

$$\bar{E}' = g(x, y(x)) \bar{E} + [S(x)]^m \varphi(x, y(x)), \; \bar{E}(a) = 0, \tag{2.1.48}$$

called the error equation; g and φ *are defined in eqs* (2.1.30) *and* (2.1.6)
(b) *the accumulated round-off error* r_k *is bounded as in eq.* (2.1.45).
The practical application of this theorem is much similar to that of the theorem given on p. 47. Also see section 2.3 for the generation of an optimal form of the weight function $S(x)$.

2.2. THE RUNGE–KUTTA METHOD; VARIOUS VERSIONS

In this section we constantly refer to the method (2.1.2) in the local interval k and for convenience we typically omit the label k.

To start with, we recall that a single-step method based on the Taylor series expansion of the solution was already discussed in section 1.8. Its increment function is obviously

$$\Phi(x, y; h) = f(x, y) + \frac{1}{2} hf'(x, y) + \ldots + \frac{1}{m!} h^{m-1} f^{(m-1)}(x, y) \tag{2.2.1}$$

where $f', f'', \ldots,$ are total derivatives with respect to x.

It is worth mentioning here that this method is a *genuine* one-step method only in the case when the total derivatives of f are appropriately evaluated. Consider for instance only the function $f'(x, y)$ in eq. (2.2.1). Since we have

$$f' = f_x + y'f_y$$

the function f' is in general a function F of *three* variables x, y and y'. As an example, if $f(x, y) = x + y$ then $f' \equiv F(x, y, y') = 1 + y'$. It becomes a function of the *two* original variables (as it is implied in eq. (2.2.1)) only *after* we explicitly account for that $y' = f(x, y)$. In the above example this means $f'(x, y) \equiv F(x, y, x + y) = 1 + x + y$.

A similar argument can be repeated word by word for all next coming total derivatives of f to conclude that the right member of eq. (2.2.1) remains a function of the three variables x, y, h, only when the derivatives of y are explicited in terms of the equation to be solved. If however the user wants to assign to these derivatives values which result from some extra evaluations (such options will be discussed in section 2.6), then the increment function Φ will contain these derivatives as additional arguments. Although in such a case the algorithm remains a one-step algorithm, it is no longer of the form (2.1.2) and therefore many of the results derived in the previous section simply break down.

Since the exact increment function Δ is the infinite series

$$\Delta(x, y; h) = f(x, y) + \frac{1}{2} hf'(x, y) + \ldots +$$

$$+ \frac{1}{m!} h^{m-1} f^{(m-1)}(x, y) + \frac{1}{(m+1)!} h^m f^{(m)}(x, y) + \ldots, \tag{2.2.2}$$

the local truncation error of the method defined in (2.2.1) is easily found to be of the form (2.1.7), viz.

$$e(x+h) = \frac{1}{(m+1)!}\, h^{m+1} f^{(m)}(x, y(x)) + 0(h^{m+2}).$$

Thus the method described in section 1.8 can reach any order if an appropriate number of terms is retained in eq. (2.2.1). Yet, the practical construction of such an algorithm is sometimes difficult because of the need of the explicit calculation of the successive derivatives of f. It seems therefore more convenient to consider forms of Φ which avoid the explicit knowledge of the derivatives of f.

The methods commonly named Kunge-Kutta methods are just of this type; in these methods Φ is expressed in terms of the values of f only.

The increment function of the Runge-Kutta methods is assumed to be a linear combination of the form

$$\Phi(x, y; h) = \sum_{i=1}^{s} w_i k_i \tag{2.2.3}$$

where w_i are some numerical weights and k_i are the values of the function f at some suitable values of the two arguments, viz.

$$\begin{aligned} k_1 &= f(x, y) \\ k_i &= f\left(x + a_i h,\ y + h \sum_{j=1}^{i-1} b_{ij} k_j\right), \quad i = 2, 3, \ldots, s. \end{aligned} \tag{2.2.4}$$

For mnemonic reasons we will characterize the method by means of the following array of parameters

$$\begin{array}{r|cccccc} a_2 & b_{21} \\ a_3 & b_{31} & b_{32} \\ a_4 & b_{41} & b_{42} & b_{43} \\ \vdots & \vdots & \vdots & \vdots \\ a_s & b_{s1} & b_{s2} & b_{s3} & \ldots & b_{s,s-1} \\ \hline m\text{th-order solution} & w_1 & w_2 & w_3 & \ldots & w_{s-1} & w_s \end{array} \tag{2.2.5}$$

For $s = 1$ the Runge-Kutta method coincides with the Euler method, eq. (1.4.1). Its order is then $m = s = 1$. For $s > 1$ the problem is to calculate the value of the parameters a_i, b_{ij} ($i = 2, 3, \ldots, s$, $j = 1, 2, \ldots$ $\ldots, s-1$) and w_i ($i = 1, 2, \ldots, s$) such that m in eq. (2.1.6) is as large as possible. As it will be seen below, for the usual choices $s = 2, 3, 4$ the equality $m = s$ holds. Note also that for $s > 4$ one has $m < s$ (see the discussion at the end of this section).

Runge-Kutta methods with $s = 2$

We put $s = 2$ in eqs (2.2.3) and (2.2.4) and search for the values of the parameters w_1, w_2, a_2 and b_{21}. To this aim k_2 is developed in powers of h

$$\begin{aligned} k_2 &= f(x + a_2h,\ y + b_{21}hk_1) = f(x + a_2h,\ y + b_{21}hf) \\ &= f + h(a_2f_x + b_{21}f\cdot f_y) + h^2\left(\frac{1}{2}a_2^2 f_{xx} + a_2b_{21}f\cdot f_{xy} + \right. \\ &\left. + \frac{1}{2}b_{21}^2 f^2 f_{yy}\right) + O(h^3) \end{aligned} \tag{2.2.6}$$

where f is the short hand notation for $f(x, y)$ and all partial derivatives are calculated at (x, y).

The increment function then reads

$$\begin{aligned} \Phi(x, y; h) &= w_1k_1 + w_2k_2 = (w_1 + w_2)f + hw_2(a_2f_x + b_{21}f\cdot f_y) \\ &+ h^2w_2\left(\frac{1}{2}a_2^2f_{xx} + a_2b_{21}f\cdot f_{xy} + \frac{1}{2}b_{21}^2 f^2 f_{yy}\right) + O(h^3). \end{aligned} \tag{2.2.7}$$

Next we write down the exact increment function (2.2.2) in terms of partial derivatives

$$\begin{aligned} \Delta(x, y; h) &= f + \frac{1}{2}h(f_x + f\cdot f_y) + \frac{1}{6}h^2(f_{xx} + 2f\cdot f_{xy} + f^2 f_{yy} \\ &+ f_x\cdot f_y + f\cdot f_y^2) + O(h^3) \end{aligned} \tag{2.2.8}$$

and finally identify the coefficients of the like powers of h in eqs (2.2.7) and (2.2.8). The coefficients of h^0 give $(w_1 + w_2)f = f$, so that

$$w_1 + w_2 = 1, \tag{2.2.9}$$

while the coefficients of h^1 give

$$w_2(a_2 f_x + b_{21} f \cdot f_y) = \frac{1}{2}(f_x + f \cdot f_y). \tag{2.2.10}$$

Since f_x and $f \cdot f_y$ are in general independent we must equalize their coefficients separately,

$$w_2 a_2 = \frac{1}{2}, \qquad w_2 b_{21} = \frac{1}{2}. \tag{2.2.11}$$

The algebraic system formed by eqs (2.2.9) and (2.2.11) consists of *three* equations for *four* unknowns and hence it is underdetermined. If we take one of these unknowns as a free parameter we obtain a unique solution for the other three unknowns. Specifically, if we take $w_2 = \lambda$ as the free parameter, then

$$w_1 = 1 - \lambda, \; a_2 = \frac{1}{2\lambda}, \quad b_{21} = \frac{1}{2\lambda}. \tag{2.1.12}$$

In conclusion the algorithm of the Runge-Kutta method with $s = 2$ is (we restore for the moment the labels)

$$\bar{y}_{k+1} = \bar{y}_k + w_1 h_k k_1 + w_2 h_k k_2 = \bar{y}_k + (1 - \lambda) h_k k_1 + \lambda h_k k_2 \tag{2.2.13 a}$$

where

$$k_1 = f(x_k, \bar{y}_k), \; k_2 = f\left(x_k + \frac{1}{2\lambda} h_k, \; \bar{y}_k + \frac{1}{2\lambda} h_k k_1\right) \tag{2.2.13 b}$$

with the *free parameter* λ.

As a matter of fact, it must be emphasized that in general the value of λ never results from an additional condition, as, for instance, by equating the coefficients of h^2 in eqs (2.2.7) and (2.2.8). In fact, when the expressions of these coefficients are compared, it is seen that they can never be made equal since, for instance, the quantity $f_x \cdot f_y$ does not occur in eq. (2.2.7) (that is its coefficient is zero) while its coefficient is 1/6 in eq. (2.2.8). If we however insist to impose the condition of equating the coefficients of $f_{xx}, f \cdot f_{xy}$ and $f^2 \cdot f_{yy}$ we obtain $w_2 a_2^2/2 = 1/6$, $w_2 a_2 b_{21} = 1/3$ and $w_2 b_{21}^2/2 = 1/6$ which gives $a_2 = 2/3$ and thus $\lambda = 3/4$. Yet, this does not imply at all that the value $\lambda = 3/4$ is better than the other choices. In fact, it may well happen that for some different value of λ the coefficient of h^2 in Φ is numerically equal with the whole coefficient of h^2 in Δ, and not with the sum of its first three terms, as implied by $\lambda = 3/4$.

On subtracting eq. (2.2.7) from eq. (2.2.8) and then multiplying by h, the local truncation error is readily found to be of the form (2.1.7) with $m = 2$ and

$$\varphi(x, y) = \left(\frac{1}{6} - \frac{1}{8\lambda}\right) f_{xx} + \left(\frac{1}{3} - \frac{1}{4\lambda}\right) f \cdot f_{xy} + \left(\frac{1}{6} - \frac{1}{8\lambda}\right) f^2 f_{yy} + \frac{1}{6}(f_x \cdot f_y + f \cdot f_y^2). \tag{2.2.14}$$

Since the value of λ is arbitrary the Runge-Kutta method with $s = 2$ is never referred to as a unique method but as a *family* of methods in which each member is identified by the value of λ. The above analysis also shows that any member of this family is of order $m = s = 2$ and that the expression of $\varphi(x, y)$ is rather complicated to be useful in practical applications, as for instance to evaluate the accumulated error by means of eq. (2.1.48). In addition, it is generally impossible to make an optimal choice for λ.

Two particular choices are however preferred, mainly because they lead to more symmetric formulas. These are $\lambda = 1$, with the algorithm

$$\bar{y}_{k+1} = \bar{y}_k + h_k f(x_k + \frac{1}{2} h_k, \bar{y}_k + \frac{1}{2} h_k f(x_k, \bar{y}_k)), \tag{2.2.15}$$

known as the *improved polygon method* or the *modified Euler method*, see for instance [1 p. 67], [2 p. 53], and $\lambda = 1/2$, viz.:

$$\bar{y}_{k+1} = \bar{y}_k + \frac{1}{2} h_k [f(x_k, \bar{y}_k) + f(x_k + h_k, \bar{y}_k + h_k f(x_k, \bar{y}_k))], \tag{2.2.16}$$

sometimes called the *improved Euler method* [2 p. 54], or the *Heun* method [3].

Runge-Kutta methods with $s = 3$

The function Φ is here of the form

$$\Phi(x, y; h) = w_1 k_1 + w_2 k_2 + w_3 k_3 \tag{2.2.17}$$

with k_1, k_2 and k_3 given by eq. (2.2.4).

To find the associated parameters $w_1, w_2, w_3, a_2, a_3, b_{21}, b_{31}$ and b_{32} we apply a similar procedure as before, namely use the expansion in powers of h for k_2 and k_3, introduce these in eq. (2.2.17) to finally identify the coefficients of like powers of h in the resultant expansions of Φ, and of Δ, eq. (2.2.2). The series for k_2 is already given in eq. (2.2.6). As for k_3 we have

$$\begin{aligned} k_3 = f + h[a_3 f_x + (b_{31} + b_{32}) f \cdot f_y] + h^2 \Bigg[\frac{1}{2} a_3^2 f_{xx} + a_3(b_{31} + b_{32}) f \cdot f_{xy} \\ + \frac{1}{2} (b_{31} + b_{32})^2 f^2 f_{yy} + b_{32}(a_2 f_x + b_{21} f \cdot f_y) f_y \Bigg] + 0(h^3). \end{aligned} \tag{2.2.18}$$

On equating the coefficients of h^0, h^1 and h^2 in Φ and Δ we get the following set of algebraic equations

$$\begin{aligned} & w_1 + w_2 + w_3 = 1, & & a_2 w_2 + a_3 w_3 = \frac{1}{2}, \\ & b_{21} w_2 + (b_{31} + b_{32}) w_3 = \frac{1}{2}, & & \frac{1}{2} a_2^2 w_2 + \frac{1}{2} a_3^2 w_3 = \frac{1}{6}, \\ & a_2 b_{21} w_2 + a_3(b_{31} + b_{32}) w_3 = \frac{1}{3}, & & \frac{1}{2} b_{21}^2 w_2 + \frac{1}{2} (b_{31}+b_{32})^2 w_3 = \frac{1}{6}, \\ & a_2 b_{32} w_3 = \frac{1}{6}, & & b_{21} b_{32} w_3 = \frac{1}{6} \end{aligned} \tag{2.2.19}$$

The two equations in the last line give $a_2 = b_{21}$. This is further introduced in the equations of the previous line to obtain that $a_3 = b_{31} + b_{32}$. The remaining four equations plus eqs. $a_2 = b_{21}$ and $a_3 = b_{31} + b_{32}$ form an algebraic system of *six* equations for *eight* unknowns w_1, w_2, w_3, a_2, a_3, b_{21}, b_{31} and b_{32}. If *two* of them are fixed the other *six* result uniquely.

As for the local truncation error, it is determined by the standard procedure namely subtract Φ from Δ and then multiply the result by h. The explicit calculation is a bit longer than in the previous case and we only quote the result: the local truncation error is of the form (2.1.7) with $m = s = 3$ and

$$\varphi(x, y) = \frac{1}{4!}[(1 - 4(a_2^3 w_2 + a_3^3 w_3)f''' + (1 - 12a_2^2 b_{32} w_3)f_y \cdot f''$$

$$+ (3 - 24a_2 a_3 b_{32} w_3)f' \cdot (f_y)' + f' \cdot f_y] \tag{2.2.20}$$

where again the primes denote the total derivatives with respect to x.

To summarize, we reach conclusions which closely resemble the ones drawn in the case $s = 2$, viz. (i) the algorithm is not unique; it actually depends on *two* free parameters (the previous case involved only *one* free parameter); (ii) the order of the method or, better saying, of the family of methods, is $m = s = 3$, (iii) the formula of the local truncation error is rather cumbersome to be operational from a practical point of view.

The most popular case is $a_2 = 1/2$, $a_3 = 1$. Since the above algebraic system determines uniquely the other coefficients as $w_1 = w_3 = 1/6$, $w_2 = 2/3$, $b_{21} = 1/2$, $b_{31} = -1$ and $b_{32} = 2$, the algorithm is:

$$\bar{y}_{k+1} = \bar{y}_k + \frac{1}{6}h_k(k_1 + k_3) + \frac{2}{3}h_k k_2 \tag{2.2.21 a}$$

with

$$k_1 = f(x_k, \bar{y}_k), \quad k_2 = f\left(x_k + \frac{1}{2}h_k, \bar{y}_k + \frac{1}{2}h_k k_1\right),$$

$$k_3 = f(x_k + h_k, \bar{y}_k + h_k(-k_1 + 2k_2)), \tag{2.2.21 b}$$

and this is known as the Kutta method.

Runge-Kutta methods with $s = 4$

Their algorithm is of the form (2.1.2) where Φ is a sum of four terms

$$\Phi(x, y; h) = w_1k_1 + w_2k_2 + w_3k_3 + w_4k_4$$

with k_1, k_2, k_3 and k_4 defined in eq. (2.2.4). The procedure to derive the values of the thirteen parameters $w_1, w_2, w_3, w_4, a_2, a_3, a_4, b_{21}, b_{31}, b_{32}, b_{41}, b_{42}, b_{43}$ is similar with that used above in cases $s = 2$ and $s = 3$. The only difference is that now the identification of the coefficients of the like powers of h in Φ and Δ also includes terms with h^3 to get the algorithm, and with h^4 to evaluate the local truncation error. For this reason the calculation becomes too long to be repeated here; for details we refer to [4]. We only quote the main conclusions which are very similar to the ones resulting in the previous cases: (i) the algorithm is not unique; it actually depends on *three* free parameters; (ii) the order of each member of this family of methods is $m = s = 4$; (iii) the formula of the local truncation error is not operational from a practical point of view.

The most popular versions at $s = 4$ are as follows:

(i) The *standard* Runge-Kutta method [5]. The array (2.2.5) of the coefficients of this version is

$$\begin{array}{r|llll} 1/2 & 1/2 & & & \\ 1/2 & 0 & 1/2 & & \\ 1 & 0 & 0 & 1 & \\ \hline \text{4th-order solution} & 1/6 & 1/3 & 1/3 & 1/6 \end{array} \qquad (2.2.22)$$

so that the algorithm is

$$\bar{y}_{k+1} = \bar{y}_k + \frac{1}{6} h_k[k_1 + k_4 + 2(k_2 + k_3)] \qquad (2.2.23\text{ a})$$

with

$$k_1 = f(x_k, \bar{y}_k),\ k_2 = f\left(x_k + \frac{1}{2} h_k, \bar{y}_k + \frac{1}{2} h_k k_1\right),$$

$$k_3 = f\left(x_k + \frac{1}{2} h_k, \bar{y}_k + \frac{1}{2} h_k k_2\right),\quad k_4 = f(x_k + h_k, \bar{y}_k + h_k k_3). \tag{2.2.23 b}$$

The local truncation error is of the form (2.1.7) where $m = s = 4$ and

$$\varphi(x, y) = -\frac{1}{2880} \cdot f^{(4)} + \frac{1}{576} f''' f_y - \frac{1}{288} (f_y^2 - f_{xy} - f_{yy} \cdot f) f''$$

$$-\frac{1}{192} (2f_{xy} f_y + 3f_{yy} f_y f - 2f_y^3 + f_{yy} f_x) f'. \tag{2.2.24}$$

(ii) The *Gill method* [6], with the following array of the parameters

$$\begin{array}{r|llll} 1/2 & 1/2 & & & \\ 1/2 & 1/\sqrt{2}-1/2 & 1-1/\sqrt{2} & & \\ 1 & 0 & -1/\sqrt{2} & 1+1/\sqrt{2} & \\ \hline \text{4th-order solution} & 1/6 & (1-1/\sqrt{2})/3 & (1+1/\sqrt{2})/3 & 1/6 \end{array} \tag{2.2.25}$$

(iii) The so called *Kutta's three-eighth's rule* [7 p. 22] with the parameters

$$\begin{array}{r|llll} 1/3 & 1/3 & & & \\ 2/3 & -1/3 & 1 & & \\ 1 & 1 & -1 & 1 & \\ \hline \text{4th-order solution} & 1/8 & 3/8 & 3/8 & 1/8 \end{array} \tag{2.2.26}$$

As a matter of fact, there are just these three versions which are compared in [8] for efficiency. The main result of this analysis is that the version (ii) is superior with respect to the memory requirements (it actually requires the smallest number of memory allocations) and also with respect to the influence of the round-off errors. Both advantages are however found to be counterbalanced by the computational time/step which is somewhat larger than for the other two methods.

The literature devoted to the Runge-Kutta methods is extremely rich and many other versions were proposed and tested along the time, see for instance [9] and references therein. The essential conclusion of such investigations resembles in spirit the conclusion of the older paper [8] in the sense that in the class of given s it can never be claimed that one and the same version is superior to the other versions in all respects. Because of this the decision of what version should be actually adopted remains to a significant extent a matter of the programmer's taste. The programmers of the IBM company, for instance, give preference to the Gill version, see [10].

To end this section we try to answer the natural question: "Why only the versions with $s = 2, 3$ and 4 are typically met in applications?" The answer basically hangs in the result that the order m of each version with $s > 4$ is *smaller* than s [11]. As an example, the Runge-Kutta algorithms with $s = 5$ are found to be methods of the fourth order, i.e. of the same order as the very methods with $s = 4$. It is then senseless to use an algorithm with $s = 5$ (which has to compute the function f *five* times in each elementary interval of the partition) to obtain an approximation to the solution which is of the same order as the one that results from an algorithm with $s = 4$ (i.e. which requires only *four* function evaluations per interval). Yet, the versions with $s > 4$ become of real use when the user is also interested to measure the errors, see [9] and also the next section.

2.3. STEP-SIZE ADJUSTMENT FOR GENUINE ONE-STEP METHODS

The theorem formulated at the end of section 2.1 allows us to appraise the errors, provided the following data are known:

(i) the order m of the method and the expression of the principal error function $\varphi(x, y)$;

(ii) the function $g(x) = f_y(x, y(x))$;

(iii) the true solution of the problem, since it figures as an argument in both φ and f_y;

(iv) the function $S(x)$ and the value of h; the latter must be sufficiently small, see condition (iii) in the theorem.

If these data are known and if the influence of the round-off errors is negligible (the last condition can be easily fulfilled in practice by using a suitable word length in the computation) the accumulated error is given by eqs (2.1.34) and (2.1.48).

The requirement (iii), paradoxical as it may seem at first sight, turns out to be the most trivial of the four. If in place of the true solution $y(x)$ the approximate values $\bar{y}_k$ are substituted in eq. (2.1.48), the right hand side of eq. (2.1.48) and thus the solution $\bar{E}(x)$ is falsified by an amount $0(h^m)$ which may be well absorbed into the correction term $0(h_k^2)$ in eq. (2.1.35) which is supposed to be negligible.

The requirement (ii) typically raises no problem while the requirement (i) does. In fact, the expression of $\varphi(x, y)$ is generally too complicate to become operational in practice, see for instance eq. (2.2.20). The practicality of the whole approach then essentially depends on the possibility of finding a simpler way to evaluate $\varphi(x, y)$ *.

The requirement (iv) is probably the most intricate one. To get the point we notice that in the aforementioned theorem the function $S(x)$ is supposed as given and the theorem tells us that the error depends on the chosen $S(x)$. Since different shapes of $S(x)$ will lead to different errors, it follows that an efficient numerical approach of differential equations by one-step methods must also consider the additional problem of generating an optimal $S(x)$, that is a shape which secures that the curve $\bar{E}(x)$ is as close as possible to the x axis.

The solution of this problem meets serious difficulties from both theoretical and practical points of view. For this reason it is usually replaced by another problem which is less ambitious but more workable. It consists of finding the function $S(x)$ which is *locally optimal* in the sense that the partition resultant from the variable stepsize rule $h_V(x; h) = S(x)h$ ensures that the *local* errors, or functions of local errors, are *uniformly* distributed on $[a, b]$ for any $h \leqslant \bar{h}$. For instance we may require that the quantities which must be uniformly distributed are either the very local truncation errors, which means

$$|e_1| = |e_2| = |e_3| = \dots \tag{2.3.1}$$

or the *relative* local errors,

$$|e_1/\bar{y}_1| = |e_2/\bar{y}_2| = |e_3/\bar{y}_3| = \dots \tag{2.3.2}$$

* A problem of the same type also occurred in the case of multistep methods. The principal error function is there $\varphi(x, y) = Cy^{(m+1)}(x)$, see eq. (1.10.3), and a convenient procedure to evaluate $y^{(m+1)}(x)$ is Milne's approach discussed in section 1.9.

or the so-called *local errors per unit interval* (the local error per unit interval $\bar{e}_{k+1}$ is defined as $\bar{e}_{k+1} = e_{k+1}/h_k$),

$$|\bar{e}_1| = |\bar{e}_2| = |\bar{e}_3| = \dots \tag{2.3.3}$$

Clearly, these three choices are not the only possible ones but they are the most popular in practice. As for which of them must be adopted, the decision merely depends on the differential equation to be solved. As an example, if the amplitude of the solution $y(x)$ is not expected to vary significantly on $[a, b]$ then the first or the third choice is recommended; contrariwise the second choice is suggested. No matter which is the case, the construction of $S(x)$ requires a way to evaluate the local errors and again the evaluation of the principal error function $\varphi(x, y)$ is playing an essential role.

The simplest procedure to compute $\varphi(x, y)$ is based on the following theorem due to Gorn and Moore [12] (see also [1, p. 82]).

THEOREM. *We consider the one-step algorithm* (2.1.2) *of order m in the current step k and for convenience suppress the label. We also assume that x, $y(x)$ and h are given and compute $\bar{y}(x + h)$ by eq.* (2.1.2)

$$\bar{y}(x + h) = y(x) + h\Phi(x, y(x); h). \tag{2.3.4}$$

Next divide the interval $[x, x + h]$ into two successive intervals of the step size $\frac{1}{2}h$ each and apply the same algorithm twice:

$$\bar{y}\left(x + \frac{1}{2}h\right) = y(x) + \frac{1}{2}h\Phi\left(x, y(x); \frac{1}{2}h\right), \tag{2.3.5 a}$$

$$\bar{\bar{y}}(x + h) = \bar{y}\left(x + \frac{1}{2}h\right) + \frac{1}{2}h\Phi\left(x + \frac{1}{2}h, \bar{y}\left(x + \frac{1}{2}h\right); \frac{1}{2}h\right) \tag{2.3.5 b}$$

Then we have the result

$$\delta(x + h) \equiv \bar{\bar{y}}(x + h) - \bar{y}(x + h) = h^{m+1}\varphi(x, y(x))(1 - 2^{-m}) + O(h^{m+2}) \tag{2.3.6}$$

Proof. On using eq. (2.3.5 a), eq. (2.3.5 b) becomes

$$\bar{\bar{y}}(x+h) = y(x) + \frac{1}{2}h\Phi\left(x, y(x); \frac{1}{2}h\right) + \frac{1}{2}h\Phi\left(x + \frac{1}{2}h, y(x) + \right.$$

$$\left. + \frac{1}{2}h\Phi\left(x, y(x); \frac{1}{2}h\right); \frac{1}{2}h\right). \tag{2.3.7}$$

By virtue of eq. (2.1.6) the sum of the first two terms in the right hand side is

$$y(x) + \frac{1}{2}h\Phi\left(x, y(x); \frac{1}{2}h\right) = y(x) + \frac{1}{2}h\Delta\left(x, y(x); \frac{1}{2}h\right) -$$

$$-\left(\frac{1}{2}h\right)^{m+1}\varphi(x, y(x)) + O(h^{m+2}) \tag{2.3.8}$$

$$= y\left(x + \frac{1}{2}h\right) - \left(\frac{1}{2}h\right)^{m+1}\varphi(x, y(x)) + O(h^{m+2}).$$

On this basis, the third term can be written as

$$\frac{1}{2}h\Phi\left(x + \frac{1}{2}h, y(x) + \frac{1}{2}h\Phi\left(x, y(x); \frac{1}{2}h\right); \frac{1}{2}h\right)$$

$$= \frac{1}{2}h\Phi\left(x + \frac{1}{2}h, y\left(x + \frac{1}{2}h\right) - \left(\frac{1}{2}h\right)^{m+1}\varphi(x, y(x)) + \right.$$

$$\left. + O(h^{m+2}); \frac{1}{2}h\right) = \frac{1}{2}h\left[\Phi\left(x + \frac{1}{2}h, y\left(x + \frac{1}{2}h\right); \frac{1}{2}h\right) + \right.$$

$$\left. + O(h^{m+1})\right] = \frac{1}{2}h\left[\Delta\left(x + \frac{1}{2}h, y\left(x + \frac{1}{2}h\right); \frac{1}{2}h\right) - \right.$$

$$\left. - \left(\frac{1}{2}h\right)^{m}\varphi\left(x + \frac{1}{2}h, y\left(x + \frac{1}{2}h\right)\right) + O(h^{m+1})\right]$$

$$= \frac{1}{2}h\Delta\left(x + \frac{1}{2}h, y\left(x + \frac{1}{2}h\right); \frac{1}{2}h\right) -$$

$$-\left(\frac{1}{2}h\right)^{m+1}\varphi(x, y(x)) + O(h^{m+2}). \tag{2.3.9}$$

(In the last line we used that $\varphi(x+h/2, y(x+h/2)) = \varphi(x,y) + O(h)$.) On collecting the results (2.3.8) and (2.3.9) in which we use that by the very definition of Δ, eq. (2.1.4), we have

$$\frac{1}{2}h\Delta\left(x, y(x); \frac{1}{2}h\right) = y\left(x+\frac{1}{2}h\right) - y(x) \tag{2.3.10 a}$$

$$\frac{1}{2}h\Delta\left(x+\frac{1}{2}h, y\left(x+\frac{1}{2}h\right); \frac{1}{2}h\right) = y(x+h) - y\left(x+\frac{1}{2}h\right) \tag{2.3.10 b}$$

so that

$$y(x+h) - y(x) \equiv h\Delta(x+h, y(x+h)) =$$

$$= \frac{1}{2}\Delta\left((x, y(x); \frac{1}{2}h\right) + \frac{1}{2}h\Delta\left(x+\frac{1}{2}h, y\left(x+\frac{1}{2}h\right); \frac{1}{2}h\right) \tag{2.3.11}$$

eq. (2.3.7) becomes

$$\bar{\bar{y}}(x+h) = y(x+h) - 2\left(\frac{1}{2}h\right)^{m+1}\varphi(x, y(x)) + O(h^{m+2}) \tag{2.3.12}$$

We also use the fact that

$$\bar{y}(x+h) = y(x+h) - h^{m+1}\varphi(x, y(x)) + O(h^{m+2}) \tag{2.3.13}$$

which results from eq. (2.1.7).

On finally subtracting eq. (2.3.13) from eq. (2.3.12) we get eq. (2.3.6), Q.E.D.

This theorem indicates that to evaluate $\varphi(x, y(x))$ we must first apply the algorithm (2.1.2) in the current interval of step size h to compute $\bar{y}(x+h)$ and next apply it twice at the step size $h/2$ to compute $\bar{\bar{y}}(x+h)$. If h is small the correction term $O(h^{m+2})$ can be neglected and eq. (2.3.6) gives the value of $\varphi(x, y(x))$. As a matter of fact, the user never disposes of the value of the exact $y(x)$; he only knows its approximate value $\bar{y}(x)$ and then the procedure actually calculates the value of $\varphi(x, \bar{y}(x))$. Yet, since $\varphi(x, y(x)) = \varphi(x, \bar{y}(x)) + E\varphi_y(x, y^*)$ where $E \equiv y(x) - \bar{y}(x)$ and $y^* \in (y(x), \bar{y}(x))$, and since E is of order h^m, it can be concluded that $\varphi(x, \bar{y}(x))$ is a good approximation of the desired $\varphi(x, \bar{y}(x))$.

As for the very local truncation error, it is evaluated by means of eq. (2.1.7) in which only the leading term is retained,

$$e(x + h) \simeq h^{m+1}\varphi(x, y(x)). \tag{2.3.14}$$

On this basis we can further proceed to the construction of the locally optimal $S(x)$. For illustration we take the condition (2.3.1) and fix some error level $\varepsilon > 0$. The partition which ensures that $|e_k| \simeq \varepsilon$ for any $k = 1, 2, 3, \ldots$ is constructed by iteration.
To evaluate the first step size h_0, for instance, we assign some test value $h_0^{(0)} > 0$ and evaluate the local error $|e_1^{(0)}| = |e(x_0 + h_0^{(0)})|$ by eq. (2.3.14)

$$|e_1^{(0)}| = h_0^{(0)m+1}|\varphi(x_0, y_0)|. \tag{2.3.15}$$

Since the desired h_0 must satisfy the condition

$$|e_1| \simeq h_0^{m+1}|\varphi(x_0, y_0)| = \varepsilon \tag{2.3.16}$$

we get

$$|\varphi(x_0, y_0)| = \frac{\varepsilon}{h_0^{m+1}} \tag{2.3.17}$$

On introducing this in eq. (2.3.15) we get

$$h_0 \simeq h_0^{(0)}(|e_1^{(0)}|/\varepsilon)^{1/(m+1)}, \tag{2.3.18}$$

so that the suggested iteration procedure is

$$|e_1^{(i)}| = |e(x_0 + h_0^{(i)})|, \; h_0^{(i+1)} = h_0^{(i)}(|e_1^{(i)}/\varepsilon)^{1/(m+1)}, \; i = 0, 1, 2, \ldots \tag{2.3.19}$$

It is stopped when it is found that $h_0^{(i+1)}$ and $h_0^{(i)}$ coincide in the preset number of figures. The desired h_0 is finally indentified as $h_0^{(i+1)}$ and the position of x_1 is located at $x_0 + h_0$.

The iteration procedure remains the same for all next coming steps, viz.

$$|e_{k+1}^{(i)}| = |e(x_k+h_k^{(i)})|, \; h_k^{(i+1)}=h_k^{(i)}(|e_k^{(i)}|/\varepsilon)^{1/(m+1)}, \; i=0, 1, 2, \ldots \quad (2.3.20)$$

In many practical cases it happens that the successive step sizes vary rather smoothly, and for this reason the starting value $h_k^{(0)}$ may be taken simply as the stepsize of the previous interval, $h^{(0)} = h_{k-1}$. It is also interesting to note that since the last step size h_{n-1} must also fulfil the condition $b = x_{n-1} + h_{n-1}$, it is generally shorter than as it would result from the iteration process alone.

Once $h_0, h_1, h_2, \ldots, h_{n-1}$ are determined, let us denote

$$h_{\max} = \max \{h_0, h_1, \ldots, h_{n-1}\} \quad (2.3.21)$$

to define the following piecewise constant function $S(x)$

$$S(x) = \begin{cases} h_0/h_{\max} & \text{for } a \leqslant x < x_1 \\ h_1/h_{\max} & \text{for } x_1 \leqslant x < x_2 \\ \vdots & \\ h_{n-1}/h_{\max} & \text{for } x_{n-1} \leqslant x \leqslant x_n = b. \end{cases} \quad (2.3.22)$$

The above construction guarantees that the variable step size rule $h_V(x; h_{\max}) = S(x)h_{\max}$ generates a partition in which the local truncation errors are uniformly distributed, that is

$$|e_{k+1}| \simeq |h_k^{m+1} \varphi(x_k, y(x_k))| = h_{\max}^{m+1}[S(x_k)]^{m+1}|\varphi(x_k, y(x_k))| \simeq \varepsilon \quad (2.3.23)$$

for any $k = 0, 1, 2, \ldots, n-2$; as explained above the last local error e_n may happen to be much smaller in absolute value than ε but this is immaterial from a practical point of view.

If the intervals $[x_k, x_{k+1} = x_k + h_k]$, $k = 0, 1, 2, \ldots, n-1$ are narrow enough so that $\varphi(x, y(x))$ can be assumed as approximately constant over each such interval the function

$$F^q(x) \equiv [S(x)]^q \varphi(x, y(x)) \quad (2.3.24)$$

is approximately constant on $[a, x_{n-1}]$ for $q = m + 1$ and the value of this constant is $\varepsilon/h_{\max}^{m+1}$. It follows that any finer partition of the form $h_V(x; h) = S(x)\, h$ with $S(x)$ defined above and $h = h_{\max}/p$, $p = 2, 3, 4, \ldots$

(such a partition means the division of each original interval $[x_k, x_{k+1}]$ in p equal subintervals) automatically preserves the property that the local truncation errors remain almost uniformly distributed,

$$|e_{k+1}| \simeq \frac{1}{p^{m+1}} \varepsilon \tag{2.3.25}$$

for $k = 0, 1, 2, \ldots, p(n-1) - 1$. (The values of $|e_{k+1}|$ for $k = p(n-1)$, $p(n-1) + 1, \ldots, pn - 1$ are bounded at a lower level.)

The practical consequence of this remark is that, if increased accuracy of the results is required, there is no need to construct the function $S(x)$ anew. Specifically, the previously evaluated $S(x)$ is preserved and only the value of h is diminished as indicated above. In this way a substantial amount of computational effort is saved.

As an illustration, let us suppose that the user must compute the solution of a given problem with an absolute error $\Delta = 10^{-4}$ and to this aim he uses a one-step method of order $m = 4$. It is also supposed that the user disposes of some qualitative information on the solution, for instance that the solution amplitude does not exhibit large variation on $[a, b]$, so that the criterion (2.3.1) is justified. We further assume that the averaged computation time/step (i.e. the time required to advance the solution one step) is of τ seconds. To evaluate the local error by means of eq. (2.3.6) the computer will consume 3τ seconds since it actually uses the one-step algorithm three times, eqs (2.3.4), (2.3.5 a) and (2.3.5 b). We now assume that the user chooses $\varepsilon = 10^{-2}$ for the local error level and to compute each step size of the partition by means of the iteration process (2.3.20) he needs an average of two iterations. Then, if n is the resultant number of steps of the partition the total effort is $\tau_1 = 6n\tau$ and at this cost the computer has furnished both the approximate solution $\bar{y}_k$ and the principal error function $\varphi(x_k, \bar{y}_k)$ at all values of k. The next sequence consists of the computation of the error equation (2.1.48) and this is performed very fast indeed since the partition is already fixed. Once the values of the magnified errors $\bar{E}_1, \bar{E}_2, \ldots \bar{E}_n$ are found, they are compared and let $\bar{E}_{\max} = \max\{|\bar{E}_1|, |\bar{E}_2|, \ldots, |\bar{E}_4|\} = 2.5\ 10^{-2}$. Since the desired accuracy is $\Delta = 10^{-4}$, eq. (2.1.34) predicts that the computation must be repeated with the new step size rule $h_V(x; h) = S(x)h$ where $h = h_{\max}(\Delta/|_{\max}E|)^{1/4} \simeq \frac{1}{4} h_{\max}$. Then the user can repeat the calculation on this new partition (the new partition has $4n$ intervals) and the point is that no additional error control is necessary in this final run. Altogether the total computing time to obtain the solution with the required accuracy is reduced to $6n\tau + 4n\tau = 10n\tau$ seconds.

A comment is now in order with respect to the connection between the *locally* optimal $S(x)$ (the only one which can be effectively constructed) and the desired *globally* optimal $S(x)$. Clearly the former, which was constructed before on the basis of condition (2.3.1), does not imply at all that it is at the same time globally optimal. As a matter of fact, it is often claimed that if the condition based on the local error per unit interval control, eq. (2.3.3), is used then the resultant $S(x)$ represents a better simulation of the desired globally optimal step distribution function. The basic argument of this claim is that condition (2.3.1) ensures that the function $F^{m+1}(x)$ is kept under control while it is function $F^m(x)$ which governs the error equation (2.1.48), and clearly the latter is kept under control only by means of condition (2.3.3). Yet, in practice, there is not much difference between the results of the two types of error control, see for instance [9] and references therein.

The computation of $\varphi(x, y)$ by means of eq. (2.3.6) is general in the sense that it can be applied for any genuine one-step method. For particular methods attempts have been made to derive specific, faster procedures and as an illustration we examine briefly the case of Runge-Kutta methods of order $m = 4$.

As discussed at the end of section 2.2 such methods result from $s \geqslant 4$. It is also mentioned there that when only the solution is computed the methods with $s = 4$ are clearly the fastest since they require only four function evaluations at each step.

When an appraisal of the local error is also needed, and this is carried out by means of the formula (2.3.6), then it is seen at once that this task becomes more costly than the very computation of the solution. In fact, while the computation of only the solution of eq. (2.1.1) by an s-stage Runge-Kutta method requires s function evaluations at each step, the application of formulae (2.3.5 a) and (2.3.5 b) requires $2s - 1$ additional function evaluations. (Specifically, s function evaluations are needed to compute Φ which occurs in eq. (2.3.5 a) and the same is true for Φ in eq. (2.3.5 b) but since $k_1 = f(x_k, \bar{y}_k)$ is common in eqs (2.3.4) and (2.3.5 a), one function evaluation is actually saved.)

The idea for deriving faster procedures is suggested by the fact that, as also discussed in section 2.2, the s-stage Runge-Kutta methods form a family and each member of the family is specified by the numerical values assigned to the free parameters. It is just this sort of undeterminacy that invites us to search whether sets of parameters exist such that the local truncation error can be expressed in an advantageous form, namely in a form which contains the same s functions $k_1, k_2, \ldots, k_s$ as it is anyway necessary to advance the very solution.

Specifically, particular sets of parameters are searched for such that the local error allows to be expressed as a linear combination of these k_j, $j = 1, 2, \ldots s$, viz.

$$e_{k+1}(= h_k^{m+1}\varphi(x_k, y_k) + 0(h_k^{m+2})) = \varphi_1 k_1 + \varphi_2 k_2 + \ldots + \varphi_s k_s + 0(h_k^{m+2}) \tag{2.3.26}$$

with *constant* weights $\varphi_1, \varphi_2, \ldots, \varphi_s$.

As announced, we specialize only to the case $m = 4$ and the result of such an investigation is that for $s = 4$ there is no solution (i.e. no version with $m = s = 4$ exists such that the error is expressed as in eq. (2.3.26)) but solutions do exist when we allow $s > 4$. Two examples follow.

The first example is the method of Merson [13] and it belongs to the family of methods with $s = 5$.

Its array of coefficients is

$$\begin{array}{r|rrrrr} 1/3 & 1/3 & & & & \\ 1/3 & 1/6 & 1/6 & & & \\ 1/2 & 1/8 & 0 & 3/8 & & \\ 1 & 1/2 & 0 & -3/2 & 2 & \\ \hline \text{4th-order solution} & 1/6 & 0 & 0 & 2/3 & 1/6 \end{array} \tag{2.3.27}$$

and the values of φ_i are: $\varphi_1 = -1/15$, $\varphi_2 = 0$, $\varphi_3 = 3/10$, $\varphi_4 = -4/15$, $\varphi_5 = 1/30$.

It is however important to emphasize that this method is applicable only when $f(x, y)$ is linear in x and y,

$$f(x, y) = Ax + By + C \tag{2.3.28}$$

in the sense that only in *this* case the method is of fourth order and the representation (2.3.26) is valid; otherwise the method is only of the *third* order.

The second example is the method of Fehlberg [14]. It corresponds to $s = 6$ but there is *no* restriction on the expression of $f(x, y)$. Its array of coefficients is

$$
\begin{array}{c|cccccc}
\frac{2}{9} & \frac{2}{9} & & & & & \\[6pt]
\frac{1}{3} & \frac{1}{12} & \frac{1}{4} & & & & \\[6pt]
\frac{3}{4} & \frac{69}{128} & -\frac{243}{128} & \frac{135}{64} & & & \\[6pt]
1 & -\frac{17}{12} & \frac{27}{4} & -\frac{27}{5} & \frac{16}{15} & & \\[6pt]
\frac{5}{6} & \frac{65}{432} & -\frac{5}{16} & \frac{13}{16} & \frac{4}{27} & \frac{5}{144} & \\[6pt]
\hline
\text{4th-order solution} & \frac{1}{9} & 0 & \frac{9}{20} & \frac{16}{45} & \frac{1}{12} & 0
\end{array}
\tag{2.3.29}
$$

and the values of φ_i are: $\varphi_1 = -1/150$, $\varphi_2 = 0$, $\varphi_3 = 3/100$, $\varphi_4 = -16/75$, $\varphi_5 = -1/20$, $\varphi_6 = 6/25$.

The advantage of using such versions is obvious. If, for instance, we use the standard Runge-Kutta method the fourth-order solution requires only four function evaluations/step but to measure the error by means of eq. (2.3.6) we need in addition seven function evaluations. In contrast, if the Fehlberg scheme is used we obtain both the fourth-order solution and the error at a total expense of only six function evaluations per step.

For more detailed examination of such techniques and for their implementations as efficient computer codes see [9].

2.4. OTHER GENUINE ONE-STEP METHODS

For the beginning we present two approaches which are not exactly *new* numerical methods. They are only some convenient adaptations of the previously discussed one-step methods. The goal of their inclusion here is mainly to illustrate that some degree of inventivity is always useful in specific applications.

We first consider the Taylor series method. Its application crucially depends on the ability to evaluate the successive derivatives of the function f. The efficiency of the method is therefore significantly increased if the user is able to accompany the algorithm by an automatic derivative generating procedure. Of course this is not possible for arbitrary f, but there are cases when this can be done without difficulty.

Such a case is discussed by Wilson [15] who assumes that $f(x, y)$ is linear in y,

$$f(x, y) = P_1(x)y + Q_1(x) \tag{2.4.1}$$

to obtain a recurrence relation for its derivatives. Indeed, $f^{(i-1)}(x, y)$ is assumed of the form

$$f^{(i-1)}(x, y) = P_i(x)y + Q_i(x), \quad i = 2, 3, 4, \dots \tag{2.4.2}$$

and this is differentiated once more:

$$\begin{aligned} f^{(i)}(x, y) &= P_i'y + P_iy' + Q_i' = P_i'y + P_i(P_1y + Q_1) + Q_i' \\ &= (P_i' + P_iP_1)y + P_iQ_1 + Q_i' \equiv P_{i+1}(x)y + Q_{i+1}(x), \end{aligned}$$

to obtain that the following recurrence relations are taking place

$$P_{i+1} = P_i' + P_iP_1, \; Q_{i+1} = Q_i' + P_iQ_1, \quad i = 1, 2, 3, \dots \tag{2.4.3}$$

Insomuch as the expressions of P_1 and Q_1 are simple enough (polynomials, for instance) this procedure is very simple indeed and the approximate solution of order m is

$$\begin{aligned} \bar{y}_{k+1} = \bar{y}_k\left(1 + h_kP_1(x_k) + \frac{1}{2!}h_k^2P_2(x_k) + \dots + \frac{1}{m!}h_k^mP_m(x_k)\right) + \\ + h_kQ_1(x_k) + \frac{1}{2!}h_k^2Q_2(x_k) + \dots + \frac{1}{m!}h_k^mQ_m(x_k) \cdot \end{aligned} \tag{2.4.4}$$

The procedure equally suggests a simple, direct error control. Suppose that equal stepsizes h are used and replace h by $-h$ in eq. (2.4.4). The right side of eq. (2.4.4) then furnishes the value of the approximate solution at the symmetric point x_{k-1}, a value which is already known from some previous application of the algorithm. The comparison of the two numbers then allows a direct appraisal of the error. If the error exceeds the acceptable level more terms are added in the Taylor series until the desired agreement is reached.

A similar way can be also adopted to solve numerically differential equations of higher order. Consider for instance the second-order equation

$$y'' = xy, \quad x \geqslant 0, \tag{2.4.5}$$

with the initial conditions $y(0) = 1$, $y'(0) = 0$, a case discussed in [7, p. 16]. We differentiate this equation $i - 2$ times to get

$$y^{(i)} = xy^{(i-2)} + (i-2)y^{(i-3)}, \quad i = 3, 4, 5, \ldots \tag{2.4.6}$$

which is a recurrence relation which enables the evaluation of $y^{(i)}$ in terms of $y^{(i-2)}$ and $y^{(i-3)}$.

The second approach we discuss here is based on the idea that sometimes conjugate use of several methods may exhibit better efficiency than the separate use of the components.

We actually follow a paper by Stoller and Morrison [16] in which a hybrid between the Runge-Kutta method and the Gauss quadrature formula is proposed.

We recall that the Gauss formula allows the integration of a function $g(x)$ on some $[a, b]$ by expressing the integral as a linear combination of the values of g at certain properly selected points $\bar{x}_s$, viz.

$$\int_a^b g(x)dx \simeq \sum_{s=1}^{p} a_s g(\bar{x}_s). \tag{2.4.7}$$

The description of such a p-point Gauss formula, which essentially means the recipe to calculate the positions $\bar{x}_s$ and the weights a_s, $s = 1, 2, \ldots, p$, can be found in any standard course of numerical analysis, see for instance [17]. On using the two-point Gauss formula on the elementary interval $[x_k, x_{k+1}]$ we have

$$y(x_{k+1}) - y(x_k) = \int_{x_k}^{x_{k+1}} y'(x)dx \simeq a_1 y'(\bar{x}_1) + a_2 y'(\bar{x}_2). \tag{2.4.8}$$

Thus, the procedure to solve the differential equation $y' = f(x, y)$ on $[x_k, x_{k+1}]$ with the initial condition $y(x_k) = \bar{y}_k$ consists of first solving the differential equation by some Runge-Kutta method in the successive intervals $[x_k, \bar{x}_1]$ and $[\bar{x}_1, \bar{x}_2]$. We thus obtain $\bar{y}(\bar{x}_1)$ and $\bar{y}(\bar{x}_2)$ respectively. The derivatives of the solution are determined by the very equation, $\bar{y}'(\bar{x}_1) = f(\bar{x}_1, \bar{y}(\bar{x}_1))$ and $\bar{y}'(\bar{x}_2) = f(\bar{x}_2, \bar{y}(\bar{x}_2))$. These are finally introduced in eq. (2.4.8) to obtain the value of $\bar{y}_{k+1}$.

The error in the result is not easy to determine but some empirical tests reported by the authors (they used the standard fourth-order version, eq. (2.2.22)) show a significant improvement in efficiency over the straightforward Runge-Kutta method.

The same type of procedure is also discussed by Hammer and Hollingsworth [18].

The application of the one-step methods to solve *systems* of differential equations is straightforward in the sense that the algorithms remain the same; the only difference is that y and f are now vectors. (The situation is thus entirely similar to the one encountered in the case of multistep methods, see section 1.8.) Due to this there is no problem to also solve differential equations of higher order since these can be always written as systems of first-order equations. Yet, specific procedures have been devised for the solutions of such equations of special form, in particular for the second-order equations in which the first derivative is absent.

We quote here the method of Nyström [19] which is applied to systems of second-order equations of special form, viz.

$$\mathbf{y}'' = \mathbf{f}(\mathbf{y}), \quad x \in [a, b], \quad \mathbf{y}(a) = \mathbf{y}_0, \mathbf{y}'(a) = \mathbf{y}_0'. \tag{2.4.9}$$

Note that $\mathbf{f}$ here depends on $\mathbf{y}$ but not on $\mathbf{y}'$ or on x. The last requirement, which seems so restrictive at first sight, turns out to be not restrictive at all since we can always rebaptise the argument x as one component of the vector $\mathbf{y}$. Consider for illustration the scalar problem

$$y'' = f(x, y), \quad x \in [a, b], y(a) = y_0', \quad y'(a) = y_0'.$$

On introducing the two-dimensional vectors

$$\mathbf{y} = \begin{pmatrix} x \\ y \end{pmatrix}, \quad \mathbf{f}(\mathbf{y}) = \begin{pmatrix} 0 \\ f(x, y) \end{pmatrix}$$

eq. $y'' = f(x, y)$ becomes of the very form $\mathbf{y}'' = \mathbf{f}(\mathbf{y})$ and the initial conditions are $\mathbf{y}_0 = \begin{pmatrix} a \\ y_0 \end{pmatrix}$, $\mathbf{y}_0' = \begin{pmatrix} 1 \\ y_0' \end{pmatrix}$.

The Nyström method exhibits striking similarities with the Runge-Kutta method. In each elementary interval $[x_k, x_{k+1} = x_k + h_k]$ it allows the computation of $\mathbf{y}_{k+1}$ in terms of given $\mathbf{y}_k$, $\mathbf{y}_k'$ and h_k as follows

$$\bar{\mathbf{y}}_{k+1} = \bar{\mathbf{y}}_k + h_k \mathbf{\Phi}(\bar{\mathbf{y}}_k, \bar{\mathbf{y}}_k'; h_k), \tag{2.4.10 a}$$

$$\bar{\mathbf{y}}_{k+1}' = \bar{\mathbf{y}}_k' + h_k \mathbf{\Psi}(\bar{\mathbf{y}}_k, \bar{\mathbf{y}}_k'; h_k), \tag{2.4.10 b}$$

where the two increment function vectors $\mathbf{\Phi}$ and $\mathbf{\Psi}$ are of the form

$$\mathbf{\Phi}(\mathbf{y}, \mathbf{y}'; h) = \mathbf{y}' + h \sum_{i=1}^{s} a_i \mathbf{k}_i, \tag{2.4.11 a}$$

$$\mathbf{\Psi}(\mathbf{y}, \mathbf{y}'; h) = \sum_{i=1}^{s} b_i \mathbf{k}_i. \tag{2.4.11 b}$$

Here a_i and b_i are some numerical weights and $\mathbf{k}_i$ are the values of the vector $\mathbf{f}(\mathbf{y})$ at some suitable values of the argument $\mathbf{y}$, viz.

$$\mathbf{k}_1 = \mathbf{f}(\mathbf{y})$$

$$\mathbf{k}_i = \mathbf{f}\left(\mathbf{y} + h u_i \mathbf{y}' + h^2 \sum_{j=1}^{i-1} w_{ij} \mathbf{k}_j\right), \quad i = 2, 3, \ldots, s. \tag{2.4.12}$$

To compactly characterize the method, the following array of coefficients is used

$$\begin{array}{r|llllll}
u_2 & w_{21} & & & & \\
u_3 & w_{31} & w_{32} & & & \\
u_4 & w_{41} & w_{42} & w_{43} & & \\
\vdots & \vdots & & & & \\
u_s & w_{s1} & w_{s2} & w_{s3} \cdots & w_{s,s-1} & \\
\hline
m\text{th-order solution} & a_1 & a_2 & a_3 \ldots & a_{s-1} & a_s \\
 & b_1 & b_2 & b_3 \ldots & b_{s-1} & b_s
\end{array} \tag{2.4.13}$$

The meaning of the order for this method is consistent with the previous definition adopted in this book. Specifically, the value of m guarantees that for eq. (2.4.9) the components of the two accumulated errors at some fixed $X \in (a, b]$,

$$\mathbf{E}(X) = \mathbf{y}(X) - \bar{\mathbf{y}}(X) \text{ and } \mathbf{E}'(X) = \mathbf{y}'(X) - \bar{\mathbf{y}}'(X)$$

tend to zero as h^m when the maximal step size h of the partition of $[a, X]$ is decreased.

The investigation of Nyström shows that for $s = 2, 3, 4$ the equality $m = s + 1$ holds, see also [1, pp. 171—173]. The three versions are as follows

$s = 2$

			(2.4.14)
2/3	1/3		
3rd-order solution	1/4	1/4	
	1/4	3/4	

$s = 3$

				(2.4.15)
1/2	1/8			
1	0	1/2		
4th-order solution	1/6	1/3	0	
	1/6	2/3	1/6	

$s = 4$

					(2.4.16)
2/5	2/25				
2/3	2/9	0			
4/5	4/25	4/25	0		
5th-order solution	23/192	25/64	−9/64	25/192	
	23/192	125/192	−27/64	125/192	

2.5. CHEBYSHEV SERIES METHOD

The most attractive property of the set of Chebyshev polynomials is that it provides particularly accurate approximations of the functions. For this reason the approximate solution of a differential equation is suggested to be determined just by expressing it as a Chebyshev series. Before examining this procedure in some detail (the procedure was originally considered by Lanczos [20] and Clenshaw [21]) we briefly recall some basic formulas concerning the Chebyshev polynomials. Consider the function $T_n(x)$ (integer n) defined on $x \in [-1, 1]$ as

$$T_n(x) \equiv \cos [n(\cos^{-1}x)]. \tag{2.5.1}$$

Note that this definition implies $T_{-n}(x) = T_n(x)$.
Using the trigonometrical identity

$$\cos[(n+1)\theta] + \cos[(n-1)\theta] = 2 \cos n\theta \cos \theta \tag{2.5.2}$$

and denoting $\theta = \cos^{-1}x$ we get after some rearrangement

$$T_{n+1}(x) = 2xT_n(x) - T_{n-1}(x). \tag{2.5.3}$$

Since from eq. (2.5.1) it results that $T_0(x) = 1$ and $T_1(x) = x$ it follows by induction on eq. (2.5.3) that T_n, $n \geqslant 0$, is a polynomial of degree n. It is called a Chebyshev polynomial. From the recurrence relation (2.5.3) we actually find that

$$T_2(x) = 2x^2 - 1, \; T_3(x) = 4x^3 - 3x, \; T_4(x) = 8x^4 - 8x^2 + 1, \tag{2.5.4}$$

etc. We also note the converse relations

$$\begin{gathered} 1 = T_0(x), \quad x = T_1(x), \quad x^2 = \frac{1}{2}(T_0(x) + T_1(x)), \\ x^3 = \frac{1}{4}(3T_1(x) + T_3(x)), \; x^4 = \frac{1}{8}(3T_0(x) + 4T_2(x) + T_4(x)), \end{gathered} \tag{2.5.5}$$

and so on.

In connection to the use of Chebyshev polynomials for solving differential equations the following four identities are of major importance

$$x^s T_r(x) = 2^{-s} \sum_{j=0}^{s} \binom{s}{j} T_{r-s+2j}(x) \tag{2.5.6}$$

$$T_r(x)T_s(x) = \frac{1}{2}[T_{r+s}(x) + T_{r-s}(x)] \tag{2.5.7}$$

$$\int^x T_r(x')\mathrm{d}x' = \begin{cases} \dfrac{1}{2}\left[\dfrac{T_{r+1}(x)}{r+1} - \dfrac{T_{r-1}(x)}{r-1}\right] & r \neq 1 \\ \\ \dfrac{1}{4}T_2(x) & r = 1 \end{cases} \tag{2.5.8}$$

$$T_r'(x) = \begin{cases} 0 & r = 0 \\ \\ 2r\left(\dfrac{1}{2}T_0 + T_2 + \ldots + T_{r-1}\right) & r \text{ odd} \\ \\ 2r(T_1 + T_3 + \ldots + T_{r-1}) & r \text{ even.} \end{cases} \tag{2.5.9}$$

We are now ready to tackle the very problem, that is

$$y' = f(x, y),\ x \in [a, b],\ y(a) = y_0. \tag{2.5.10}$$

As usually we introduce a partition $x_0 = a$, $x_1 = x_0 + h_0, \ldots, x_{k+1} = x_k + h_k, \ldots, x_n = b$, and focus on the current interval $[x_k, x_{k+1}]$ of length h_k.

To use Chebyshev polynomials we must first transform the current interval into the standard one $[-1, 1]$. This can be done easily by means of the linear transformation

$$X = \frac{1}{h_k}[2x - (x_k + x_{k+1})]. \tag{2.5.11}$$

Using it and introducing $Y(X) = y(x)$ the original equation to be solved on $[x_k, x_{k+1}]$, viz.

$$y' = f(x, y),\ x \in [x_k, x_{k+1}],\ y(x_k) = \bar{y}_k \tag{2.5.12}$$

becomes

$$Y' = F(X, Y),\ X \in [-1, 1],\ Y(-1) = \bar{y}_k, \tag{2.5.13}$$

with

$$F(X, Y) = \frac{1}{2} h_k f(x, y). \tag{2.5.14}$$

In practice the application of the method based on Chebyshev series is restricted to the differential equations in which the function $f(x, y)$ is linear in y and with polynomial coefficients, viz.

$$f(x, y) = p(x)y + q(x) \tag{2.5.15}$$

where $p(x)$ and $q(x)$ are polynomials in x; if so, its transform $F(X, Y)$ has the same structure,

$$F(X, Y) = P(X)Y + Q(X) \tag{2.5.16}$$

as it can be checked easily.

There are several possible ways to solve the problem (2.5.13) through Chebyshev series. One version, also discussed in [17, p. 349], is as follows;

We start from the Chebyshev series for $Y(X)$,

$$Y(X) = \sum_{r=0}^{\infty}{}' a_r T_r(X) = \frac{1}{2} a_0 T_0(X) + a_1 T_1(X) + a_2 T_2(X) + \dots \tag{2.5.17}$$

(Σ' denotes summation with the first term halved; such a convention is very popular in connection with the operations on Chebyshev series since it typically leads to formulas with more symmetry.) On integrating $Y' = F(X, Y)$ it results

$$Y(X) = \int P(X') Y(X') \mathrm{d}X' + \int Q(X') \mathrm{d}X' + C \tag{2.5.18}$$

where C is an arbitrary constant. We now express $P(X)$ and $Q(X)$ in terms of Chebyshev polynomials to construct the Chebyshev series for the expression in the right member of eq. (2.5.18),

$$\int P(X') Y(X') \mathrm{d}X' + \int Q(X') \mathrm{d}X' + C = \sum_{r=0}^{\infty}{}' c_r T_r(X) \tag{2.5.19}$$

Clearly, each coefficient c_r is some linear combination of the unknown a_r; the coefficient c_0 also depends on the arbitrary constant C. Eq. (2.5.18) implies $a_r = c_r$, $r = 0, 1, 2, \ldots$ and we thus obtain an infinite set of linear algebraic equations in the a_r. We usually solve the first few of these equations (except for $r = 0$ as this involves the arbitrary constant) and so find an approximation of the general solution of the differential equation; this is finally fixed in terms of the preset initial condition in problem (2.5.13).

To illustrate the process we take the following example

$$Y' = -XY + X, \quad X \in [-1, 1], \quad Y(-1) = 0. \tag{2.5.20}$$

On integrating, the equation becomes

$$Y = -\int X'Y(X')\mathrm{d}X' + \frac{1}{2}X^2 + C \tag{2.5.21}$$

and if series (2.5.17) is used for Y then from eqs (2.5.6) and (2.5.8) it results that

$$-\int X'\,Y(X')\,\mathrm{d}X' = -\frac{1}{4}(a_1 - a_3)T_1(X) - \frac{1}{4}\sum_{r=2}^{\infty}\frac{1}{r}(a_{r-2} - a_{r+2})T_r(X). \tag{2.5.22}$$

If we also notice that

$$\frac{1}{2}X^2 + C = \left(\frac{1}{4} + C\right) + \frac{1}{4}T_2(X) \tag{2.5.23}$$

the right member of eq. (2.5.21) becomes

$$-\int X'Y(X')\mathrm{d}X' + \frac{1}{2}X^2 + C = \sum_{r=1}^{\infty}{}' c_rT_r(X) \tag{2.5.24}$$

with

$$
\begin{aligned}
c_0 &= \frac{1}{2} + 2C \\
c_1 &= -\frac{1}{4}(a_1 - a_3), \\
c_2 &= -\frac{1}{4}\cdot\frac{1}{2}(a_0 - a_4) + \frac{1}{4}, \\
c_r &= -\frac{1}{4r}(a_{r-2} - a_{r+2}), \quad r = 3, 4, 5, \ldots
\end{aligned}
\tag{2.5.25}
$$

On equating coefficients a_r and c_r, $r = 1, 2, 3, \ldots$ we have in order

$$
\begin{aligned}
a_1 &= -\frac{1}{4}(a_1 - a_3), \\
a_2 &= -\frac{1}{8}(a_0 - a_4) + \frac{1}{4}, \\
a_r &= -\frac{1}{4r}(a_{r-2} - a_{r+2}) \quad r = 3, 4, 5, \ldots
\end{aligned}
\tag{2.5.26}
$$

This is an infinite set of linear algebraic equations for the a_r which, if solved, yields the exact general solution of the given differential equation. Yet, an infinite system of equations can never be solved in practice, therefore some truncation is necessary. To be specific we look for an approximation $\overline{Y}(X)$ of $Y(X)$ which consists of a prescribed finite number of terms in the series (2.5.17), five say, viz.

$$
\overline{Y}(X) = \frac{1}{2}\bar{a}T_0 + \bar{a}_1T_1 + \bar{a}_2T_2 + \bar{a}_3T_3 + \bar{a}_4T_4. \tag{2.5.27}
$$

We thus require that $\bar{a}_r$, $r = 1, 2, 3, 4$ satisfy the first four equations in the set (2.5.26) and assign $\bar{a}_5 = \bar{a}_6 = \ldots = 0$. The system to be solved then is

$$
\begin{aligned}
&5\bar{a}_1 - \bar{a}_3 = 0 \\
&\bar{a}_0 + 8\bar{a}_2 - \bar{a}_4 = 2 \\
&\bar{a}_1 + 12\bar{a}_3 = 0 \\
&\bar{a}_2 + 16\bar{a}_4 = 0
\end{aligned}
\tag{2.5.28}
$$

in which $\bar{a}_0$ is kept as a free parameter.

The first and the third equation collectively give $\bar{a}_1 = \bar{a}_3 = 0$ while the remaining two equations give $\bar{a}_4 = -\dfrac{1}{16}\bar{a}_2 = \dfrac{\bar{a}_0 - 2}{129}$. Thus the approximate *general* solution of the differential equation is

$$\bar{Y}(X) = \frac{1}{2}\bar{a}_0 T_0(X) - \frac{16(\bar{a}_0 - 2)}{129} T_2(X) + \frac{\bar{a}_0 - 2}{129} T_4(X). \tag{2.5.29}$$

The value of $\bar{a}_0$ is fixed in terms of the initial condition; since $T_0(-1) = T_2(-1) = T_4(-1) = 1$ we must have

$$\bar{Y}(-1) = 0 = \frac{1}{2}\bar{a}_0 - \frac{16(\bar{a}_0 - 2)}{129} + \frac{\bar{a}_0 - 2}{129} \tag{2.5.30}$$

whence

$$\bar{a}_0 = -\frac{20}{33}. \tag{2.5.31}$$

Thus the approximate solution of the problem (2.5.20) is

$$\bar{Y}(X) = -\frac{10}{33} + \frac{32}{99} T_2(X) - \frac{2}{99} T_4(X) = -\frac{16}{99}(4 - 5X^2 + X^4). \tag{2.5.32}$$

(For completeness note that the exact solution is $X(Y) = 1 - \exp[(1 - X^2)/2]$, and on $[-1, 1]$ the maximum value of the error $Y(X) - \bar{Y}(X)$ occurs at $X = 0$ when $Y(0) - \bar{Y}(0) \simeq 2.2\ 10^{-3}$.)

One interesting feature of the method is that it allows a direct, specific backward error analysis and we illustrate this on the same example. To measure the extent in which $\bar{Y}(X)$ approximates the exact solution $Y(X)$ means essentially to appraize the deviation of the $\bar{a}_r$ from the true a_r. The latter actually suggests checking how exactly the $\bar{a}_r$ verify the original infinite set (2.5.26) for the a_r. We see that the values: $\bar{a}_0$ arbitrary, $\bar{a}_1 = 0$, $\bar{a}_2 = -16(\bar{a}_0 - 2)/129$, $\bar{a}_3 = 0$, $\bar{a}_4 = (\bar{a}_0 - 2)/129$, and $\bar{a}_r = 0$ for $r = 5, 6, 7, \ldots$, satisfy exactly all equations of the system (2.5.26) except that for $r = 6$ which must be replaced by

$$\bar{a}_6 = -\frac{1}{4.6}(\bar{a}_4 - \bar{a}_8) + \frac{1}{4.6}\bar{a}_4. \tag{2.5.33}$$

This equation would result if a term $\bar{a}_4 T_6(X)/24$ were added to the right member of eq. (2.5.21). Thus $\bar{Y}(X)$ (which also obeys the initial condition $\bar{Y}(-1) = 0$, hence $\bar{a}_0 = -20/33$ and $\bar{a}_4 = -2/99$) is found to satisfy the equation

$$\bar{Y}(X) = -\int X'\bar{Y}(X')\mathrm{d}X' + \frac{1}{2}X^2 - \frac{1}{1188}T_6(X) + C. \tag{2.5.34}$$

It is further differentiated (to write $T_6'(X)$ use eq. (2.5.9)):

$$\bar{Y}'(X) = -X\bar{Y}(X) + X - \frac{1}{99}[T_1(X) + T_3(X) + T_5(X)] \tag{2.5.35}$$

and subtracted from the original equation (2.5.20). We thus find that the local error $e(X) = Y(X) - \bar{Y}(X)$ satisfies

$$e' = -Xe + \frac{1}{99}[T_1(X) + T_3(X) + T_5(X)],\ X \in [-1, 1],\ e(-1) = 0. \tag{2.5.36}$$

It is interesting to note that although this error equation, when solved, will produce numerically the error, it will never yield the same sort of information as we typically acquire in the case of other numerical methods. The concept of the *order* of a numerical method, for instance, is outside the spirit of the Chebyshev series method.

The Chebyshev series method can be easily extended to solve linear equations of higher order and with polynomial coefficients. For illustration we consider the case of a second-order equation. Let

$$Y'' = R(X)Y' + P(X)Y + Q(X),\ X \in [-1, 1] \tag{2.5.37}$$

where R, P and Q are polynomials and the initial conditions are fixed, viz. $Y(-1) = \bar{y}$, $Y'(-1) = \bar{y}'$.
We integrate eq. (2.5.37) twice to obtain

$$Y = \int^{X} dx' RY + \int^{X} dx'' \int^{x''} dx'(P - R')Y + \int^{X} dx'' \int^{x''} dx' Q + C_1 X + C_2 \tag{2.5.38}$$

where C_1 and C_2 are arbitrary constants, and proceed as before to obtain a system of linear equations in the Chebyshev coefficients. Note that this time we avoid to use the equations equating the coefficients of both T_0 and T_1 since these involve the arbitrary constants. These equations are replaced by the two initial conditions in the same way as it was done in eq. (2.5.30) for the above example.

To end the section we recall that the realisation of the Chebyshev series method considered here represents only one possible way; for other versions see, for instance, [20], [21] and [7, Chapter 10].

2.6. SPECIAL ONE-STEP METHODS

To start with, we recall that the one-step methods are defined as methods which compute the solution at each new mesh point x_{k+1} by using information about the solution from only the very last point x_k. As for the amount of such an information the class of the one-step methods is divided into two sub-classes, namely genuine and special one-step methods.

For the initial value problem (2.1.1) the genuine one-step methods are characterized by the fact that the only required information at x_k is the value of the solution while the special one-step methods require some additional data, as for instance the values of the successive derivatives of the solution.

There are several versions of special one-step methods. We first discuss a *closed* version.

This version consists essentially of the standard Taylor series method plus some extra procedure to iteratively estimate the remainder. In each interval of the partition the (approximate) solution is written as an mth-degree polynomial whose parameters are calculated on the basis of the initial conditions, in the first interval, and from continuity conditions in the other intervals.

Specifically, in the first interval the approximate solution of problem (2.1.1) is assumed to be of the form

$$\bar{y}(x) = y_0 + \frac{x - x_0}{1!} y_0' + \frac{(x - x_0)^2}{2!} y_0'' + \\ + \dots + \frac{(x - x_0)^{m-1}}{(m-1)!} y_0^{(m-1)} + \frac{(x - x_0)^m}{m!} C_0 \tag{2.6.1}$$

where $y_0', y_0'', \dots, y_0^{(m-1)}$ are determined from the initial condition in the same way as in the standard Taylor series method while C_0 is a number not yet determined.

To determine C_0 we require that $\bar{y}(x)$ satisfies the given differential equation at $x = x_1$,

$$\bar{y}_1' = f(x_1, \bar{y}_1). \tag{2.6.2}$$

Upon denoting

$$U = y_0' + \frac{h_0}{1!} y_0'' + \dots + \frac{h_0^{m-2}}{(m-2)!} y_0^{(m-1)} \tag{2.6.3 a}$$

$$V = y_0 + \frac{h_0}{1!} y_0' + \dots + \frac{h_0^{m-1}}{(m \dot{-} 1)!} y_0^{(m-1)}, \tag{2.6.3 b}$$

$h_0 = x_1 - x_0$, eq. (2.6.2) becomes

$$C_0 = \frac{(m-1)!}{h_0^{m-1}} \left[f\left(x_1, V + \frac{h_0^m}{m!} C_0 \right) - U \right] \tag{2.6.4}$$

which is an implicit algebraic equation for C_0 wich is solved by iteration as explained in section 1.9.

Once C_0 is evaluated the approximate solution $\bar{y}(x)$ is uniquely determined for any $x \in [x_0, x_1]$.

To find out the solution on the next interval we first compute the values at x_1 of the solution (2.6.1) *and* of its $m-1$ derivatives,

$$\bar{y}_1 = y_0 + \frac{h_0}{1!} y_0' + \frac{h_0^2}{2!} y_0'' + \ldots + \frac{h_0^{m-1}}{(m-1)!} y_0^{(m-1)} + \frac{h_0^m}{m!} C_0,$$

$$\bar{y}_1' = y_0' + \frac{h_0}{1!} y_0'' + \frac{h_0^2}{2!} y_0''' + \ldots + \frac{h_0^{m-2}}{(m-2)!} y_0^{(m-1)} + \frac{h_0^{m-1}}{(m-1)!} C_0$$

$$\vdots \qquad (2.6.5)$$

$$\bar{y}_1^{(m-1)} = y_0^{(m-1)} + h_0 C_0.$$

The solution on $[x_1, x_2]$ is written in a form similar to eq. (2.6.1)

$$\bar{y}(x) = \bar{y}_1 + \frac{x - x_1}{1!} \bar{y}_1' + \frac{(x - x_1)^2}{2!} \bar{y}_1'' + \ldots +$$

$$+ \frac{(x - x_1)^{m-1}}{(m-1)!} y_1^{(m-1)} + \frac{(x - x_1)^m}{m!} C_1 \qquad (2.6.6)$$

and to evaluate the constant C_1 we require that $\bar{y}(x)$ satisfies the given differential equation at $x = x_2$,

$$\bar{y}_2' = f(x_2, \bar{y}_2) \qquad (2.6.7)$$

which equally leads to an implicit equation for C_1.

The procedure is repeated in the same way on each of the next coming intervals. It is seen that to advance the solution each step the procedure requires not only the value of the solution at the lower end of the elementary interval but also the values of its $m-1$ derivatives: thus it is a *special* one-step method. It is equally seen that the procedure leads to the solution of an implicit algebraic equation; for this reason it is called *closed.* The approximate solution produced by this method is a function $\bar{y}(x)$, $x \in [a, b]$, which is a *piecewise* polynomial of degree m and with *jumps* of the mth derivative at the mesh points.

Such functions are called *spline* functions and for this reason the method is also called a *spline function method* to solve differential equations. As a matter of fact, the method was originally derived just in such a context by Loscalzo and Talbot [22].

The analysis undertaken in [23] also shows the important result that this method is convergent only if $m \leqslant 3$. Shortly the argument is as follows.

We assume for simplicity that the mesh points are equally spaced at the step size h.

The discussed method yields the solution as a spline function $\bar{y}(x) \in C^{m-1}[a, b]$ which satisfies the equation

$$\bar{y}'(x_i) = f(x_i, \bar{y}(x_i)), \; x_0 = a, \; x_i = x_0 + hi, \; i = 1, 2, 3, \ldots \tag{2.6.8}$$

Let us now take a sequence of $m + 1$ successive mesh points (remember that m is the degree of the piecewise polynomial) $x_k, x_{k+1}, \ldots, x_{k+m}$. Then the values of $\bar{y}$ and $\bar{y}'$ at these points are linear dependent. We write this fact in a particularly transparent way, viz.

$$\begin{aligned} & a_m^{(m)}\bar{y}(x_{k+m}) + a_{m-1}^{(m)}\,\bar{y}(x_{k+m-1}) + \ldots + a_0^{(m)}\bar{y}(x_k) \\ & = h(b_m^{(m)}\,\bar{y}'(x_{k+m}) + b_{m-1}^{(m)}\bar{y}'(x_{k+m-1}) + \ldots + b_0^{(m)}\bar{y}'(x_k)) \end{aligned} \tag{2.6.9}$$

where $a_s^{(m)}$ and $b_s^{(m)}$, $s = 0, 1, 2, \ldots, m$ are some constants. The values of the coefficients $a^{(m)}$ and $b^{(m)}$ for $m = 2, 3$, and 4 are collected in Table 9.

The similarity with the general multistep algorithm (1.6.1) is striking. Moreover, on comparing the coefficients in Table 9 with the coefficients of the closed multistep methods discussed in section 1.5 we find out that the spline function method with $m = 2$ is equivalent to the trapezoidal rule (1.5.2), the method with $m = 3$ is equivalent to the method of Milne (1.5.5) but the method with $m = 4$ has no analog.

Table 9

Coefficients $a_s^{(m)}$ and $b_s^{(m)}$, $s = 0, 1, 2, \ldots, m$, for $m = 2, 3$ and 4

m \ s	$a_s^{(m)}$ 0	1	2	3	$b_s^{(m)}$ 0	1	2	3
2	−1	1			1/2	1/2		
3	−1	0	1		1/3	4/3	1/3	
4	−1	−3	3	1	1/2	11/4	11/4	1/4

The last remark is fully normal because this method is unstable, and therefore it is not convergent. (Recall that a multistep method is convergent if and only if it is both consistent and stable.)

In fact, the characteristic equation of the multistep method with the coefficients written in the last row of Table 9 is

$$-d^3 - 3d^2 + 3d + 1 = 0$$

with the roots $d_1 = 1$, $d_2 = -2 + \sqrt{3}$ and $d_3 = -2 - \sqrt{3}$. The absolute value of d_3 clearly exceeds 1.

It can be equally proved that the method remains unstable for any $m > 4$.

The remark that the two applicable particularizations of the spline function method are equivalent to some multistep methods raises the question of which of these kinds of methods is to be preferred in practice.

The spline function method has the advantage that it gives the solution at *any* point in $[a, b]$ while the corresponding multistep algorithm gives it only at the *mesh points*. On the other hand, the first method involves more arithmetic operations at each step than the latter, and therefore it is slower. Altogether, the answer depends on the specific criteria of the user.

It is also worthy to be mentioned that reason for the fact that the above method ceases to be convergent for $m > 3$ is that *too many conditions of continuity* have been imposed on the approximate solution. It is then expected that if the number of conditions of continuity is decreased the resultant method remains convergent for higher values of m as well. This idea was explored by Hulme [24] who formulated an open version of the special one-step algorithms.

The basic assumption remains as above, namely that the approximate solution is given by a piecewise polynomial of degree m.

In the first interval of the partition $\bar{y}(x)$ is written as

$$\bar{y}(x) = y_0 + \frac{x - x_0}{1!} y_0' + \frac{(x - x_0)^2}{2!} y_0'' + \ldots$$

$$+ \frac{(x - x_0)^{m-1}}{(m-1)!} y_0^{(m-1)} + \frac{(x - x_0)^m}{m!} y_0^{(m)}. \qquad (2.6.10)$$

(Compare with eq. (2.6.1) to see that the only difference is that C_0 is now replaced by $y_0^{(m)}$. The latter is computed as in the Taylor series method.) Naturally, $\bar{y}(x)$ defined in eq. (2.6.1) is a better approximation of the exact solution than $\bar{y}(x)$ defined in eq. (2.6.10) but the price paid for it is the occurrence of an implicit algebraic equation. Let us now fix some integer value $0 \leqslant p < m$ and calculate the values at x_1 of $\bar{y}(x)$ given by eq. (2.6.10) and of its first p derivatives

$$\begin{aligned} \bar{y}_1 &= y_0 + \frac{h_0}{1!} y_0' + \frac{h_0^2}{2!} y_0'' + \ldots + \frac{h_0^{m-1}}{(m-1)!} y_0^{(m-1)} + \frac{h_0^m}{m!} y_0^{(m)} \\ \bar{y}_1' &= y_0' + \frac{h_0}{1!} y_0'' + \ldots + \frac{h_0^{m-2}}{(m-2)!} y_0^{(m-1)} + \frac{h_0^{m-1}}{m!} y_0^{(m)} \\ &\vdots \\ \bar{y}_1^{(p)} &= y_1^{(p)} + \frac{h_0}{1!} y_1^{(p+1)} \ldots \frac{h_0^{m-p-1}}{(m-p-1)!} y_0^{(m-1)} + \frac{h_0^{m-p}}{(m-p)!} y_0^{(m)}. \end{aligned} \tag{2.6.11}$$

The derivatives of order $p+1, p+2, \ldots, m$ are computed by a different procedure, namely in terms of the original differential equation

$$\bar{y}_1^{(j)} = f^{(j-1)}(x_1, \bar{y}(x_1)), \qquad j = p+1, p+2, \ldots, m. \tag{2.6.12}$$

Once the numbers $\bar{y}_1$ and $\bar{y}_1^{(j)}$, $j = 1, 2, \ldots, m$ have been determined as shown, the approximate solution on $[x_1, x_2]$ is written as

$$\begin{aligned} \bar{y}(x) = \bar{y}_1 &+ \frac{x - x_1}{1!} \bar{y}_1' + \frac{(x - x_1)^2}{2!} \bar{y}_1'' + \ldots \\ &+ \frac{(x - x_1)^{m-1}}{(m-1)!} \bar{y}_1^{(m-1)} + \frac{(x - x_0)^m}{m!} \bar{y}_1^{(m)}. \end{aligned} \tag{2.6.13}$$

In general, given the approximate solution $\bar{y}(x)$ in the interval $[x_{k-1}, x_k]$,

$$\begin{aligned} \bar{y}(x) = \bar{y}_{k-1} &+ \frac{x - x_{k-1}}{1!} \bar{y}_{k-1}' + \frac{(x - x_{k-1})^2}{2!} \bar{y}_{k-1}'' + \ldots \\ &+ \frac{(x - x_{k-1})^{m-1}}{2!} \bar{y}_{k-1}^{(m-1)} + \frac{(x - x_{k-1})^m}{m!} \bar{y}_{k-1}^{(m)} \end{aligned} \tag{2.6.14}$$

the approximate solution in the next interval $[x_k, x_{k+1}]$ is constructed as follows

$$\bar{y}(x) = \bar{y}_k + \frac{x - x_k}{1!}\bar{y}'_k + \frac{(x - x_k)^2}{2!}\bar{y}''_k + \dots$$
$$+ \frac{(x - x_k)^{m-1}}{(m-1)!}\bar{y}_k^{(m-1)} + \frac{(x - x_k)^m}{m!}\bar{y}_k^{(m)} \tag{2.6.15}$$

where the numbers $\bar{y}_k, \bar{y}'_k, \dots, \bar{y}_k^{(p)}$ are the values at $x=x_k$ of $\bar{y}(x)$ given in eq. (1.6.14) and of its first p derivatives, while the numbers $\bar{y}_k^{(p+1)}, \bar{y}_k^{(p+2)}, \dots, \bar{y}_k^{(m)}$ are calculated by

$$\bar{y}_k^{(j)} = f^{(j-1)}(x_k, \bar{y}(x_k)), \qquad j = p+1, p+2, \dots, m. \tag{2.6.16}$$

(The argument $\bar{y}(x_k)$ in this equation is obviously that defined by eq. (2.6.14).)

We see that this method yields an approximate solution $\bar{y}(x)$, $x \in [a, b]$ which is equally a piecewise mth-degree polynomial as it was in the case of the spline function method. The difference is that while in the case of the spline function method only $\bar{y}^{(m)}(x)$ exhibited jumps at the mesh points, in the present case all $\bar{y}^{(p+1)}(x), \bar{y}^{(p+2)}(x), \dots \bar{y}^{(m)}(x)$ exhibit jumps. It is also significant to notice that this algorithm is open since it does not involve the solution of an implicit algebraic equation.

Hulme's analysis also shows that this algorithm is convergent for $m \geqslant 1$ and any $p = 0, 1, 2, \dots, m-1$; its order is $m - p$.

To end this section we present a hybrid between the two previous special one-step methods for the case of linear second-order differential equations. Consider the initial value problem

$$y'' + a(x)y' + b(x)y + c(x) = 0, \qquad x \in [a, b], y(a) = y_0, y'(a) = y'_0, \tag{2.6.17}$$

which was also treated in section 1.13.

To find the algorithm, eq. (2.6.17) is first written as a system of two first-order equations

$$y' = z \tag{2.6.18 a}$$

$$z' = -a(x)z - b(x)y - c(x) \tag{2.6.18 b}$$

with the initial conditions $y(a) = y_0$, $z(a) = y'_0$.

We focus on the interval $[x_k, x_{k+1}]$ and assume that we know the values of the approximate solutions $\bar{y}(x)$ and $\bar{z}(x)$ at $x = x_k$, say $\bar{y}_k$ and $\bar{z}_k$. The problem is then to compute $\bar{y}(x_k + h_k)$ and $\bar{z}(x_k + h_k)$.

We assume that $\bar{y}(x)$ and $\bar{z}(x)$ are polynomials of the third degree

$$\bar{y}(x) = \bar{y}_k + (x - x_k)\,\bar{y}'_k + \frac{1}{2}(x - x_k)^2\,\bar{y}''_k + (x - x_k)^3\,R \quad (2.6.19\text{ a})$$

$$\bar{z}(x) = \bar{z}_k + (x - x_k)\,\bar{z}'_k + \frac{1}{2}(x - x_k)^2\,\bar{z}''_k + (x - x_k)^3\,T \quad (2.6.19\text{ b})$$

As stated, we know $\bar{y}_k$ and $\bar{z}_k$; the values of $\bar{y}'_k$ and $\bar{z}'_k$ are calculated at once from the very equations (2.6.18 a and b) taken at $x = x_k$. The value of $\bar{y}''_k$ results from the original equation (2.6.17) while to determine z'' we differentiate eq. (2.6.18 b) with respect to x,

$$z'' = -a'z - b'y - c' - az' - by' =$$

$$= -(a' + b - a^2)z - (b' - ab)y - c' + ac \quad (2.6.20)$$

and put $x = x_k$ in its right member.

Thus, upon denoting

$$Y(x - x_k) = \bar{y}_k + (x - x_k)\bar{y}'_k + \frac{1}{2}(x - x_k)^2\bar{y}''_k, \quad (2.6.21\text{ a})$$

$$Z(x - x_k) = \bar{z}_k + (x - x_k)\,\bar{z}'_k + \frac{1}{2}(x - x_k)^2\bar{z}''_k, \quad (2.6.21\text{ b})$$

eqs (2.6.19 a) and (2.6.19 b) become

$$\bar{y}(x) = Y(x - x_k) + (x - x_k)^3R, \quad (2.6.22\text{ a})$$

$$\bar{z}(x) = Z(x - x_k) + (x - x_k)^3T. \quad (2.6.22\text{ b})$$

In short, to obtain $Y(x - x_k)$ and $Z(x - x_k)$ we used exactly the strategy of the open version corresponding to $p = 0$ and $m = 2$. To determine the constants R and T we adopt a strategy which reminds the closed version. Eqs (2.6.22 a) and (2.6.22 b) taken at $x = x_{k+1} = x_k + h$ actually give

$$Y'(h) + 3h^2R = Z(h) + h^3T \tag{2.6.23 a}$$

and

$$Z'(h) + 3h^2T = -a_{k+1}(Z(h) + h^3T) - b_{k+1}(Y(h) + h^3R) - c_{k+1} \tag{2.6.23 b}$$

with $Y'(h) = \bar{y}_k' + h\bar{y}_k''$ and $Z'(h) = \bar{z}_k' + h\bar{z}_k''$.

This algebraic system is solved for R and T with the solution

$$R = \{h^2[3(3+ha_{k+1}) + b_{k+1}h^2]\}^{-1}[hu - (3 + ha_{k+1})(-Z(h) + Y'(h))] \tag{2.6.24 a}$$

$$T = \{h^2[3(3 + ha_{k+1}) + b_{k+1}h^2]\}^{-1}[3u + b_{k+1}h(-Z(h) + Y'(h))] \tag{2.6.24 b}$$

where

$$u = a_{k+1}Z(h) - b_{k+1}Y(h) - c_{k+1} - Z'(h). \tag{2.6.25}$$

The pair of approximate solutions $\bar{y}(x)$ and $\bar{z}(x)$ is thus fully determined on $[x_k, x_{k+1}]$.

This hybrid method is easily extended to the case of systems of linear equations, i.e. when a and b are N by N matrices of functions while c, y and y' are N dimensional vectors. The only point is that in this case we have to be careful with the *order* in which we perform the matrix operations, particularly the matrix multiplications. As at matter of fact, the expressions (2.6.24 a), (2.6.24 b) and (2.6.25) were so written just to account for this circumstance.

Note finally that, judged in terms of efficiency, the hybrid method used at constant stepsizes is found to be approximately as accurate as the Numerov method. Its capability of using variable step sizes and of yielding the solution at any point of $[a, b]$ (not only at the mesh points as with the Numerov method) are advantages which should not be underestimated.

REFERENCES

1. P. Henrici, *Discrete Variable Methods in Ordinary Differential Equations*, John Wiley and Sons, New York and London, 1964.
2. L. Collatz, *The Numerical Treatment of Differential Equations*, Springer, Berlin, 1960.
3. K. Heun, *Neue Methode zur approximativen Integration der Differentialgleichungen einer unabhängigen Veränderlichen*, Z. Math. Physik, **45**, 23 (1900).
4. Z. Kopal, *Numerical Analysis*, John Wiley and Sons, New York and London, 1955.
5. C. Runge, *Über die numerische Auflosung von Differentialgleichungen*, Math. Ann., **46**, 167 (1895); W. Kutta, *Beitrag zur näherungsweisen Integration totaler Differentialgleichungen*, Z. Math. Phys., **46**, 435 (1901).
6. S. Gill, *A process for the step-by-step integration of differential equations in an automatic digital computing machine*, Proc. Camb. Phil. Soc., **47**, 96 (1951).
7. L. Fox (editor), *Numerical Solution of Ordinary and Partial Differential Equations*, Pergamon Press, Oxford–London–New York–Paris, 1962.
8. D. W. Martin, *Runge-Kutta methods for integrating differential equations on high-speed digital computers*, Computer J. **1**, 118 (1958).
9. L. F. Shampine and H. A. Watts, *Practical solution of ordinary differential equations*, Sandia Laboratories Report SAND-76-0585, Albuquerque, New Mexico, 1976; see also *The art of writing a Runge-Kutta code, Part I* in Mathematical Software III (J. R. Rice, Ed.) Academic Press, New York, 1977, p. 257, and ibid. Part. II, Appl. Math. Comput., *5*, 93 (1979).
10. SSP: System/360 Scientific Subroutine Package, IBM document GH20-0205-4 (1970).
11. J. C. Butcher, *On Runge-Kutta processes of high order*, J. Austral. Math. Soc., **4**, 179 (1964), and *On the attainable order of Runge-Kutta methods*, Math. Comp., **19**, 488 (1965).
12. S. Gorn and R. Moore, *Automatic error control – the initial value problem in ordinary differential equations*, Report 893, Ballistic Research Labs., Aberdeen Proving Ground, Md., 1953.
13. R. H. Merson, *An operational method for the study of integration processes*, Proceedings of conference on data processing and automatic computing machines at Weapons Research Establishment, Salisbury, South Australia, 1957.
14. E. Fehlberg, *Low order classical Runge-Kutta formulas with stepsize control and their application to some heat transfer problems*, NASA Technical Report NASA TR R-315, 1969; also Computing **6**, 61 (1970).
15. E. M. Wilson, *A note on the numerical integration of differential equations* Quart. *J.* Mech., **2**, 208 (1949).

16. L. Stoller and D. Morrison, *A method for the numerical integration of ordinary differential equations*, Math. Tab., Wash., **12**, 269 (1958).
17. G. M. Phillips and P. J. Taylor, *Theory and applications of numerical analysis*, Academic Press, London—New York—San Francisco, 1973.
18. P. C. Hammer and T. W. Hollingsworth, *Trapezoidal methods of approximating solutions of differential equations*, Math. Tables Aids Comput., **9**, 92 (1955).
19. E. J. Nyström, *Über die numerische Integration von Differentialgleichungen*, Acta Soc. Sci. Fenn., **50**, No. 13, 1 (1925).
20. C. Lanczos, *Trigonometric interpolation of empirical and analytical data*, J. Math. Phys., **17**, 123 (1938); *Applied Analysis*, Pitman, London, 1957.
21. C. W. Clenshaw, *The numerical solution of linear differential equations in Chebyshev series*, Proc. Camb. Phil. Soc., **53**, 134 (1957).
22. F. R. Loscalzo and T. D. Talbot, *Spline function approximations for solutions of ordinary differential equations*, Bull. Amer. Math. Soc., **73**, 438 (1967) and SIAM J. Numer. Anal. **4**, 433 (1967).
23. F. R. Loscalzo, in *Theory and applications of spline functions* (T.N.E. Greville editor), Academic Press, New York, 1969.
24. B. L. Hulme, *Piecewise polynomial Taylor methods for initial value problems*, Numer. Math., **17**, 367 (1971).

Chapter 3

Piecewise perturbation methods

The methods described in the previous two chapters are classical methods and they are investigated in detail in the numerical analysis. Most of them are general in the sense that they can be applied to any system of first-order equations, with no serious restriction upon the number of equations in the system or upon the expression of f. Moreover, since an Nth-order differential equation can be written as a system of N first-order equations, it can be also said that such methods equally solve numerically the Cauchy problem for differential equations of higher order.

We also presented methods which can be applied to only some special forms of differential equations as, for instance, the methods of Cowell, of de Vogelaere, or of Nyström (in section 2.4), which all refer to equations of the second order. By their very motivation, the methods namely devised for special differential equations must be (and actually so they are) more convenient than the general methods applied to the same problems.

The solution of *linear* second-order differential equations exhibits large practical interest since such equations describe a lot of important physical phenomena and physicists significantly contributed with new special numerical methods. The contribution of physicists was mainly to activate for numerical purposes of some ideas which originate in branches of mathematics which are closer to them as, for instance, the mathematical physics. Such an idea is that of *perturbation approximation* and, as it will be seen in this chapter, it leads to numerical methods of higher efficiency.

Before going to devise perturbation algorithms we will take a simple example to illustrate some of the typical difficulties which occur when solving linear second-order equations by classical methods.

3.1. DIFFICULTIES WITH THE CLASSICAL METHODS

We take the innocent-looking initial value problem

$$y'' = -a^2 y, \quad x \geqslant 0, \quad y(0) = 1, \quad y'(0) = 0, \; a \geqslant 0 \tag{3.1.1}$$

with the exact solution $y = \cos(ax)$.
We will solve this equation by two numerical methods: the Runge-Kutta method with $s = 2$ and $\lambda = 1/2$ (a one-step method, see section 2.2 in particular eq. (2.2.16)) and the Numerov method (a special multistep method, see section 1.13) and assume that the partition is equidistant for both methods

$$x_0 = a, \quad x_i = x_0 + ih, \quad i = 1, 2, 3, \ldots$$

We particularly focus on the dependence, at fixed h, of the accuracy of the numerical solution vs. the value of the parameter a and first consider the *Runge-Kutta* method.

To apply it we must first transform the given second-order equation into a system of two first-order equations.

It can be checked easily that the associated system is

$$\mathbf{y}' = \mathbf{f}(x, \mathbf{y}), \quad \mathbf{y} \equiv \begin{pmatrix} y^1 \\ y^2 \end{pmatrix}, \quad \mathbf{f}(x, \mathbf{y}) = \begin{pmatrix} f^1 \equiv y^2 \\ f^2 \equiv -a^2 y^1 \end{pmatrix}, \quad x \geqslant 0 \tag{3.1.2}$$

with the initial condition

$$\mathbf{y}(0) = \begin{pmatrix} y^1(0) = 1 \\ y^2(0) = 0 \end{pmatrix}. \tag{3.1.3}$$

(We have followed the general procedure presented in "Preliminaries" in connection with eq. (15).)

We now seek for the coefficients $\mathbf{k}_1$ and $\mathbf{k}_2$ of the Runge-Kutta method

$$\mathbf{k}_1 \equiv \begin{pmatrix} k_1^1 \\ k_1^2 \end{pmatrix} = \begin{pmatrix} f^1(x_k, \bar{\mathbf{y}}_k) = \bar{y}_k^2 \\ f^2(x_k, \bar{\mathbf{y}}_k) = -a^2 \bar{y}_k^1 \end{pmatrix} \tag{3.1.4 a}$$

$$\mathbf{k}_2 \equiv \begin{pmatrix} k_2^1 \\ k_2^2 \end{pmatrix} = \begin{pmatrix} f^1(x_k + h, \bar{\mathbf{y}}_k + h\mathbf{k}_1) = \bar{y}_k^2 + hk_1^2 = \bar{y}_k^2 - ha^2 \bar{y}_k^1 \\ f^2(x_k + h, \bar{\mathbf{y}}_k + h\mathbf{k}_1) = -a^2(\bar{y}_k^1 + hk_1^1) = -a^2(\bar{y}_k^1 + h\bar{y}_k^2) \end{pmatrix}. \tag{3.1.4 b}$$

The Runge-Kutta algorithm (2.2.16),

$$\bar{\mathbf{y}}_{k+1} = \bar{\mathbf{y}}_k + \frac{1}{2} h(\mathbf{k}_1 + \mathbf{k}_2), \tag{3.1.5}$$

then reads

$$\begin{pmatrix} \bar{y}^1_{k+1} \\ \bar{y}^2_{k+1} \end{pmatrix} = \begin{pmatrix} \bar{y}^1_k + \frac{1}{2} h(k^1_1 + k^1_2) \\ \bar{y}^2_k + \frac{1}{2} h(k^2_1 + k^2_2) \end{pmatrix} = \begin{pmatrix} \left(1 - \frac{1}{2} h^2a^2\right) \bar{y}^1_k + h\bar{y}^2_k \\ - ha^2\bar{y}^1_k + \left(1 - \frac{1}{2} h^2a^2\right) \bar{y}^2_k \end{pmatrix} \tag{3.1.6}$$

or, in compressed matrix form,

$$\bar{\mathbf{y}}_{k+1} = \mathbf{M}(h)\,\bar{\mathbf{y}}_k \tag{3.1.7}$$

where

$$\mathbf{M}(h) = \begin{pmatrix} 1 - \frac{1}{2} h^2a^2 & h \\ - ha^2 & 1 - \frac{1}{2} h^2a^2 \end{pmatrix}. \tag{3.1.8}$$

The exact solution of eq. (3.1.2) can be written in a similar form

$$\mathbf{y}(x_{k+1}) = \mathbf{N}(h)\,\mathbf{y}(x_k) \tag{3.1.9}$$

with

$$\mathbf{N}(h) = \begin{pmatrix} \cos(ah) & \frac{1}{a} \sin(ah) \\ - a\sin(ah) & \cos(ah) \end{pmatrix} \tag{3.1.10}$$

as it can be readily checked for.

To evaluate the local truncation error of algorithm (3.1.7) we put there

$$\bar{\mathbf{y}}_k = \mathbf{y}(x_k) = \begin{pmatrix} \cos(ax_k) \\ -a\sin(ax_k) \end{pmatrix} \tag{3.1.11}$$

and subtract eq. (3.1.7) from eq. (3.1.9) to get

$$\mathbf{e}_{k+1} \equiv \mathbf{y}(x_{k+1}) - \bar{\mathbf{y}}_{k+1} = (\mathbf{N}(h) - \mathbf{M}(h))\,\mathbf{y}(x_k) \tag{3.1.12}$$

which on components becomes

$$e^1_{k+1} = y(x_{k+1}) - \bar{y}_{k+1} = \left[\cos(ah) - \left(1 - \frac{1}{2}h^2a^2\right)\right]\cos(ax_k) -$$

$$- [\sin(ah) - ah]\sin(ax_k) \tag{3.1.13 a}$$

and

$$e^2_{k+1} = y'(x_{k+1}) - \bar{y}'_{k+1} = a\,[-\sin(ah) + ah]\cos(ax_k) -$$

$$- a\left[\cos(ah) - \left(1 - \frac{1}{2}a^2h^2\right)\right]\sin(ax_k). \tag{3.1.13 b}$$

We see that the Runge-Kutta solution means essentially a piecewise polynomial approximation of the exact solution of the given problem. In fact, eq. (3.1.13 a) shows that the Runge-Kutta solution is as accurate as is the parabola

$$\left(1 - \frac{1}{2}a^2\delta^2\right)\cos(ax_k) - a\delta\sin(ax_k), \qquad \delta \in [0, h] \tag{3.1.14}$$

in comparison with the exact solution

$$\cos(ax) = \cos[a(x_k + \delta)] = \cos(a\delta)\cos(ax_k) - \sin(a\delta)\sin(ax_k). \tag{3.1.15}$$

Clearly, for fixed h the quality of this approximation impairs with increasing a.

We now consider the *Numerov* method.

This algorithm is a particular form of algorithm (1.11.18) and in order to help understanding what we are going to say in connection to the application of this algorithm to the sample problem (3.1.1) we first slightly modify the procedure of deriving the consistency conditions (1.11.22).

We start with the same associated functional as in eq. (1.11.20),

$$L(z(x); h) = a_i z(x + ih) + a_{i-1} z(x + (i-1)h) + \dots + a_0 z(x) -$$
$$- h^2(b_i z''(x + ih) + b_{i-1} z''(x + (i-1)h) + \dots + b_0 z''(x)) \quad (3.1.16)$$

and search for the conditions to be satisfied by the coefficients a_s and b_s $s = 0, 1, 2, \dots, i$, so that equation

$$L(z(x); h) = 0 \quad (3.1.17)$$

is identically satisfied for any x and h if $z(x)$ is a *power function*, $z(x) = x^p$, $p = 0, 1, 2, 3, \dots$

For $p = 0$, that is $z(x) = 1$, eq. (3.1.17) gives at once

$$a_0 + a_1 + a_2 + \dots + a_i = 0. \quad (3.1.18)$$

For $p = 1$, i.e. $z(x) = x$ we have

$$(a_i + a_{i-1} + \dots + a_0)\, x + (ia_i + (i-1)\, a_{i-1} + \dots + a_1)\, h = 0$$

which, by virtue of eq. (3.1.18), reduces to

$$a_1 + 2a_2 + \dots + ia_i = 0. \quad (3.1.19)$$

Going on to higher values of p we get in the same way that

$$\frac{1}{p!}(a_1 + 1^p a_2 + \dots + i^p a_i) - \frac{1}{(p-2)!}(0^{p-2} b_0 + 1^{p-2} b_1 +$$
$$+ \dots + i^{p-2} b_i) = 0. \quad (3.1.20)$$

Since these equations can be written compactly as $C_p = 0$, $p = 0, 1, 2, \dots$ with C_p given by (1.11.22) it results that the consistency of algorithm (1.11.18) can be formulated in the following alternative way: the algorithm is consistent and of order m if it is identically satisfied in the interval $[x_k, x_{k+i}]$ by any power function x^p, $p = 0, 1, 2, \ldots,$ $m + 1$. Since the algorithm is also linear something more can be said, namely that the algorithm is consistent and of order m if it is identically satisfied by any linear combination of the quoted power functions, that is by any $(m + 1)$th degree polynomial

$$y(x) = A_0 + A_1 x + A_2 x^2 + \ldots + A_{m+1} x^{m+1}. \tag{3.1.21}$$

We now come back to the Numerov algorithm. Since this is a *two*-step algorithm of order $m = 4$ it results that it will be appropriate for use only when in each two-step interval the exact solution of the differential equation to be solved is reasonably approximated by a fifth-degree polynomial.

For the sample problem (3.1.1), where the algorithm is

$$\bar{y}_{k+1} + 2c\bar{y}_k + \bar{y}_{k-1} = 0, \quad 2c = \frac{\dfrac{5}{6}\, a^2h^2 - 2}{1 + \dfrac{1}{12}\, a^2h^2}, \; k = 1, 2, 3, \ldots, \tag{3.1.22}$$

the numerical results will then be satisfactory only when the exact solution $\cos(ax) = \cos\,[a(x_k + \delta)]$ is well approximated on $[x_{k-1}, x_{k+1}]$ (that is $\delta \in [-h, h]$) by a fifth-degree polynomial in δ. Of course, if h is fixed the quality of this approximation impairs with increasing a, and this fact is reflected also on the expression of the local truncation error. In fact, on using the trigonometric identities

$$y(x_{k\pm 1}) = \cos\,[a(x_k \pm h)] = \cos\,(ax_k)\cos\,(ah) \mp \sin\,(ax_k)\sin\,(ah)$$

and on assuming that the entrance data in algorithm (3.1.22) are the exact values of the solutions i.e. $\bar{y}_k = y(x_k) = \cos\,(ax_k)$ and $\bar{y}_{k-1} =$ $= y(x_{k-1}) = \cos\,[a(x_k - h)]$ we get

$$\begin{aligned} e_{k+1} &\equiv y(x_{k+1}) - \bar{y}_{k+1} = y(x_{k+1}) + 2cy(x_k) + y(x_{k-1}) = \\ &= 2\cos\,(ax_k)\,[\cos\,(ah) + c] \end{aligned} \tag{3.1.23}$$

The dependence of the local truncation error in the solution against a for the Runge-Kutta method (eq. (3.1.13 a)) and for the Numerov method is illustrated in Table 10 for $x_k = 2\pi$ and $h = 1/8$.

Table 10

Variation with a of the local truncation error for the Runge-Kutta and Numerov methods for the sample problem (3. 1. 1); $x_k = 2\pi$ and $h = 1/8$

a	Runge-Kutta	Numerov
0	0	0
1	$-1.02\ 10^{-5}$	$1.59\ 10^{-8}$
2	$-1.62\ 10^{-4}$	$1.01\ 10^{-6}$
3	$-8.20\ 10^{-4}$	$1.14\ 10^{-5}$
4	$-2.58\ 10^{-3}$	$6.31\ 10^{-5}$
6	$-1.29\ 10^{-2}$	$6.91\ 10^{-4}$
8	$-4.03\ 10^{-2}$	$3.68\ 10^{-3}$
10	$-9.66\ 10^{-2}$	$1.31\ 10^{-2}$

These data confirm that when h is fixed the accuracy of both methods decreases when a increases and the reason is that, as stated, both methods are based on local *polynomial* approximations of the solution, approximations which by their very nature impair with increasing a. For the same reason if we require equal accuracy at any a the only way to reach this is by corresponding adjustment of the value of the step size. As an illustration, if we want to solve eq. (3.1.1) on the interval [0, 100] for $a = 100$ at the *same* accuracy as it is reached for $a = 1$ with $h = 1/8$ (note that the accuracy depends on the product ah) then we need a step size $h_1 = 1/(8\,a) = 1/800$. Thus the computer will carry out 800 iterations when $a = 1$ but 80,000 iterations when $a = 100$, i.e. it will spend a hundred times more computation time. Moreover, although the round-off errors accumulated after 800 iterations admittedly do not damage the result too much, they surely do so after 80,000 iterations so that the very program must be rewritten in double precision with the corresponding increase in both memory allocation and computation time.

In short, all difficulties come out from one and the same source namely that the two considered methods were devised on the belief that polynomials are good piecewise approximations to the solutions of differential equations and the point is that this belief is *common* for all methods of the previous two chapters. However, such a belief is actually correct only in the *limit of small h*, and this very restriction is the reason why the classical methods become hardly operational (sometimes just nonoperational) in many practical cases.

It is then suggested to try devising methods in which the solution is approximated by other types of functions than the polynomials. Even more radically, one may try devising methods which simply avoid the idea of approximating the solution directly.

Mathematical physics offers a different type of approximation. It is the perturbation approximation. In its spirit the given equation is replaced by another differential equation (we call it a *reference* differential equation), which can be solved *exactly;* the deviation of the solution of the reference equation from the solution of the original equation is further estimated by means of the perturbation theory.

Clearly, in this way the emphasis is translated from the approximation of *the solution* (the basic attitude when deriving classical methods) to the approximation of *the equation*. The latter approach is more natural than the former also because it simply accounts for *known* data (namely the expression of the equation) in place of unknown quantities as is the solution of the equation.

To derive numerical perturbation methods means essentially to follow the idea of the perturbation approximation, which in mathematical physics is typically applied globally, i.e. on the whole domain of the equation at once, see e.g. [1] but apply it in the spirit of the numerical analysis, that is *piecewise*. The condition that the reference equation must admit analytic solution seriously restricts the class of differential equations for which piecewise perturbation methods can be formulated. Fortunately the linear equations belong to this class. We particularly discuss the case of *second*-order linear differential equations, a case often encountered in problems of physics and engineering. As a matter of fact, the first proposals for piecewise perturbation techniques occurred just in such a context [2—4].

Each piecewise perturbation method is defined in terms of the recipe used for the piecewise approximation of the coefficients of the given equation. If, for instance, the coefficients are approximated by piecewise constants the method is referred to as a *constant* perturbation method while if piecewise straight lines are used the method is called a *line* perturbation method.

Such methods will be considered in detail in the next sections of this chapter. As a by-product of these investigations some hints will be obtained concerning classes of functions which must replace the polynomials to lead to algorithms of classical form but with increased accuracy, see section 3.8.

3.2. A CONSTANT PERTURBATION METHOD FOR LINEAR NON-HOMOGENEOUS SECOND-ORDER EQUATIONS

To illustrate the process of devising piecewise perturbation methods we consider the initial value problem for a linear nonhomogeneous second-order equation, that is

$$y'' + a(x)y' + b(x)y + c(x) = 0, \quad x \in [a, b],\ y(a) = y_0,\ y'(a) = y_0' \tag{3.2.1}$$

where $a(x)$, $b(x)$ and $c(x)$ are given, well-behaved functions, and construct a simple, unsophisticated method for this problem. As usually, we introduce a partition of $[a, b]$ through the mesh points $x_0 = a < < x_1 < x_2 < \ldots < x_n = b$. This partition is in general nonequidistant.

In accordance with the discussion made at the end of the previous section, in each interval $[x_k, x_{k+1}]$, $k = 0, 1, 2, \ldots$. the given equation is approximated by an equation which can be solved analytically. The approximation refers to the coefficient functions $a(x)$, $b(x)$ and $c(x)$ and the simplest choice is through constants. We thus introduce the constant approximations

$$a(x) \to \tilde{a}_k,\ b(x) \to \tilde{b}_k,\ c(x) \to \tilde{c}_k, \text{ for } x \in [x_k, x_{k+1}]$$

where the constants $\tilde{a}_k$, $\tilde{b}_k$ and $\tilde{c}_k$ are obtained in terms of the values of $a(x)$, $b(x)$ and $c(x)$, by some suitable recipe of approximation, for instance

$$\tilde{a}_k = \frac{1}{2}(a(x_k) + a(x_{k+1})), \quad \tilde{b}_k = \frac{1}{2}(b(x_k) + b(x_{k+1})),$$

$$\tilde{c}_k = \frac{1}{2}(c(x_k) + c(x_{k+1})). \tag{3.2.2}$$

In this way the following approximating equation is obtained

$$\tilde{y}'' + \tilde{a}_k\tilde{y}' + \tilde{b}_k\tilde{y} + \tilde{c}_k = 0, \quad x \in [x_k, x_{k+1}], \tag{3.2.3}$$

an equation which can be solved by hand.

To form its general solution we construct separately the general solution of the associated homogeneous equation and a particular solution of the given equation.

The associated homogeneous equation is

$$z'' + \tilde{a}_k z' + \tilde{b}_k z = 0, \quad x \in [x_k, x_{k+1}]. \tag{3.2.4}$$

We assume that $z(x)$ is of the form $z(x) = \exp[\omega(x - x_k)]$ and determine the constant ω by selfconsistency. We thus introduce this $z(x)$ in eq. (3.2.4) and find out an algebraic equation for ω,

$$\omega^2 + \tilde{a}_k \omega + \tilde{b}_k = 0 \tag{3.2.5}$$

which has the solutions

$$\omega^{\pm} = \frac{1}{2} [- \tilde{a}_k \pm (\tilde{a}_k^2 - 4\tilde{b}_k)^{1/2}]. \tag{3.2.6}$$

When $\tilde{a}_k^2 \neq 4\tilde{b}_k$ ω^+ and ω^- do not coincide. Then the two linear independent solutions of eq. (3.2.4) are $z^{\pm}(x) = \exp[\omega^{\pm}(x - x_k)]$ and thus the general solution is their linear combination, viz.

$$z(x) = A_k \exp[\omega^+(x - x_k)] + B_k \exp[\omega^-(x - x_k)] \tag{3.2.7}$$

where A_k and B_k are arbitrary constants.

A particular solution of eq. (3.2.3) is

$$\tilde{y}^p = - \tilde{c}_k / \tilde{b}_k \equiv U_k \tag{3.2.8}$$

if $b_k \neq 0$ and

$$\tilde{y}^p = - \frac{\tilde{c}_k}{\tilde{a}_k} (x - x_k) \tag{3.2.9}$$

if $\tilde{b}_k = 0$ and $\tilde{a}_k \neq 0$.

If $\tilde{a}_k = \tilde{b}_k = 0$ the *general* solution of eq. (3.2.3) is simply

$$\tilde{y} = - \frac{1}{2} \tilde{c}_k (x - x_k)^2 + C(x - x_k) + D$$

with C and D arbitrary constants.

In the following we assume that $\tilde{a}_k \neq 0$, $\tilde{b}_k \neq 0$ and $\tilde{a}_k^2 \neq 4\tilde{b}_k$. If so, the general solution of eq. (3.2.3) is the sum of the solution $z(x)$ given in eq. (3.2.7) and the particular solution $\tilde{y}^p$ in form (3.2.8),

$$\tilde{y}(x) = A_k \exp[\omega^+(x - x_k)] + B_k \exp[\omega^-(x - x_k)] + U_k. \quad (3.2.10)$$

Its first-order derivative is

$$\tilde{y}'(x) = A_k\omega^+ \exp[\omega^+(x - x_k)] + B_k\omega^- \exp[\omega^-(x - x_k)]. \quad (3.2.11)$$

The values of A_k and B_k are determined by the initial conditions $\tilde{y}(x_k) = \tilde{y}_k$ and $\tilde{y}'(x_k) = \tilde{y}'_k$. Eqs (3.2.10) and (3.2.11) thus read

$$\begin{aligned} \tilde{y}_k &= A_k + B_k + U_k \\ \tilde{y}'_k &= \omega^+ A_k + \omega^- B_k \end{aligned} \quad (3.2.12)$$

with the solution

$$A_k = \frac{1}{\omega^+ - \omega^-}(-\omega^- \tilde{y}_k + \omega^- U_k + \tilde{y}'_k),$$

$$B_k = \frac{1}{\omega^+ - \omega^-}(\omega^+ \tilde{y}_k - \omega^+ U_k - \tilde{y}'_k). \quad (3.2.13)$$

Eqs (3.2.10), (3.2.11) and (3.2.13) define the desired algorithm [5] in each elementary interval $[x_k, x_{k+1}]$.

The solution is advanced as in any one-step method. In the first interval, i.e. $k = 0$, we first construct $\tilde{a}_0$, $\tilde{b}_0$ and $\tilde{c}_0$ then ω^+, ω^- and U_0. These and the given values of y_0 and y'_0 are used to determine A_0 and B_0. The solution (3.2.10) and its derivative (3.2.11) are thus uniquely determined on $[x_0, x_1]$ and their values at $x = x_1$ are taken as initial conditions on the next interval $k = 1$ where the procedure is repeated the same way.

This algorithm exhibits some salient advantages over the classical algorithms and all of them come from the fact that the only approximation involved refers to the input, known functions $a(x)$, $b(x)$ and $c(x)$.

First of all, the quality of the approximate solution can be appreciated just from the very beginning, in terms of the deviations $a(x) - \tilde{a}_k$, $b(x) - \tilde{b}_k$ and $c(x) - \tilde{c}_k$. In particular the user has the possibility to conveniently adjust the partition so that these deviations are kept under a certain level. For contrast we recall that, when the classical methods are applied, the quality of the approximate solution can be judged only after several test runs were carried out.

Second, this algorithm admits in general a rather wide range for the values of h at which it gives results of acceptable accuracy. For comparison we recall that the classical methods are applicable only when h is so small that the piecewise polynomial approximation to the solution is a good approximation.

As a matter of fact, the concept of *order* of a numerical method, which is so popular for classical methods, only partially characterizes the quality of a piecewise perturbation algorithm. To explain this point we note that the introduction of the concept of the order of a numerical method (we have repeatedly considered this issue in the previous chapters) is based on the same belief as the very methods devised there and for this reason the characterization of a classical method by its order refers to the same restrained range of values of h at which the method is used in practice. On the other hand, as noted, the range is significantly larger for piecewise perturbation methods and thus if orders are assigned to such methods, these will be useful to judge the methods for accuracy only in an asymptotically narrow zone of the practically available range of h.

For illustration we consider the problem

$$y'' - 4xy' + (4x^2 + \alpha^2 - 2)\, y + \alpha^2 \exp(x^2) = 0, \quad y_0 = 1, \quad y' = \beta, \tag{3.2.14}$$

$x \in [0, 1.5]$, for different values of $\alpha \in [1, 25]$ and $\beta \in [0, 25]$. Its exact solution is known

$$y(x) = \left[1 + \frac{\beta}{\alpha} \sin(\alpha x)\right] \exp(x^2). \tag{3.2.15}$$

We compare the results from this method to the ones furnished by the classical fourth-order Runge-Kutta-Gill method, eq. (2.2.25). The run with the two methods at one and the same $h = 0.1$ clearly confirms the points mentioned above. It is seen that the accuracy of the solution given by the second method fluctuates with α and β. It is satisfactory for $\alpha \leqslant 15$ and any β as well as for $\alpha > 15$ and $\beta = 0$ but when $\alpha > 15$ and $\beta \neq 0$ the accuracy impairs with increasing α and/or β.

Such a behaviour of the accuracy is easily understood since the shape of the true solution strongly depends on α and β. When $\beta = 0$, for instance, the solution is smooth enough to be satisfactorily approximated by a fifth-degree polynomial in each interval of width $h = 0.1$. When α and β increase such approximation is obviously less and less accurate.

In contrast, the accuracy of the results furnished by the constant perturbation method remains practically the same at any α and β. On Fig. 1 the results given by the two methods at the mesh points are collected for $\alpha = \beta = 20$. The constant perturbation method was also used to compute values of $\tilde{y}(x)$ inbetween the mesh points and

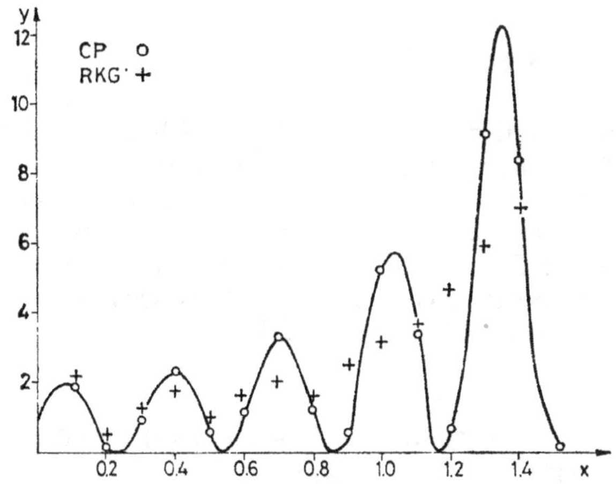

Fig. 1 — Numerical results from constant perturbation (CP) and Runge-Kutta-Gill (RKG) methods at $h = 0.1$ for problem (3.2.14).

the result exhibits a relative error of approximately one percent. For this reason on Fig. 1 the curves $y(x)$ and $\tilde{y}(x)$ coincide.

The order of the discussed piecewise perturbation method is $m = 2$. (We omit the proof since it is actually very simple.) While this would guarantee that at small h, $h = 0.01$ say, the accuracy will be better with the Runge-Kutta-Gill method than with the piecewise perturbation method, the knowledge of the order alone disregards the other, equally important piece of information namely that the latter method is valid over a larger range in h than the former, particularly at $h = 0.1$. It can thus be said that to better characterize a piecewise perturbation method the knowledge of the order must be supplemented with some specific information but the problem of how mathematically expressing such information is still open. Some steps towards this goal are reported only for constant perturbation methods applied to homogeneous equations, see section 3.4.

A natural question is whether approximations better than constants exist for $a(x)$, $b(x)$ and $c(x)$ such that the reference equation can still be solved analytically. The answer is positive and for instance we can take linear approximation for $a(x)$

$$a(x) \to \tilde{a}_k(x) = \tilde{a}^1 (x - x_k) + \tilde{a}^2;$$

$b(x)$ and $c(x)$ remain approximated by constants $\tilde{b}_k$ and $\tilde{c}_k$. The resultant reference equation

$$\tilde{y}'' + (\tilde{a}_k^1(x - x_k) + \tilde{a}_k^2)\, \tilde{y}' + \tilde{b}_k \tilde{y} + \tilde{c}_k = 0 \tag{3.2.16}$$

admits analytic solutions in terms of the Whittacker functions, see for instance [6]. Although this $\tilde{y}(x)$ will be certainly closer to the exact solution $y(x)$ than is $\tilde{y}(x)$ furnished by eq. (3.2.10), such approach does not seem to be more convenient in practice since the evaluation of the Whittacker functions on computer is not only more difficult but also much more time consuming than the evaluation of the exponential functions required by eq. (3.2.10).

3.3. GENERAL PIECEWISE PERTURBATION APPROACH

In the following we constantly refer to linear second-order homogeneous equations of particular form, namely a form in which the first derivative is absent. The problem to be solved numerically is thus

$$y'' = g(x)y, \quad x \in [a, b], \quad y(a) = y_0, \quad y'(a) = y_0', \tag{3.3.1}$$

where $g(x)$ is some given bounded real function.

Note in passing that algorithms for such an equation are equally applicable for linear homogeneous second-order equations of a general form

$$z'' + a(x)\, z' + b(x)\, z = 0 \tag{3.3.2}$$

since z is given by

$$z(x) = \exp\left[-\frac{1}{2} \int^{x} a(x')\, dx' \right] y(x) \tag{3.3.3}$$

where y satisfies $y'' = g(x)y$ with

$$g(x) = \frac{1}{2}a'(x) + \frac{1}{4}a^2(x) - b(x). \tag{3.3.4}$$

As usually, we introduce a partition of $[a, b]$, viz. $x_0 = a < x_1 < < x_2 < \ldots < x_n = b$. There is no special restriction upon the manner of distributing the mesh points except that if $g(x)$ is discontinuous at some points then these points should be taken as mesh points. We focus on the current interval $[x_k, x_{k+1}]$ of length h_k and look for a piecewise perturbation algorithm which propagates the solution from x_k all the way up to x_{k+1}. We introduce the current variable $\delta = x - x_k$, $\delta \in [0, h_k]$ and denote $X = x_k$, $f(\delta) \equiv g(X + \delta)$ and $h = h_k$. The one-step problem then is

$$y''(X + \delta) = f(\delta)\, y(X + \delta), \quad \delta \in [0, h] \tag{3.3.5}$$

with some known initial conditions $y(X) = A, y'(X) = B$.

Let us denote by $u(\delta)$ and $v(\delta)$ the two solutions of eq. (3.3.5) which satisfy some particular initial conditions, viz.

$$u'' = f(\delta)u, \quad u(0) = 1, \quad u'(0) = 0 \tag{3.3.6}$$

and

$$v'' = f(\delta)v, \quad v(0) = 0, \quad v'(0) = 1. \tag{3.3.7}$$

By the very construction, functions u and v are linear independent and their wronskian (we recall that the wronskian is defined as $W[u, v] = uv' - u'v$) is equal to 1.

It follows that the solution of eq. (3.3.5) with initial conditions A and B can be conveniently written in matrix form as follows

$$\begin{pmatrix} y(X + \delta) \\ y'(X + \delta) \end{pmatrix} = \begin{pmatrix} u(\delta) & v(\delta) \\ u'(\delta) & v'(\delta) \end{pmatrix} \begin{pmatrix} A \\ B \end{pmatrix}. \tag{3.3.8}$$

It is thus seen that the role of u and v is of propagating the exact solution from X to the desired $X + \delta$ and for this reason u and v are called exact propagators.

We associate to eq. (3.3.5) an equation of the same form

$$\tilde{y}''(X + \delta) = \tilde{f}(\delta)\tilde{y}(X + \delta), \quad \delta \in [0, h] \tag{3.3.9}$$

where the expression of $\tilde{f}(\delta)$ is chosen in such a way that this equation has known analytic solutions.

We are particularly interested in the propagators $\tilde{u}(\delta)$ and $\tilde{v}(\delta)$ associated to eq. (3.3.9), that is solutions of problems

$$\tilde{u}'' = \tilde{f}(\delta)\tilde{u}, \quad \tilde{u}(0) = 1, \quad \tilde{u}'(0) = 0 \tag{3.3.10}$$

and

$$\tilde{v}'' = \tilde{f}(\delta)\tilde{v}, \quad \tilde{v}(0) = 0, \quad \tilde{v}'(0) = 1, \tag{3.3.11}$$

respectively.

Function $\tilde{f}$, eq. (3.3.9) and its propagators $\tilde{u}$ and $\tilde{v}$ will be called hereafter as *reference* function, equation and propagators, respectively and our purpose actually is to construct the *unknown* propagators u and v in terms of the *known* reference propagators $\tilde{u}$ and $\tilde{v}$.

Perturbation theory is applied in this aim and its idea mainly consists of introducing a parametrically dependent function $\mathscr{F}(\delta;\lambda)$, $\lambda \in [0, 1]$ which reproduces the given $f(\delta)$ and the reference $\tilde{f}(\delta)$ when λ assumes its extreme values viz. $\mathscr{F}(\delta;0) = \tilde{f}(\delta)$, $\mathscr{F}(\delta;1) = f(\delta)$. Once $\mathscr{F}(\delta;\lambda)$ constructed the propagators u and v, which we want to evaluate, and $\tilde{u}$ and $\tilde{v}$, which we already know, are the particular cases $\lambda = 1$ and $\lambda = 0$ of the propagators $u(\delta;\lambda)$ and $v(\delta;\lambda)$ of the differential equation

$$y''(X+\delta;\lambda) = \mathscr{F}(\delta;\lambda)\, y(X+\delta;\lambda), \qquad \delta \in [0, h]. \tag{3.3.12}$$

The simplest form of $\mathscr{F}(\delta;\lambda)$ is

$$\mathscr{F}(\delta;\lambda) = \tilde{f}(\delta) + \lambda \Delta f(\delta) \tag{3.3.13}$$

where the deviation

$$\Delta f(\delta) \equiv f(\delta) - \tilde{f}(\delta) \tag{3.3.14}$$

is called perturbation.

The propagators $u(\delta;\lambda)$ and $v(\delta;\lambda)$ are searched for as power series in the coupling parameter λ

$$u(\delta;\lambda) = \sum_{q=0}^{\infty} \lambda^q u_q(\delta), \qquad v(\delta;\lambda) = \sum_{q=0}^{\infty} \lambda^q v_q(\delta). \tag{3.3.15}$$

Both propagators will be generically denoted by p; Eqs (3.3.15) are thus collectively written as

$$p(\delta;\lambda) = \sum_{q=0}^{\infty} \lambda^q p_q(\delta) \tag{3.3.16}$$

where $p = u$ if $p(0;\lambda) = 1$, $p'(0;\lambda) = 0$ and $p = v$ if $p(0;\lambda) = 0$, $p'(0;\lambda) = 1$.

To calculate p_q we introduce $p(\delta;\lambda)$ into eq. (3.3.12),

$$p''(\delta;\lambda) = (\tilde{f}(\delta) + \lambda \Delta f(\delta))\, p(\delta;\lambda) \tag{3.3.17}$$

and organize the terms in powers of λ

$$\lambda^0[p_0'' - \tilde{f}(\delta)p_0] + \sum_{q=1}^{\infty} \lambda^q[p_q'' - \tilde{f}(\delta)\, p_q - \Delta f(\delta)\, p_{q-1}] = 0. \tag{3.3.18}$$

This is identically satisfied for any $\lambda \in [0, 1]$ only if the δ dependent weight of λ^q vanishes for any $q = 0, 1, 2, \ldots$, that is

$$p_0'' = \tilde{f}(\delta)p_0 \tag{3.3.19 a}$$

$$p_q'' = \tilde{f}(\delta)\, p_q + \Delta f(\delta)\, p_{q-1}(\delta), \qquad q = 1, 2, 3, \ldots \tag{3.3.19 b}$$

To find out the suitable initial conditions for these differential equations we start from the remark that if we put $\lambda = 0$ in eq. (3.3.16) we get $p(\delta; 0) = p_0(\delta)$. Since, on the other hand, we also know that $p(\delta; 0) = \tilde{p}(\delta)$ we get $\tilde{p}(\delta) = p_0(\delta)$. For this reason the initial values of the differences $p(\delta;\lambda) - p_0(\delta)$ and $p'(\delta;\lambda) - p_0'(\delta)$ are forced to vanish,

$$\sum_{q=1}^{\infty} \lambda^q p_q(0) = 0, \qquad \sum_{q=1}^{\infty} \lambda^q p_q'(0) = 0, \qquad \lambda \in [0, 1] \tag{3.3.20}$$

whence

$$p_q(0) = p_q'(0) = 0 \tag{3.3.21}$$

for any $q = 1, 2, 3, \ldots$.

In short, we have obtained the following

THEOREM. *Eq.* (3.3.5) *with the initial conditions A and B has solution* (3.3.8) *where the propagators u and v are written as perturbation series*

$$p(\delta) = p_0(\delta) + p_1(\delta) + p_2(\delta) + \dots \tag{3.3.22}$$

(p stands for u or v); the zeroth order propagator $p_0(\delta)$ *is exactly the reference propagator* $\tilde{p}(\delta)$ *while the correction* $p_q(\delta)$ $(q = 1, 2, 3, \dots)$ *is the solution of the problem*

$$p_q'' = \tilde{f}(\delta) p_q + \Delta f(\delta)\, p_{q-1}(\delta), \quad p_q(0) = p_q'(0) = 0. \tag{3.3.23}$$

In practice only a finite number of terms is retained in the sum (3.3.22). If p_Q is the last term retained, that is $p(\delta)$ is approximated by

$$\bar{p}(\delta) = \tilde{p}(\delta) + p_1(\delta) + \dots + p_Q(\delta) \tag{3.3.24}$$

the resultant algorithm

$$\begin{pmatrix} \bar{y}(X+\delta) \\ \bar{y}'(X+\delta) \end{pmatrix} = \begin{pmatrix} \bar{u}(\delta) & \bar{v}(\delta) \\ \bar{u}'(\delta) & \bar{v}'(\delta) \end{pmatrix} \begin{pmatrix} A \\ B \end{pmatrix} \tag{3.3.25}$$

is called a Qth order perturbation algorithm with respect to the reference function $\tilde{f}(\delta)$.

The quality of this algorithm depends on the approximation of the given $f(\delta)$ through the reference function $\tilde{f}(\delta)$ and also on the number of perturbative corrections retained.

Fit of the reference function

The function $\tilde{f}(\delta)$ is typically assumed of polynomial form,

$$\tilde{f}(\delta) = \tilde{f}_0 + \tilde{f}_1 \delta + \tilde{f}_2 \delta^2 + \dots + \tilde{f}_t \delta^t \tag{3.3.26}$$

and to calculate its coefficients we choose to impose the condition that the function

$$I^2(\tilde{f}_0, \tilde{f}_1, \dots, \tilde{f}_t) \equiv \int_0^h [\Delta f(\delta)]^2 \, d\delta \tag{3.3.27}$$

is minimized, a condition which is known as the least squares fit. As a matter of fact, this is not the only possible fit. Other conditions may be for instance that either

$$\max_{0 \leqslant \delta \leqslant h} |\Delta f(\delta)| \tag{3.3.28}$$

or

$$I^1(\tilde{f}_0, \tilde{f}_1, \ldots, \tilde{f}_t) \equiv \int_0^h |\Delta f(\delta)|\, d\delta \tag{3.3.29}$$

is minimized. It will be however seen that the least squares fit is closest to the spirit of perturbation methods; see next section.

To explicitly evaluate the $\tilde{f}_n$ we set to zero the first-order partial derivative of I^2 with respect to each $\tilde{f}_n$ and obtain the following system of linear equations

$$\sum_{n=0}^{t} I_{mn} \tilde{f}_n = J_n, \qquad m = 0, 1, 2, \ldots, t \tag{3.3.30}$$

where

$$I_{mn} = \int_0^h \delta^{m+n}\, d\delta = \frac{1}{(m+n+1)} h^{m+n+1}, \qquad J_n = \int_0^h \delta^n f(\delta)\, d\delta, \tag{3.3.31}$$

a system which can be solved numerically by some standard subroutine. (True, the system is ill-conditioned and as t increases the accuracy of the resultant $\tilde{f}_0, \tilde{f}_1, \ldots, \tilde{f}_t$ tends to deteriorate rapidly due to the round-off errors but in our case t remains sufficiently small (up to 2 or so) so that no special problem is arising.)

A point which will be exploited below, in particular when we investigate the convergence properties of the algorithm (3.3.25), is that when h tends to zero $\tilde{f}(\delta)$ tends to $f(0)$. For illustration we take the constant $\tilde{f}(\delta)$ that is $t = 0$ in eq. (3.3.26). We get

$$\tilde{f}(\delta) \equiv \tilde{f}_0 = J_0(h) / I_{00}(h) = \frac{1}{h} \int_0^h f(\delta)\, d\delta \tag{3.3.32}$$

where, on applying the l'Hopital rule, it results that $\lim_{h \to 0} \tilde{f}(\delta) = f(0)$.

Convergence

Algorithm (3.3.25) is a genuine one-step method so that to investigate its convergence it suffices to put it in form (2.1.2) and to verify that the conditions of the theorem given on p. 118 are satisfied. We actually use the vector extension of that theorem. It is as follows.

THEOREM. *Consider the genuine one-step algorithm*

$$\bar{\mathbf{y}}(X+h) = \bar{\mathbf{y}}(X) + h\,\mathbf{\Phi}(X, \bar{\mathbf{y}}(X); h) \tag{3.3.33}$$

for the system

$$\mathbf{y}'(x) = \mathbf{f}(x, \mathbf{y}), \quad x \in [a, b], \quad \mathbf{y}(a) = \mathbf{y}_0 \tag{3.3.34}$$

If the increment function vector $\mathbf{\Phi}(x, \mathbf{y}; h)$ *is continuous in all three variables* $x \in [a, b]$, $\mathbf{y}(\|\mathbf{y}\| \in (-\infty, +\infty))$ *and* $h \geqslant 0$ *and also a constant* L *exists such that*

$$\|\mathbf{\Phi}(x, \mathbf{y}; h) - \mathbf{\Phi}(x, \mathbf{y}^*; h)\| \leqslant L\|\mathbf{y} - \mathbf{y}^*\| \tag{3.3.35}$$

then a necessary and sufficient condition for the convergence of algorithm (3.3.33) *is that*

$$\mathbf{\Phi}(x, \mathbf{y}; 0) = \mathbf{f}(x, \mathbf{y}). \tag{3.3.36}$$

Eq. (3.3.1) can be equivalently written as a system of two first-order equations

$$\begin{pmatrix} y \\ y' \end{pmatrix}' = \begin{pmatrix} 0 & 1 \\ g(x) & 0 \end{pmatrix} \begin{pmatrix} y \\ y' \end{pmatrix} \tag{3.3.37}$$

and the increment function vector $\mathbf{\Phi}$ of algorithm (3.3.25) is

$$\mathbf{\Phi}(X, \mathbf{y}(X); h) = \frac{1}{h} \begin{pmatrix} \bar{u}(h) - 1 & \bar{v}(h) \\ \bar{u}'(h) & \bar{v}'(h) - 1 \end{pmatrix} \begin{pmatrix} y(X) \\ y'(X) \end{pmatrix} \tag{3.3.38}$$

for $h > 0$. We choose to define the value of $\mathbf{\Phi}$ at $h = 0$ as follows

$$\mathbf{\Phi}(X, \mathbf{y}(X); 0) = \begin{pmatrix} 0 & 1 \\ g(X) & 0 \end{pmatrix} \begin{pmatrix} y(X) \\ y'(X) \end{pmatrix}. \tag{3.3.39}$$

Clearly $\mathbf{\Phi}$ defined for $h \geqslant 0$ by eqs (3.3.38) and (3.3.39) is continuous in X and $\mathbf{y}$, satisfies condition (3.3.35) and also automatically satisfies eq. (3.3.36). It only remains to be shown that $\mathbf{\Phi}$ is continuous in h, in particular when h tends to zero. We explicitly write down the dependence on h of the reference function and denote

$$\tilde{F}(\delta; h) = \begin{cases} \tilde{f}(\delta) & \text{if } 0 \leqslant \delta \leqslant h, \quad h > 0 \\ f(0) & \text{if } \delta = h = 0 \end{cases} \tag{3.3.40}$$

As mentioned, $\tilde{F}(\delta; h)$ is continuous in h, in particular at $h = 0$. Hence the perturbation

$$\Delta F(\delta; h) = f(\delta) - \tilde{F}(\delta; h) \tag{3.3.41}$$

is also continuous in h and $\Delta F(0, 0) = 0$.

Now calculate the limit of $\mathbf{\Phi}$ when h tends to zero; we use again the l'Hopital rule,

$$\begin{aligned} \lim_{h\to 0} \mathbf{\Phi}(X, \mathbf{y}(X); h) &= \lim_{h\to 0} \begin{pmatrix} \bar{u}'(h) & \bar{v}'(h) \\ \bar{u}''(h) & \bar{v}''(h) \end{pmatrix} \begin{pmatrix} y(X) \\ y'(X) \end{pmatrix} \\ &= \begin{pmatrix} \bar{u}'(0) & \bar{v}'(0) \\ \bar{u}''(0) & \bar{v}''(0) \end{pmatrix} \begin{pmatrix} y(X) \\ y'(X) \end{pmatrix} \end{aligned} \tag{3.3.42}$$

where the last member of the chain is justified by virtue of the fact that $\tilde{F}(\delta; h)$ and $\Delta F(\delta; h)$ are continuous in h. If we further use the fact that $\bar{u}'(0) = 0$ and $\bar{v}'(0) = 1$ (due to the initial conditions) and that $\bar{u}''(0)$ and $\bar{v}''(0)$ satisfy on components eq. (3.3.23) with $\tilde{f}(0) = f(0) = g(X)$ and $\Delta f(0) = 0$ it results that $\bar{u}''(0) = g(X)\bar{u}(0) = g(X)$ and $\bar{v}''(0) = g(X)\bar{v}(0) = 0$. Altogether $\mathbf{\Phi}$ is continuous in h and since this was the only point which required special investigation the final conclusion is that algorithm (3.3.25) with fit choice (3.3.26) for $\tilde{f}(\delta)$ is convergent.

Error analysis

To measure the error associated to algorithm (3.3.25) it is sufficient to consider the errors in propagators and in their first derivative

$$\begin{aligned} \Delta p(\delta) &= p_{Q+1}(\delta) + p_{Q+2}(\delta) + \dots \\ \Delta p'(\delta) &= p'_{Q+1}(\delta) + p'_{Q+2}(\delta) + \dots \end{aligned} \tag{3.3.43}$$

with $p = u, v$.

The *local truncation error*, for instance, which remains defined the same way as for classical methods namely as the set

$$\begin{aligned} e(X+h) &\equiv y(X+h) - \bar{y}(X+h) \\ e'(X+h) &\equiv y'(X+h) - \bar{y}'(X+h) \end{aligned} \tag{3.3.44 a}$$

calculated under the assumption that the initial conditions on the considered one-step interval $[X, X+h]$ are the exact $y(X)$ and $y'(X)$, reads

$$\begin{pmatrix} e(X+h) \\ e'(X+h) \end{pmatrix} = \begin{pmatrix} \Delta u(h) & \Delta v(h) \\ \Delta u'(h) & \Delta v'(h) \end{pmatrix} \begin{pmatrix} y(X) \\ y'(X) \end{pmatrix}. \tag{3.3.44 b}$$

The *accumulated* truncation error can also be evaluated in terms of Δp and $\Delta p'$ and to see this we first recall that in each step of the partition the exact solution of problem (3.3.5) satisfies eq. (3.3.8) with $A = y(X)$ and $B = y'(X)$ while what the considered perturbation method tentatively solves is the problem

$$z''(X+\delta) = f(\delta)\, z(X+\delta), \qquad \delta \in [0, h] \tag{3.3.45}$$

with the initial conditions $z(X) = \bar{y}(X)$, $z'(X) = \bar{y}'(X)$ where $\bar{y}(X)$ and $\bar{y}'(X)$ are values resulted from the repeated application of the same algorithm from $x = a$ up to $x = X$. Then the deviations $E(X) = y(X) - \bar{y}(X)$ and $E'(X) = y'(X) - \bar{y}'(X)$ collectively represent the accumulated error at $x = X$.

On subtracting eq. (3.3.45) from eq. (3.3.5) we find that the function $\varepsilon(X+\delta) \equiv y(X+\delta) - z(X+\delta)$ is a solution of the problem

$$\begin{aligned} &\varepsilon''(X+\delta) = f(\delta)\, \varepsilon(X+\delta), \quad \delta \in [0, h], \\ &\varepsilon(X) = E(X), \ \varepsilon'(X) = E'(X). \end{aligned} \tag{3.3.46}$$

The considered algorithm does not solve problem (3.3.45) exactly; it only yields some approximations $\bar{y}(X+\delta)$ and $\bar{y}'(X+\delta)$ through eq. (3.3.25) with $A = \bar{y}(X)$, $B = \bar{y}'(X)$. Then, if we denote

$$\begin{aligned} \Delta z(X+h) &= z(X+h) - \bar{y}(X+h), \\ \Delta z'(X+h) &= z'(X+h) - \bar{y}'(X+h) \end{aligned} \tag{3.3.47}$$

the two components of the accumulated error at $X+h$ read

$$E(X+h) \equiv y(X+h) - \bar{y}(X+h) = \varepsilon(X+h) + \Delta z(X+h)$$
$$E'(X+h) \equiv y'(X+h) - \bar{y}'(X+h) = \varepsilon'(X+h) + \Delta z'(X+h). \quad (3.3.48)$$

Deviations $\Delta z(X+h)$ and $\Delta z'(X+h)$ satisfy an analog of eq. (3.3.44 b) viz.

$$\begin{pmatrix} \Delta z(X+h) \\ \Delta z'(X+h) \end{pmatrix} = \begin{pmatrix} \Delta u(h) & \Delta v(h) \\ \Delta u'(h) & \Delta v'(h) \end{pmatrix} \begin{pmatrix} \bar{y}(X) \\ \bar{y}'(X) \end{pmatrix}, \quad (3.3.49)$$

and thus they are fully determined in terms of Δp and $\Delta p'$. Let us restore for the moment the interval labels: $x_s = X$, $x = x_s + \delta$, $x \in [x_s, x_{s+1} = x_s + h_s]$, $g(x) = f(\delta)$, $\varepsilon_s(x) = \varepsilon(X+\delta)$, $E_s = E(X)$, $E'_s = E'(X)$, $\Delta z_{s+1} = \Delta z(X+h)$, $\Delta z'_{s+1} = \Delta z'(X+h)$. We are thus in the conditions of lemma given on p. 88, with $d_{s+1} = -\Delta z_{s+1}$, $d'_{s+1} = -\Delta z'_{s+1}$ and $E_0 = E'_0 = 0$.
Eq. (1.12.23) gives

$$\varepsilon_{n-1}(x_n = b) = \sum_{s=1}^{n-1} [F(x_s)\,\Delta z'_s - F'(x_s)\,\Delta z_s]$$

so that

$$E_{n-1}(x_n = b) \equiv \varepsilon_{n-1}(b) + \Delta z_n = \sum_{s=1}^{n} (F(x_s)\Delta' z_s - F'(x_s)\Delta z_s). \quad (3.3.50)$$

This formula gives the way to calculate the accumulated truncation errors at the very end point $x_n = b$ in terms of the known Δz_s and $\Delta z'_s$, $s = 1, 2, \ldots, n$ and of the solution of the original equation $F'' = g(x)F$ with the initial conditions taken at the *same* point where the accumulated error is measured, that is at $x_n = b$.

This result can be easily adapted to evaluate the accumulated error in the solution at any mesh point x_k. It namely shows that

$$E_{k-1}(x_k) = y(x_k) - \bar{y}(x_k) \quad (3.3.51)$$

is given by

$$E_{k-1}(x_k) = \sum_{s=1}^{k} [F(x_s)\, \Delta z_s' - F'(x_s)\, \Delta z_s] \tag{3.3.52}$$

where $F(x)$ is the solution of the original equation with the initial conditions $F(x_k) = 0,\ F'(x_k) = -1$.

If eq. (1.12.24) is worked out the same way a similar formula for the accumulated error in the derivative of the solution at x_k is attained namely that

$$E'_{k-1}(x_k) = y'(x_k) - \bar{y}'(x_k) \tag{3.3.53}$$

is given by

$$E'_{k-1}(x_k) = \sum_{s=1}^{k} [D(x_s)\, \Delta z_s' - D'(x_s)\, \Delta z_s] \tag{3.3.54}$$

where $D(x)$ is the solution of $D'' = g(x)D$ with the initial conditions $D(x_k) = 1,\ D'(x_k) = 0$.

Although eqs (3.3.52) and (3.3.54) are exact their practical use relies on two approximations: (i) only the first term of the perturbation series for the propagator errors $\Delta p(h)$ and $\Delta p'(h)$, eq. (3.3.43), is retained; (ii) $F(x_s)$, $F'(x_s)$, $D(x_s)$ and $D'(x_s)$ are replaced by the approximate values $\bar{F}(x_s)$, $\bar{F}'(x_s)$, $\bar{D}(x_s)$ and $\bar{D}'(x_s)$ obtained by integrating eq. $y'' = g(x)y$ *backward* through the same piecewise perturbation method.

As a matter of fact the backward integration raises no special problem. Suppose that we want to integrate eq. (3.3.5) when the initial conditions are imposed at $X + h$. We premultiply both members of eq. (3.3.8) by matrix

$$\begin{pmatrix} v'(\delta) & -v(\delta) \\ -u'(\delta) & u(\delta) \end{pmatrix}.$$

By virtue of the matrix identity

$$\begin{pmatrix} v'(\delta) & -v(\delta) \\ -u'(\delta) & u(\delta) \end{pmatrix} \begin{pmatrix} u(\delta) & v(\delta) \\ u'(\delta) & v'(\delta) \end{pmatrix} = \begin{pmatrix} 1 & 0 \\ 0 & 1 \end{pmatrix} \tag{3.3.55}$$

(this comes directly from $W[u, v] = 1$) we get

$$\begin{pmatrix} y(X) \\ y'(X) \end{pmatrix} = \begin{pmatrix} v'(\delta) & -v(\delta) \\ -u'(\delta) & u(\delta) \end{pmatrix} \begin{pmatrix} y(X+\delta) \\ y'(X+\delta) \end{pmatrix} \tag{3.3.56}$$

so that to obtain $y(X)$ and $y'(X)$ in terms of $y(X+h)$ and $y'(X+h)$ we have to put here $\delta = h$. The algorithm for backward integration thus remains of the very type (3.3.25) with the only modification that the approximate propagators and their derivatives have different positions and signs in the propagation matrix.

The investigation of the *accumulated round-off errors* for algorithm (3.3.25) deserves no special attention since it is fully covered by the general analysis given in section 2.1 for genuine one-step methods.

Construction of the perturbative corrections

There are several ways to construct the perturbative corrections p_q, $q = 1, 2, 3, \ldots$. One of them, perhaps the simplest, starts from the assumption that each p_q is a linear combination of the reference propagators and of their first-order derivative,

$$p_q(\delta) = a_q(\delta)\tilde{u}(\delta) + b_q(\delta)\tilde{v}(\delta) + c_q(\delta)\tilde{u}'(\delta) + d_q(\delta)\tilde{v}'(\delta) \tag{3.3.57}$$

where a_q, b_q, c_q and d_q are functions to be determined. To this end we first explicitly evaluate p_q' and p_q''

$$\begin{aligned} p_q' &= a_q'\tilde{u} + a_q\tilde{u}' + b_q'\tilde{v} + b_q\tilde{v}' + c_q'\tilde{u}' + c_q\tilde{u}'' + d_q'\tilde{v}' + d_q\tilde{v}'' \\ &= a_q'\tilde{u} + (a_q + c_q')\tilde{u}' + (b_q + d_q')\tilde{v}' + b_q'\tilde{v} + c_q\tilde{f}\tilde{u} + d_q\tilde{f}\,\tilde{v} \\ &= (a_q' + c_q\tilde{f})\tilde{u} + (b_q' + d_q\tilde{f})\tilde{v} + (a_q + c_q')\tilde{u}' + (b_q + d_q')\tilde{v}', \end{aligned} \tag{3.3.58}$$

$$\begin{aligned} p_q'' = {} & [a_q'' + (2c_q' + a_q)\tilde{f} + c_q\tilde{f}']\tilde{u} + [b_q'' + (2d_q' + b_q)\tilde{f} + d_q\tilde{f}']\tilde{v} + \\ & + (c_q'' + 2a_q' + c_q\tilde{f})\tilde{u}' + (d_q'' + 2b_q' + d_q\tilde{f})\tilde{v}' \end{aligned} \tag{3.3.59}$$

and next conveniently write the right member of eq. (3.3.23) as

$$\begin{aligned} \tilde{f}p_q + \Delta f p_{q-1} = {} & (a_q\tilde{f} + \Delta f a_{q-1})\tilde{u} + (b_q\tilde{f} + \Delta f b_{q-1})\tilde{v} + \\ & + (c_q\tilde{f} + \Delta f c_{q-1})\tilde{u}' + (d_q\tilde{f} + \Delta f d_{q-1})\tilde{v}'. \end{aligned} \tag{3.3.60}$$

We now identify the coefficients of $\tilde{u}, \tilde{v}, \tilde{u}'$ and $\tilde{v}'$ in eqs (3.3.59) and (3.3.60) to get the following system of four differential equations

$$
\begin{aligned}
&a_q'' + 2c_q'\tilde{f} + c_q\tilde{f}' = \Delta f a_{q-1},\\
&b_q'' + 2d_q'\tilde{f} + d_q\tilde{f}' = \Delta f b_{q-1},\\
&c_q'' + 2a_q' = \Delta f c_{q-1},\\
&d_q'' + 2b_q' = \Delta f d_{q-1}.
\end{aligned}
\tag{3.3.61}
$$

To evaluate the initial conditions we explicitly take into account that $\tilde{u}(0) = \tilde{v}'(0) = 1$ and $\tilde{u}'(0) = \tilde{v}(0) = 1$. Eqs (3.3.57) and (3.3.58) thus give

$$a_q(0) + d_q(0) = 0, \; a_q'(0) + c_q(0)\tilde{f}(0) + b_q(0) + d_q'(0) = 0 \tag{3.3.62}$$

The solution goes on sequentially with increasing q. For $q = 1$ we have $a_0 = 1$, $b_0 = c_0 = d_0 = 0$ if $p = u$ and $a_0 = c_0 = d_0 = 0$ but $b_0 = 1$ if $p = v$. We then solve the system for a_1, b_1, c_1 and d_1; these are next introduced in the right members of equations (3.3.61) at $q = 2$, and so on.

What is particularly important to note is that the system (3.3.61—62) consists of *four* differential equations with only *two* initial conditions and hence its solution is not unique. This circumstance is not unpleasant at all, it is just very convenient. In fact it leaves open the possibility of further searching for a_q, b_q, c_q and d_q among functions of preset, easily computable forms.

The simplest is the polynomial form but of course not any $\tilde{f}(\delta)$ will allow expressing a_q, b_q, c_q and d_q as polynomials. Reasonable candidates for being reference functions in a perturbative scheme will therefore be the expressions of $\tilde{f}(\delta)$ which, in addition to generating analytic expressions for the reference propagators, also allow expressing the corrections p_q, $q = 1, 2, 3, \ldots$ as in eq. (3.3.57) with *polynomial* coefficients.

The second requirement restricts the class of candidates to only a few members among whom constant and linear reference functions are the most popular. These will be treated separately in the next two sections.

3.4. CONSTANT PERTURBATION METHODS

To solve the one-step problem

$$y''(X+\delta) = f(\delta)y(X+\delta), \ \delta \in [0,h], \ y(X) = A, \ y'(X) = B \tag{3.4.1}$$

where $f(\delta)$ is real, bounded and continuous, we use the reference equation

$$\tilde{y}''(X+\delta) = \tilde{f}\tilde{y}(X+\delta), \ \delta \in [0,h] \tag{3.4.2}$$

where $\tilde{f}$ is a real constant

$$\tilde{f} = F \equiv \frac{1}{h}\int_0^h f(\delta)\, d\delta, \tag{3.4.3}$$

see eq. (3.3.32).
When $F \neq 0$ the general solution of the reference equation is

$$\tilde{y}(X+\delta) = A_1 \exp[F^{1/2}\delta] + A_2 \exp[-F^{1/2}\delta] \tag{3.4.4}$$

with A_1 and A_2 arbitrary constants. The propagators $\tilde{u}(\delta)$ and $\tilde{v}(\delta)$ are the particular solutions which correspond to $A_1 = A_2 = 1/2$ in $\tilde{u}(\delta)$ and to $A_1 = -A_2 = 1/(2F^{1/2})$ in $\tilde{v}(\delta)$.
When $F = 0$ the general solution is

$$\tilde{y}(X+\delta) = A_1 + A_2\delta$$

and thus $\tilde{u}(\delta) = 1$ and $\tilde{v}(\delta) = \delta$.
Altogether we define the functions

$$\xi(F,\delta) \equiv \begin{cases} \cos(|F|^{1/2}\delta) & \text{for } F \leqslant 0 \\ \cosh(F^{1/2}\delta) & \text{for } F > 0 \end{cases} \tag{3.4.5}$$

and

$$\eta_0(F,\delta) \equiv \begin{cases} \sin(|F|^{1/2}\delta)/|F|^{1/2} & \text{for } F < 0 \\ \delta & \text{for } F = 0 \\ \sinh(F^{1/2}\delta)/F^{1/2} & \text{for } F > 0 \end{cases} \tag{3.4.6}$$

to compactly write the reference propagators as

$$\tilde{u}(\delta) = \xi(F, \delta), \quad \tilde{v}(\delta) = \eta_0(F, \delta), \tag{3.4.7}$$

for *any* real F.
Their derivatives are

$$u'(\delta) = F\eta_0(F, \delta), \quad v'(\delta) = \xi(F, \delta). \tag{3.4.8}$$

We follow the convention made in connection with eq. (3.3.25) to call the numerical method in which the exact u, v, u' and v' are replaced by these $\tilde{u}$, $\tilde{v}$, $\tilde{u}'$ and $\tilde{v}'$ as a zeroth order constant (reference function) perturbation method, CPM(0).

Perturbation corrections

The corrections $p_1, p_2, p_3, \ldots$ can be constructed according to the procedure outlined in the previous section. It is however useful to take advantage of the fact that for the constant reference function the propagators and their derivative are connected via eq. (3.4.8). Thus, the corrections can be searched for as linear combinations of the reference propagators alone,

$$p_q(\delta) = a_q(\delta)\tilde{u}(\delta) + b_q(\delta)\tilde{v}(\delta). \tag{3.4.9}$$

On applying the same procedure as on p. 193 it readily results that a_q and b_q satisfy the system

$$\begin{aligned} a_q'' + 2b_q' &= \Delta f(\delta)a_{q-1} \\ b_q'' + 2Fa_q' &= \Delta f(\delta)b_{q-1} \end{aligned} \tag{3.4.10}$$

with the initial conditions

$$a_q(0) = 0, \; a_q'(0) + b_q(0) = 0. \tag{3.4.11}$$

We use this result only to show that if $\Delta f(\delta)$ is a polynomial in δ then a_q and b_q are also polynomials in δ. To see this we start with the remark that the coefficients in the left member of each of the two equations (3.4.10) are numbers, *not* functions of δ. For this reason a particular solution of polynomial form always exists provided the inhomogeneous term is a polynomial. In addition, due to the fact that the coefficient of a_q and of b_q is zero in each of the two equations, the free terms of these polynomial solutions are still at our disposal and they are determined uniquely by the two initial conditions, eq. (3.4.11).

It remains only to check whether the right members of the two equations are polynomials in δ and for this it is enough to consider the case $q = 1$. Indeed, for $p = u$ one has $a_0 = 1$ and $b_0 = 0$ while for $p = v$ one has $a_0 = 0$ and $b_0 = 1$. Thus, if $\Delta f(\delta)$ is a polynomial then $\Delta f a_0$ and $\Delta f b_0$ will equally be polynomials and so a_1 and b_1 will actually result. This guarantees that the same remains true for a_q and b_q with $q = 2, 3, 4, \ldots$

Eqs (3.4.10—11) can be used without difficulty to explicitly calculate the perturbation corrections but we will renounce to do so and prefer another, slightly modified procedure which takes the advantage of being more transparent for practical appraisal of the error.

We thus construct the following functions

$$\eta_1(F, \delta) = [\delta\xi(F, \delta) - \eta_0(F, \delta)]/F, \tag{3.4.12 a}$$

$$\eta_s(F, \delta) = [\delta^2\eta_{s-2}(F, \delta) - (2s-1)\eta_{s-1}(F, \delta)]/F, \quad s = 2, 3, 4, \ldots \tag{3.4.12 b}$$

(The reader will recognize that each η_s is actually some particular linear combination of the reference propagators $\tilde{u} = \xi$ and $\tilde{v} \equiv \eta_0$.)

Some properties of functions $\xi, \eta_0, \eta_1, \eta_2, \ldots$ are listed below

(i) Differentiation with respect to δ

$$\frac{\partial\xi(F,\delta)}{\partial\delta} = F\eta_0(F, \delta); \frac{\partial\eta_0(F,\delta)}{\partial\delta} = \xi(F, \delta);$$
$$\frac{\partial\eta_s(F, \delta)}{\partial\delta} = \delta\eta_{s-1}(F, \delta), \; s = 1, 2, 3, \ldots \tag{3.4.13}$$

(ii) Differentiation with respect to F

$$\frac{\partial\xi(F,\delta)}{\partial F} = \frac{1}{2}\delta\eta_0(F, \delta); \; \frac{\partial\eta_s(F,\delta)}{\partial F} = \frac{1}{2}\eta_{s+1}(F, \delta), \quad s = 0, 1, 2, \ldots \tag{3.4.14}$$

(iii) $\xi(F, \delta)$ defined in eq. (3.4.5) actually depends only on the "compressed" argument $Z \equiv F\delta^2$

$$\xi(F, \delta) \equiv \bar{\xi}(Z) = \begin{cases} \cos(|Z|^{1/2}) & \text{if } Z \leqslant 0 \\ \cosh(Z^{1/2}) & \text{if } Z > 0. \end{cases} \tag{3.4.15}$$

At the same time $\eta_0(F, \delta)$ can be written as $\eta_0(F, \delta) = \delta\bar{\eta}_0(Z)$ with

$$\bar{\eta}_0(Z) = \begin{cases} \sin(|Z|^{1/2})/|Z|^{1/2} & \text{if } Z < 0 \\ 1 & \text{if } Z = 0 \\ \sinh(Z^{1/2})/Z^{1/2} & \text{if } Z > 0. \end{cases} \tag{3.4.16}$$

It follows that any $\eta_s(F, \delta)$, $s = 1, 2, 3, \ldots$ can be written in a form in which variables δ and Z are separated, viz.

$$\eta_s(F, \delta) = \delta^{2s+1}\bar{\eta}_s(Z) \tag{3.4.17}$$

where

$$\bar{\eta}_1(Z) = [\bar{\xi}(Z) - \bar{\eta}_0(Z)]/Z \tag{3.4.18 a}$$

and

$$\bar{\eta}_s(Z) = [\bar{\eta}_{s-2}(Z) - (2s-1)\bar{\eta}_{s-1}(Z)]/Z, \quad s = 2, 3, 4, \ldots \tag{3.4.18 b}$$

(iv) The expansion of $\bar{\eta}_s(Z)$ in powers of Z reads

$$\bar{\eta}_s(Z) = 2^s \sum_{q=0}^{\infty} g_{sq} Z^q/(2q + 2s + 1)! \tag{3.4.19}$$

with

$$g_{sq} = \begin{cases} 1 & \text{if } s = 0 \\ (q+1)(q+2)\ldots(q+s) & \text{if } s > 0 \end{cases} \tag{3.4.20}$$

(v) The behaviour of $\bar{\eta}_s$ at large $|Z|$ is

$$\bar{\eta}_s(Z) \simeq \begin{cases} \bar{\xi}(Z)/Z^{(s+1)/2} & \text{for odd } s \\ \bar{\eta}_0(Z)/Z^{s/2} & \text{for even } s. \end{cases} \tag{3.4.21}$$

The remarkable significance of the functions $\xi, \eta_0, \eta_1, \eta_2, \eta_3, \ldots$ stems from the properties (iii), (iv) and (v) which collectively say that these can be written as

$$\xi(F, \delta) = \bar{\xi}(Z), \quad \eta_s(F, \delta) = \delta^{2s+1}\bar{\eta}_s(Z)$$

where the set $\bar{\xi}(Z)$, $\bar{\eta}_0(Z)$, $\bar{\eta}_1(Z), \ldots$ exhibits definite *hierarchy* with respect to the numerical importance. Thus $\bar{\xi}(Z)$ is the "biggest" member of the set and it is followed in order by $\bar{\eta}_0(Z)$, $\bar{\eta}_1(Z)$, $\bar{\eta}_2(Z)\ldots$. The degree of discrepancy significantly depends on Z and it is at its lowest level when $Z = 0$, increasing with $|Z|$. These points are well visualized on Fig. 2.

Fig. 2 also suggests a simple way to construct a reasonably close bound for $\bar{\eta}_s(Z)$, $s = 1, 2, 3, \ldots$ For negative Z it is seen that $\bar{\eta}_s(Z)$ is an oscillating function whose amplitude damps out when $Z \to -\infty$. When $|Z|$ is sufficiently large eq. (3.4.21) is valid and since $\max_{Z \leqslant 0} |\bar{\xi}(Z)| = 1$ and the amplitude of $\bar{\eta}_0(Z)$ behaves as $1/|Z|^{1/2}$ at large $-Z$, it results that the amplitude of $\bar{\eta}_s(Z)$, $s = 1, 2, 3, \ldots$ will asymptotically damp out as $1/|Z|^{(s+1)/2}$ for any s.

We thus can take some point Z^C in the asymptotic region, $Z^C = -s^2(s+1)^2$, for instance, and construct the straight line segment which connects the points of coordinates $\left(Z = 0,\ \bar{\eta}_s(0) = \dfrac{1}{(2s+1)!!}\right)$ and $(Z = Z^C,\ 1/|Z^C|^{(s+1)/2})$. For $Z < Z^C$ the asymptotic behaviour is preserved.

Altogether the following envelope bound is obtained

$$|\bar{\eta}_s(Z)| \lesssim B_s^-(Z) \equiv$$

$$\equiv \begin{cases} -Z\,\dfrac{-\dfrac{1}{(2s+1)!!}+\dfrac{1}{[s(s+1)]^{s+1}}}{s^2(s+1)^2}+\dfrac{1}{(2s+1)!!} & \text{if } -s^2(s+1)^2 \leqslant Z \leqslant 0 \\ 1/|Z|^{(s+1)/2} & \text{if } Z < -s^2(s+1)^2. \end{cases} \tag{3.4.22}$$

For positive Z all these functions increase exponentially with Z and it is therefore convenient to look for a bound for the "normalized" function $\bar{\eta}_s(Z)/\bar{\xi}(Z)$. We repeat the same procedure to find out that

$$|\bar{\eta}_s(Z)/\bar{\xi}(Z)| \lesssim B_s^+(Z) \equiv$$

$$\equiv \begin{cases} Z\,\dfrac{\dfrac{1}{[s(s+1)]^{s+1}}-\dfrac{1}{(2s+1)!!}}{s^2(s+1)^2}+\dfrac{1}{(2s+1)!!} & \text{if } 0 < Z \leqslant s^2(s+1)^2 \\ 1/Z^{(s+1)/2} & \text{if } Z > s^2(s+1)^2. \end{cases} \tag{3.4.23}$$

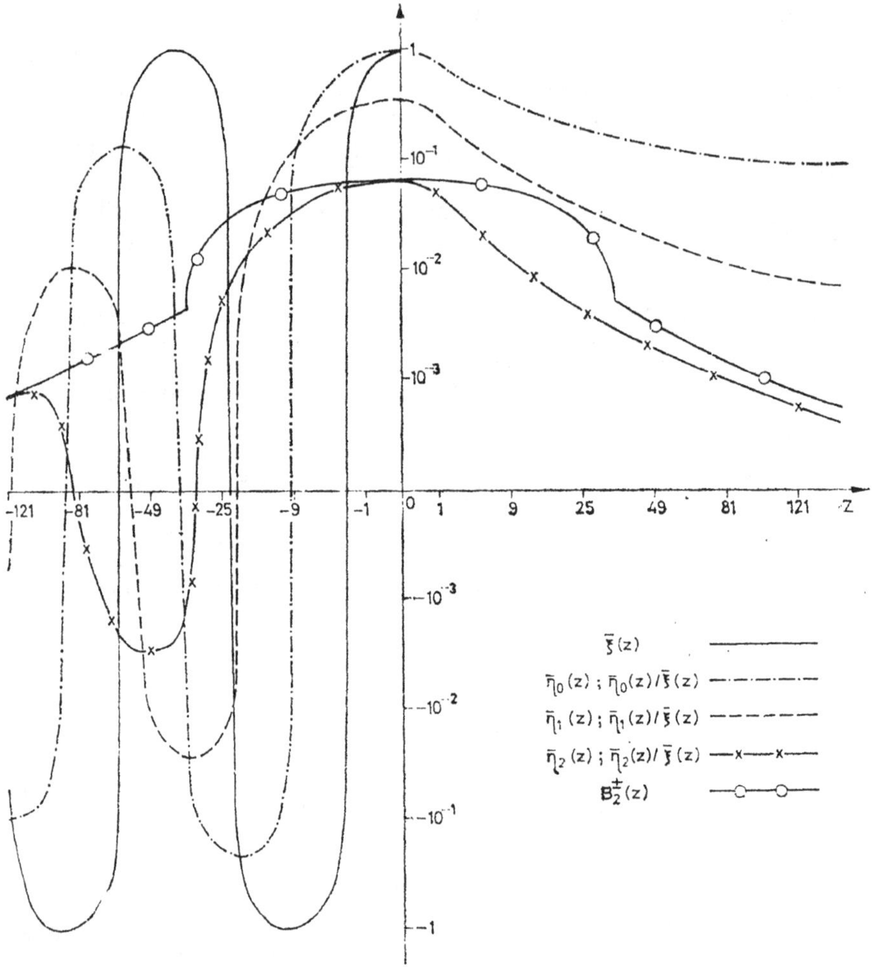

Fig. 2. — Graphs of $\bar{\xi}(Z)$, $\bar{\eta}_0(Z)$, $\bar{\eta}_1(Z)$ and $\bar{\eta}_2(Z)$ for $Z \leqslant 0$ and of the normalized $\bar{\eta}_0(Z)/\bar{\xi}(Z)$, $\bar{\eta}_1(Z)/\bar{\xi}(Z)$ and $\bar{\eta}_2(Z)/\bar{\xi}(Z)$ for $Z > 0$. The bound $B_{\overline{2}}^{\pm}(Z)$ is also plotted.

True, the quality of these bounds is not very good (for illustration the curve $B_{\overline{2}}^{\pm}(Z)$ is drawn on Fig. 2) and of course closer bounds may be also devised. Yet, the above expressions are so simple that they quickly allow a decent measure of the errors associated to CP algorithms.

To construct the qth correction p_q we assume that the inhomogeneous term in eq. (3.3.23) is a linear combination of the functions ξ and $\eta_0, \eta_1, \eta_2 \dots$

$$\Delta f(\delta)p_{q-1}(\delta) = Q(\delta)\xi(F, \delta) + P_0(\delta)\eta_0(F, \delta) + P_1(\delta)\eta_1(F, \delta) + \dots \tag{3.4.24}$$

where $Q, P_0, P_1, \dots$ are known functions, and search for p_q of the form

$$p_q(\delta) = C_0(\delta)\eta_0(F, \delta) + C_1(\delta)\eta_1(F, \delta) + C_2(\delta)\eta_2(F, \delta) + \dots \tag{3.4.25}$$

We differentiate eq. (3.4.25) twice with respect to δ and form $p_q'' - Fp_q$. One gets

$$p_q'' - Fp_q = 2C_q'\xi(F, \delta) + (C_0'' + 2C_1'\delta + 2C_1)\eta_0(F, \delta) + \dots$$

$$+ [C_s'' + 2C_{s+1}'\delta + 2(s + 1)C_{s+1}]\eta_s(F, \delta) + \dots \tag{3.4.26}$$

This must be equal to $\Delta f(\delta)p_{q-1}(\delta)$. Upon identifying the coefficients of $\xi, \eta_0, \eta_1, \eta_2, \dots$ in eqs (3.4.24) and (3.4.26) we get

$$2C_0' = Q(\delta),\ C_s'' + 2[C_{s+1}'\delta + (s + 1)C_{s+1}] = P_s(\delta),\ s = 1, 2, 3, \dots \tag{3.4.27}$$

These equations are solved by iteration: $C_0(\delta)$ is given by

$$C_0(\delta) = \frac{1}{2}\int_0^\delta Q(\delta_1)\, d\delta_1 \tag{3.4.28 a}$$

while $C_s(\delta)$, $s = 1, 2, 3, \dots$, results as

$$C_s(\delta) = \frac{1}{2}\delta^{-s}\int_0^\delta \delta_1^{s-1}[P_{s-1}(\delta_1) - C_{s-1}''(\delta_1)]\, d\delta_1, \tag{3.4.28 b}$$

and we thus have all ingredients necessary to evaluate the perturbative corrections as sums over the hierarchized set of functions $\xi, \eta_0, \eta_1, \eta_2, \dots$

It is interesting to remark that each new correction p_q starts with a term of higher order than the previous one, and to see this we examine the case $p = u$. (The case $p = v$ can be treated in a similar way.)

When $q = 1$, eq. (3.4.24) consists of a single term, namely $Q(\delta)\xi$ with $Q(\delta) = \Delta f(\delta)$, hence the resultant u_1 starts with the η_0 component. It follows that the inhomogeneous term $\Delta f u_1$ which occurs when $q = 2$ is a series which actually begins with the η_0 component and thus u_2 will result as a series which starts with the η_1 component, and so on. As for the derivative of p_k, eq. (3.4.25) gives at once that

$$p_q'(\delta) = C_0(\delta)\xi + (C_0' + \delta C_1)\eta_0 + (C_1' + \delta C_2)\eta_1 + \dots \qquad (3.4.29)$$

and thus similar connections are taking place between q and the position of the first nonvanishing term in this series.

In short we have the following result.

The first nonvanishing term in the series (3.4.25) *is the term corresponding to* η_{q-1} *in* u_q, *and to* η_q *in* v_q; *in series* (3.4.29) *it corresponds to* ξ *in* u_1', *to* η_{q-2} *in* u_q' *for* $q \geqslant 2$, *and to* η_{q-1} *in* v_q'.

A direct consequence of this result is that it provides a good basis to justify that the choice (3.4.3) for $\tilde{f}$ is an *optimal* choice.

In fact, in practical applications the user is mainly interested in the values of $\bar{y}$ and $\bar{y}'$ at $\delta = h$ since these will play the role of initial conditions in the next interval of the partition. For this reason, the optimal value of $\tilde{f}$ will be that which ensures that the very zeroth order algorithm yields solutions which are as accurate as possible at $\delta = h$. Equivalently, the optimal value of $\tilde{f}$ is the one which ensures that the values at $\delta = h$ of the first-order corrections are as small as possible and thus we have to examine the leading terms of these corrections. On using eqs (3.4.28 a) and (3.4.28 b) at $q = 1$ we find out that the leading term is $\frac{1}{2}\eta_0 \int_0^h \Delta f(\delta)\, d\delta$ in $u_1(h)$ and in $v_1'(h)$, $\frac{1}{2}\xi \int_0^h \Delta f(\delta)\, d\delta$ in $u_1'(h)$ and $\frac{1}{2h}\eta_1 \int_0^h \Delta f(\delta)\, d\delta$ in $v_1(h)$. All these vanish when

$$\int_0^h \Delta f(\delta)\, d\delta = 0, \qquad (3.4.30)$$

an equation which is just the equivalent of the least squares condition (3.3.27) with $t = 0$, thus justifying the choice (3.4.3).

A pilot reference function

In principle the above procedure can be used for any form of $\Delta f(\delta)$. Practical difficulties however arise due to the fact that the number of terms in series (3.4.24) and (3.4.25) is infinite and also because the practical evaluation of higher order terms in each correction becomes increasingly difficult. Both drawbacks are cured at once if $\Delta f(\delta)$ is a polynomial in δ. In this case these series consist of a *finite* number of terms for any q and the C_s are polynomials in δ which can be readily specified.

This suggests to add an intermediary step in the procedure which consists of approximating the given $f(\delta)$ through $f^*(\delta)$, a polynomial in δ. $f^*(\delta)$ is called a *pilot* reference function. Its degree must be high enough to ensure that $f^*(\delta)$ is a good approximation of $f(\delta)$ on $[0, h]$ but not too high since in such a case difficulties may occur in connection to the fact that the system (3.3.30—31) is ill-conditioned. The parabolic function

$$f^*(\delta) = f_0^* + f_1^*\delta + f_2^*\delta^2 \tag{3.4.31}$$

is a reasonable compromise and its coefficients f_0^*, f_1^* and f_2^* are computed by eqs (3.3.30—31) corresponding to the least squares fit.

If this is so the original $f(\delta)$ can be written as

$$f(\delta) = F + \Delta f^*(\delta) + \Delta\tilde{f}(\delta), \quad \delta \in [0, h] \tag{3.4.32}$$

where

$$F = \frac{1}{h}\int_0^h f(\delta)\, d\delta = \frac{1}{h}\int_0^h f^*(\delta)\, d\delta = f_0^* + \frac{1}{2}hf_1^* + \frac{1}{3}h^2f_2^*, \tag{3.4.33}$$

$$\Delta f^*(\delta) \equiv f(\delta) - f^*(\delta) \tag{3.4.34}$$

and

$$\Delta\tilde{f}(\delta) \equiv f^*(\delta) - F = f_1^*\left(\delta - \frac{1}{2}h\right) + f_2^*\left(\delta^2 - \frac{1}{3}h^2\right). \tag{3.4.35}$$

Δf^* and $\Delta \tilde{f}$ are called residual and pilot perturbation, respectively. Since the pilot perturbation is of polynomial form the perturbative corrections from it can be determined without difficulty in closed form. Only *these* corrections will be retained into the algorithm while the error due to the neglect of Δf^* will be considered later on. Particular attention is to be paid on the accuracy of the numerical computation of the coefficients f_0^*, f_1^* and f_2^*. As a matter of fact sufficient accuracy is obtained if the integrals J_n, $n = 0, 1, 2$ in eq. (3.3.31) are computed by either the five-point Gauss formula or the six-point Newton-Cotes formula. In the latter case f_0^*, f_1^* and f_2^* result of the form

$$f_0^* = [161 f_1 + 400 f_2 + 50 f_6 - (200 f_3 + 50 f_4 + 25 f_5)]/336 \quad (3.4.36\text{ a})$$

$$f_1^* = \frac{1}{336\,h}[950 f_4 + 1250 f_3 - (611 f_1 + 1025 f_2 + 175 f_5 + 389 f_6)] \quad (3.4.36\text{ b})$$

$$f_2^* = \frac{25}{84\,h^2}[5(f_1 + f_6) + 6(f_2 + f_5) - 11(f_3 + f_4)], \quad (3.4.36\text{ c})$$

where $f_i = f\left(\frac{1}{5}(i-1)h\right)$, $i = 1, 2, 3, 4, 5, 6$. On using these formulas the following value of F results

$$F \equiv f_0^* + \frac{1}{2}hf_1^* + \frac{1}{3}h^2 f_2^* = \frac{1}{288}[19(f_1 + f_6) + 75(f_2 + f_5) + 50(f_3 + f_4)]. \quad (3.4.37)$$

Of course quadrature procedures of higher accuracy than the quoted ones can be used as well. What is important is that, for any such choice, if the original $f(\delta)$ is differentiable sufficiently many times, the residual perturbation remains of the same form as with the six-point Newton-Cotes formula, and it is

$$\Delta f^*(\delta) = \frac{1}{120} f^{(3)} h^3 [20(\delta/h)^3 - 30(\delta/h)^2 + 12\,\delta/h - 1] + \frac{1}{840} f^{(4)} h^4 [35(\delta/h)^4 - 60(\delta/h)^2 + 32\delta/h - 3] + 0(h^5), \quad (3.4.38)$$

where $f^{(j)}$ is the value of the jth derivative of f at $\delta = 0$. Expression (3.4.38) will be used to appraise the influence of $\Delta f^*(\delta)$ on the accuracy of the algorithm.

Pilot perturbation: first- and second-order perturbation corrections

We explicitly consider the case $p = u$ for $\Delta \tilde{f}(\delta)$ given by eq. (3.4.35).

First-order correction, $q = 1$. Eq. (3.4.24) consists of a single term with

$$Q(\delta) = f_1^* \left(\delta - \frac{1}{2} h\right) + f_2^* \left(\delta^2 - \frac{1}{3} h^2\right).$$

Eq. (3.4.28 a) then gives

$$C_0(\delta) = \frac{1}{4} f_1^*(\delta^2 - h\delta) + \frac{1}{6} f_2^*(\delta^3 - h^2\delta) \tag{3.4.39}$$

while from eq. (3.4.28 b) it results that

$$C_1(\delta) = -\frac{1}{4} f_1^* - \frac{1}{4} f_2^* \delta \tag{3.4.40}$$

and $C_s(\delta) = 0$ for $s = 2, 3, 4 \ldots$

Thus the first-order correction u_1 is

$$u_1(\delta) = \left[\frac{1}{4} f_1^*(\delta^2 - h\delta) + \frac{1}{6} f_2^*(\delta^3 - h^2\delta)\right] \eta_0(F, \delta) - \\ - \frac{1}{4} (f_1^* + f_2^* \delta)\, \eta_1(F, \delta). \tag{3.4.41}$$

Its first derivative results from eq. (3.4.29)

$$u_1'(\delta) = \left[\frac{1}{4} f_1^*(\delta^2 - h\delta) + \frac{1}{6} f_2^*(\delta^3 - h^2\delta)\right] \xi(F, \delta) + \\ + \left[\frac{1}{4} f_1^*(\delta - h) + \frac{1}{12} f_2^*(3\delta^2 - 2h^2)\right] \eta_0(F, \delta) - \frac{1}{4} f_2^* \eta_1(F, \delta). \tag{3.4.42}$$

Second-order correction, $q = 2$. We multiply $\Delta\tilde{f}(\delta)$ by $u_1(\delta)$ of eq. (3.4.41). Eq. (3.4.24) thus is

$$\Delta f(\delta)\, u_1(\delta) = \left[f_1^{*2}\left(\delta^2 - \frac{3}{2}h\delta^2 + \frac{1}{2}h^2\delta\right) + \right.$$

$$+ \frac{1}{12} f_1^* f_2^* (5\delta^4 - 4h\delta^3 - 3h^2\delta^2 + 2h^3\delta) +$$

$$\left. + \frac{1}{18} f_2^{*2}(3\delta^5 - 4h^2\delta^3 + h^4\delta)\right] \eta_0(F, \delta) +$$

$$+ \left[-\frac{1}{4} f_1^{*2}\left(\delta - \frac{1}{2}h\right) - \frac{1}{4} f_1^* f_2^* \left(2\delta^2 - \frac{1}{2}h\delta - \frac{1}{3}h^2\right) - \right.$$

$$\left. - \frac{1}{12} f_2^{*2}(3\delta^3 - h^2\delta)\right] \eta_1(F, \delta). \tag{3.4.43}$$

Formulas (3.4.28 a-b) are applied to get the following result

$$u_2(\delta) = \delta(\delta - h)^2 \left[\frac{1}{32} f_1^{*2} + \frac{1}{24} f_1^* f_2^* (\delta + h) + \right.$$

$$\left. + \frac{1}{72} f_2^{*2}(\delta + h)^2\right] \eta_1(F, \delta) + \left[\frac{1}{96} f_1^{*2}(-7\delta + 6h) - \right.$$

$$\left. - \frac{1}{48} f_1^* f_2^* (6\delta^2 - 3h\delta - 2h^2) - \frac{1}{360} f_2^{*2}\delta(19\delta^2 - 15h^2)\right] \eta_2(F, \delta) +$$

$$+ \left[\frac{1}{24} f_1^* f_2^* + \frac{19}{480} f_2^{*2}\delta\right] \eta_3(F, \delta). \tag{3.4.44}$$

Its first derivative reads

$$u_2'(\delta) = \delta^2(\delta - h)^2 \left[\frac{1}{32} f_1^{*2} + \frac{1}{24} f_1^* f_2^* (\delta + h) + \right.$$

$$\left. + \frac{1}{72} f_2^{*2}(\delta + h)^2\right] \eta_0(F, \delta) + \left[\frac{1}{96} f_1^{*2}(29\delta^2 - 42h\delta + 12h^2) + \right.$$

$$+\frac{1}{48}f_1^*f_2^*(2\delta^2 - 3h\delta^2 - 2h^2\delta + 2h^3) +$$

$$+\frac{1}{360}f_2^{*2}(6\delta^4 - 15h^2\delta^2 + 5h^4)\Bigg]\,\eta_1(F,\delta) +$$

$$+\Bigg[-\frac{7}{96}f_1^{*2} - \frac{1}{48}f_1^*f_2^*(10\delta - 3h) +$$

$$+\frac{1}{480}f_2^{*2}(-57\delta^2 + 20h^2)\Bigg]\,\eta_2(F,\delta) + \frac{19}{480}f_2^{*2}\eta_2(F,\delta). \tag{3.4.45}$$

The expressions of these corrections at $\delta = h$ are much shorter

$$u_1(h) = -\frac{1}{4}(f_1^* + f_2^*h)\,\eta_1(F,h) \tag{3.4.46}$$

$$u_1'(h) = \frac{1}{4}f_2^*\left(\frac{1}{3}h^2\eta_0(F,h) - \eta_1(F,h)\right) = \frac{1}{12}f_2^*F\eta_2(F,h) \tag{3.4.47}$$

$$u_2(h) = -h\left(\frac{1}{96}f_1^{*2} + \frac{1}{48}f_1^*f_2^*h + \frac{1}{90}f_2^{*2}h^2\right)\eta_2(F,h) +$$

$$+\frac{1}{24}\left(f_1^*f_2^* + \frac{19}{20}f_2^{*2}h\right)\eta_3(F,h) \tag{3.4.48}$$

and

$$u_2'(h) = -h^2\left(\frac{1}{96}f_1^{*2} + \frac{1}{48}f_1^*f_2^*h + \frac{1}{90}f_2^{*2}h^2\right)\eta_1(F,h) -$$

$$-\left(\frac{7}{96}f_1^{*2} + \frac{7}{48}f_1^*f_2^*h + \frac{37}{480}f_2^{*2}h^2\right)\eta_2(F,h) + \frac{19}{480}f_2^{*2}\eta_3(F,h). \tag{3.4.49}$$

The same treatment can be repeated to calculate the corrections to the other propagator $p = v$. We give below only the values at $\delta = h$

$$v_1(h) = -\frac{1}{12} f_2^* \eta_2(F, h) \tag{3.4.50}$$

$$v_1'(h) = \frac{1}{4}(f_1^* + f_2^* h)\,\eta_1(F, h) \tag{3.4.51}$$

$$v_2(h) = -\left(\frac{1}{96} f_1^{*2} + \frac{1}{48} f_1^* f_2^* h + \frac{1}{90} f_2^{*2} h^2\right) \eta_3(F, h) +$$

$$+ \frac{1}{160} f_2^{*2} \eta_4(F, h) \tag{3.4.52}$$

$$v_2'(h) = -h\left(\frac{1}{96} f_1^{*2} + \frac{1}{48} f_1^* f_2^* h + \frac{1}{90} f_2^{*2} h^2\right) \eta_2(F, h) -$$

$$- \frac{1}{24}\left(f_1^* f_2^* + \frac{21}{20} f_2^{*2} h\right) \eta_3(F, h). \tag{3.4.53}$$

The algorithm in which Q corrections from the pilot perturbation (3.4.35) are included into the propagation matrix is called as a Qth order constant (reference function) perturbation method and is abbreviated as CPM(Q). The explicit formulas given above allow immediate application on computer for $Q = 0$, 1 and 2.

Error analysis

The immediate practical purpose of the error analysis of any numerical method is to generate hopefully simple error formulas which are to be used to measure the accuracy of the results in the typical case when h is sufficiently small. It is important to recall that the ground of the condition that "h is sufficiently small" is different in classical and piecewise perturbation methods. In fact, when a classical method is applied to solve eq. (3.3.1) this condition is based on the fact that the solution must be well approximated through piecewise polynomials and the latter means simply that h must be such that $|g(x)|h^2 \leqslant T$ where T is a threshold value which is of the order of unity; T slightly varies from method to method. In contrast, in case of a piecewise perturbation method the condition "h is sufficiently small" has no connection with the magnitude of $|g(x)|h^2$ but with the magnitude of the *deviation* of $g(x)$ from its piecewise polynomial approximation. For this reason it often happens that "small h" in the latter sense is actually much larger than in the former sense. We recall for illustration that for eq. (3.1.1) the Numerov method required $h = 1/8$ to produce as many as five exact figures for $a = 3$ while if CPM(0) is applied *any* h can be coined "sufficiently small" and the result is absolutely exact simply because the given $g(x) = -a^2$ and the reference constant $F = -a^2$ coincide.

A reasonable point of view is to admit that for perturbation methods the condition "h is sufficiently small" means that such h secures that the piecewise perturbation becomes so small that *only the first neglected* perturbative correction is sufficient to measure the error.

This point in mind we come back to the CPM(Q) to recall that the real perturbation here is $\Delta f^* + \Delta\tilde{f}$ where Δf^* is disregarded while Q corrections are retained from the pilot perturbation $\Delta\tilde{f}$. Thus, when h is small enough in the sense delimited above, the error is measured by the sum of the first-order correction from Δf^* and the $(Q+1)$th correction from $\Delta\tilde{f}$.

The evaluation of the first-order correction from Δf^* given by the two terms explicitly written in the right side of eq. (3.4.38) is made easily through formulas (3.4.28 a) and (3.4.28 b) and the result is

$$u_1^*(h) = -\frac{1}{240}\left(f^{(3)} + \frac{1}{2}hf^{(4)}\right)F\eta_3(F, h) \tag{3.4.54}$$

$$v_1^*(h) = -\frac{1}{3360}f^{(4)}F\eta_4(F, h) \tag{3.4.55}$$

$$u_1^{*\prime}(h) = -Fv_1^*(h),\ v_1^{*\prime}(h) = -u_1^*(h). \tag{3.4.56}$$

To set up a practically meaningful schedule for the error analysis two remarks are in order.

The first remark is that the formulas of the reference propagators and of the corrections are similar in structure. In fact, each of them consists of a sum of terms and each such term is a product of two contributions, one from $Z = Fh^2$ (this simulates $g(x)h^2$ on the current interval $x \in [X, X + h]$) and the other which comes from the very perturbation. Naturally, no contribution from the perturbation occurs in the reference propagators, while expression of $u_1(h)$, eq. (3.4.46), for instance, can be written as

$$u_1(h) = -\frac{1}{4}h^3(f_1^* + f_2^*h)\bar{\eta}_1(Z) \tag{3.4.57}$$

and it is clearly a product in which $\bar{\eta}_1(Z)$ depends on Z only while the front factor $-\frac{1}{4}h^3(f_1^* + f_2^*h)$ depends on the perturbation.

The second remark is that in many practical cases the user is faced with problems in which $g(x)$ is of the form $g(x) = V(x) - E$ where $V(x)$ is a given function and E a parameter and eq. (3.3.1) is to be solved repeatedly at different values of E. (Specifically, such a problem typically occurs when a two-point problem is formulated for equation $y'' = g(x)y$ whose eigenvalues are searched for by means of a shooting procedure; see also the next chapter.) In such a case Z depends on *both* $V(x)$ and E while the perturbation depends *only* on $V(x)$. Thus, while a classical method will require different step sizes for different values of E, the perturbation methods are expected to need roughly one and the same partition for any E, with obvious simplification of the computational effort.

Altogether a practically relevant error analysis must furnish error formulas which are valid for any Z; comparison with classical methods will be reached in the particular case of small $|Z|$.

As stated, we always assume that h is sufficiently small in the above sense.

CPM(0). For this version the analysis is simplified because some particular relationships exist between propagators and their derivatives.

In fact the error is measured by the sum of the first-order corrections from Δf^* and from $\Delta \tilde{f}$,

$$\begin{aligned} \Delta u(h) &= u_1(h) + u_1^*(h), \quad \Delta v(h) = v_1(h) + v_1^*(h) \\ \Delta u'(h) &= u_1'(h) + u_1^{*\prime}(h), \quad \Delta v'(h) = v_1'(h) + v_1^{*\prime}(h) \end{aligned} \tag{3.4.58}$$

but only Δu and Δv must be examined because $\Delta u'$ and $\Delta v'$ are of the same form, viz.

$$\Delta v'(h) = -\Delta u(h), \quad \Delta u'(h) = -F\Delta v(h). \tag{3.4.59}$$

On collecting eqs (3.4.46), (3.4.50), (3.4.54) and (3.4.55) we get formulas

$$\Delta u(h) = -\frac{1}{4}h^3(f_1^* + f_2^* h)\bar{\eta}_1(Z) - \frac{1}{240}h^5\left(f^{(3)} + \frac{1}{2}hf^{(4)}\right)Z\bar{\eta}_3(Z) \tag{3.4.60}$$

and

$$\Delta v(h) = -\frac{1}{12}h^5 f_2^* \bar{\eta}_2(Z) - \frac{1}{3360}h^7 f^{(4)} Z\bar{\eta}_1(Z) \tag{3.4.61}$$

which measure the error of CPM(0).

We now examine the behaviour of the error in various regimes of Z and first take the case of small $|Z|$. We use series expansions (3.4.19). in which we retain only the leading term

$$\Delta u(h) \simeq -\frac{1}{12}h^3(f_1^* + f_2^* h) - \frac{1}{5 \times 7!}h^7\left[f^{(3)} + \frac{1}{2}f^{(4)}h\right]F, \tag{3.4.62}$$

$$\Delta v(h) \simeq -\frac{1}{180}h^5 f_2^* - \frac{4}{35 \times 9!}h^9 f^{(4)} F. \tag{3.4.63}$$

We see that the smallest power of h in the four components of the error matrix

$$\begin{pmatrix} \Delta u(h) & \Delta v(h) \\ \Delta u'(h) & \Delta v'(h) \end{pmatrix} \tag{3.4.64}$$

is 3 and thus, since CPM(0) is, as any piecewise perturbation method, a genuine one-step method, its order is *two*.

Next take the asymptotic case of large *negative* Z and use the asymptotic behaviour (3.4.21)

$$\Delta u(h) \simeq -\frac{1}{4F}h\left[f_1^* + f_2^* h + \frac{1}{60}h^2\left(f^{(3)} + \frac{1}{2}f^{(4)}h\right)\right]\cos(|F|^{1/2}h), \tag{3.4.65}$$

$$\Delta v(h) \simeq -\frac{1}{12|F|^{1/2}F}h^2\left(f_2^* + \frac{1}{280}h^2 f^{(4)}\right)\sin(|F|^{1/2}h). \tag{3.4.66}$$

It is important to note that the regime of negative Z exhibits some peculiarity which must be accounted for adequately. In fact, due to the oscillating behaviour of the basic functions $\bar{\xi}, \bar{\eta}_0, \bar{\eta}_1, \ldots$ some values of Z exist at which the error vanishes; in the case of *large* negative Z this happens when $\cos(|Z|^{1/2}) = 0$ or when $\sin(|Z|^{1/2}) = 0$. Thus to correctly appraise the error at such critical values use must be made of second-order corrections. Yet eqs (3.4.65) and (3.4.66) can be used with sufficient confidence as a basis to generate error *bounds*. We simply take the maximal value of sin and cos to get the asymptotic bounds

$$|\Delta u(h)| \lesssim \frac{1}{4|F|} h \left| f_1^* + f_2^* h + \frac{1}{60} h^2 \left(f^{(3)} + \frac{1}{2} h f^{(4)} \right) \right| \tag{3.4.67}$$

and

$$|\Delta v(h)| \lesssim \frac{1}{12|F|^{3/2}} h^2 \left| f_2^* + \frac{1}{280} h^2 f^{(4)} \right| . \tag{3.4.68}$$

In the regime of large positive Z the basic functions $\bar{\xi}, \bar{\eta}_0, \bar{\eta}_1, \ldots$ essentially behave as exponential functions of $Z^{1/2}$ and then "normalized" absolute errors should be used to measure the accuracy of the method. We actually use eq. (3.4.23) and obtain

$$|\Delta u(h)/\bar{\xi}(Z)| \lesssim \frac{1}{4F} h \left| f_1^* + f_2^* h + \frac{1}{60} h^2 \left(f^{(3)} + \frac{1}{2} h f^{(4)} \right) \right| \tag{3.4.69}$$

and

$$|\Delta v(h)/\bar{\xi}(Z)| \lesssim \frac{1}{12F^{3/2}} h^2 \left| f_2^* + \frac{1}{280} h^2 f^{(4)} \right| . \tag{3.4.70}$$

Eqs (3.4.62-63), (3.4.67-68) and (3.4.69-70) collectively suggest that CPM(0) is complementary to the classical methods. In fact, its error components are maximal somewhere around $Z = 0$ to damp out at least as $|Z|^{-1}$ with increasing $|Z|$ thus contrasting the behaviour of the errors of classical methods which, as a rule, are minimal around $Z = 0$ to rapidly increase with $|Z|$.

The quoted dependence of the error of CPM(0) against Z is not surprising at all and to understand it qualitatively we recall that

$$f(\delta) = F + \Delta f^*(\delta) + \Delta \tilde{f}(\delta)$$

and that CPM(0) actually approximates $f(\delta)$ through F. For this reason the accuracy of the method is essentially governed by the quality of this approximation, that is by the ratio between the neglected and retained contributions $(\Delta f^*(\delta) + \Delta \tilde{f}(\delta))/F$. We thus see that for given $\Delta f^*(\delta)$ and $\Delta \tilde{f}(\delta)$ the error is expected to vary just as $|F|^{-1}$ or, equivalently, as $|Z|^{-1}$.

The most frequent case in practice is that of *intermediary* values of Z and the simplest and safest attitude to characterize the error is to refer to error bounds which are valid for any value of Z. Such expressions result by taking the absolute values of the terms in eqs (3.4.60) and (3.4.61) and using the bounds (3.4.22) and (3.4.23) for the basic functions. We thus have

$$|\Delta u(h)| \lesssim \frac{1}{4} h^3 |f_1^* + f_2^* h| B_1^-(Z) + \frac{1}{240} h^5 \left| f^{(3)} + \frac{1}{2} h f^{(4)} \right| |Z| B_3^-(Z), \tag{3.4.71}$$

$$|\Delta v(h)| \lesssim \frac{1}{12} h^5 |f_2^*| B_2^-(Z) + \frac{1}{3360} h^7 |f^{(4)}|\, |Z|\, B_4^-(Z) \tag{3.4.72}$$

when $Z < 0$ and

$$|\Delta u(h)/\bar{\xi}(Z)| \lesssim \frac{1}{4} h^3 |f_1^* + f_2^* h| B_1^+(Z) +$$

$$+ \frac{1}{240} h^5 \left| f^{(3)} + \frac{1}{2} h f^{(4)} \right| Z B_3^+(Z) \tag{3.4.73}$$

$$|\Delta v(h)/\bar{\xi}(Z)| \lesssim \frac{1}{12} h^5 |f_2^*| B_2^+(Z) + \frac{1}{3360} h^7 |f^{(4)}| Z B_4^+(Z) \tag{3.4.74}$$

when $Z > 0$.

CPM(1). Here the error is measured by the sum of the second-order corrections from $\Delta \tilde{f}$ and the first-order corrections from Δf^*. The procedure is the same as before with the only difference that now each of the four components of the error matrix must be treated separately whereas no relationship as eq. (3.4.59) exists.

As illustration we consider only $\Delta u(h)$. We use eqs (3.4.48) and (3.4.54) to get

$$\Delta u(h) = u_2(h) + u_1^*(h)$$

$$= -h^6\left(\frac{1}{96}f_1^{*2} + \frac{1}{48}f_1^*f_2^*h + \frac{1}{90}f_2^{*2}h^2\right)\bar{\eta}_2(Z) + \tag{3.4.75}$$

$$+\frac{1}{24}h^7\left(f_1^*f_2^* + \frac{19}{20}f_2^{*2}h\right)\bar{\eta}_3(Z) - \frac{1}{240}h^5\left(f^{(3)} + \frac{1}{2}hf^{(4)}\right)Z\bar{\eta}_3(Z).$$

When $|Z|$ is small this reduces to

$$\Delta u(h) \simeq -\frac{1}{6!}h^6\left(\frac{1}{2}f_1^{*2} + \frac{5}{7}f_1^*f_2^*h + \frac{11}{42}f_2^{*2}h^2\right) -$$

$$-\frac{1}{5\times 7!}h^7\left(f^{(3)} + \frac{1}{2}f^{(4)}h\right)F \tag{3.4.76}$$

while the behaviour of $\Delta u(h)$ at large negative Z reads

$$\Delta u(h) \simeq -\frac{1}{|F|^{1/2}F}h^3\left(\frac{1}{96}f_1^{*2} + \frac{1}{48}f_1^*f_2^*h + \frac{1}{90}f_2^{*2}h^2\right)\sin(|Z|^{1/2}) +$$

$$+\frac{1}{24F^2}h^3\left(f_1^*f_2^* + \frac{19}{20}f_2^{*2}h\right)\cos(|Z|^{1/2}) -$$

$$-\frac{1}{240F}h^3\left(f^{(3)} + \frac{1}{2}hf^{(4)}\right)\cos(|Z|^{1/2}). \tag{3.4.77}$$

Again critical values of Z exists in the regime $Z < 0$ so that the associated bound should be used, viz.

$$|\Delta u(h)| \lesssim \frac{1}{|F|^{3/2}}h^3\left|\frac{1}{96}f_1^{*2} + \frac{1}{48}f_1^*f_2^*h + \frac{1}{90}f_2^{*2}h^2\right| +$$

$$+\frac{1}{24F^2}h^3\left|f_1^*f_2^* + \frac{19}{20}f_2^{*2}h\right| + \frac{1}{240|F|}h^3\left|f^{(3)} + \frac{1}{2}f^{(4)}h\right| \tag{3.4.78}$$

and the same bound is valid at large *positive* Z for the normalized error $\Delta u(h)/\bar{\xi}(Z)$.

The procedure can be repeated without difficulty to Δv, $\Delta u'$ and $\Delta v'$ and on collecting these expressions we find out that, as in the previous case, the error is maximal around $Z=0$ and it decreases as $|Z|^{-1}$ when $|Z|$ is increased. When $|Z|$ is small the lowest power of h in the four components of the error matrix is 5 and thus CPM(1) is a *fourth-order* method.

An interesting feature is that the dependence on Z of the contributions from $\Delta\tilde{f}$ and from Δf^* is different. In fact, while in the case of CPM(0) both contributions exhibited essentially the same dependence at large $|Z|$, (as seen on eqs (3.4.67—70), they decrease as $|Z|^{-1}$ in $\Delta u(h)$ and as $|Z|^{-3/2}$ in $\Delta v(h)$), in the case of CPM(1) the second contribution remains the same while the first decreases faster; compare for instance eqs (3.4.67) and (3.4.78). It can thus be said that the contribution from Δf^* tends to dominate when $|Z|$ increases.

A similar error analysis of CPM(2) would first require the knowledge of the third-order corrections from $\Delta\tilde{f}$ but, although there is no difficulty in explicitly calculating them by means of eq. (3.4.28 b), the interest in obtaining such formulas is more or less academic since when $|Z|$ is increased these contributions will surely damp out faster than the second-order contributions*. (For illustration we note in passing that an explicit calculation shows that $u_3(h)$ damps out as $|F|^{-2}$). As such, the only regime where these corrections may play some significant role is that of small $|Z|$ and the point is that in this regime the corresponding formulas can be obtained through a direct, much simpler procedure. It is suggested by the remark that the case of small $|Z|$ is formally covered by putting $Z=0$ in eq. (3.4.19). Since $h \neq 0$, $Z=0$ implies $F=0$ and thus $\tilde{u}(\delta)=1$ and $\tilde{v}(\delta)=\delta$. Moreover if the original differential equation for correction (3.3.23) with $\tilde{f}(\delta)=F=0$ is used, its solution is given by quadrature

$$p_q(\delta)=\int_0^{\delta} d\delta_2 \int_0^{\delta_2} \Delta f(\delta_1) p_{q-1}(\delta_1) d\delta_1, \quad q=1,2,3,\ldots \tag{3.4.79}$$

---

* On using the results noted on p. 202, it can be shown that, in general, the $(q+1)$th correction will damp out faster than the qth correction when $|Z|$ is increased.

with $p_0(\delta) = 1$ for $p = u$ and $p_0(\delta) = \delta$ for $p = v$. Such a simple procedure gives the following expressions of the third-order corrections at $\delta = h$

$$u_3(h) = \frac{1}{9!} h^9 \left(f_1^{*3} + \frac{17}{5} f_1^{*2} f_2^* h + \frac{607}{165} f_1^* f_2^{*2} h^2 + \frac{71}{55} f_2^{*3} h^3 \right) \quad (3.4.80)$$

$$v_3(h) = \frac{1}{10!} h^{11} \left(\frac{2}{3} f_1^{*2} f_2^* + \frac{4}{33} f_1^* f_2^{*2} h + \frac{874}{1289} f_2^{*3} h^2 \right) \quad (3.4.81)$$

$$u_3'(h) = \frac{1}{9!} h^9 \left(8 f_1^{*2} f_2^* + 16 f_1^* f_2^{*2} h + \frac{64}{1289} f_2^{*3} h^2 \right) \quad (3.4.82)$$

$$v_3'(h) = -\frac{1}{9!} h^9 \left(f_1^{*3} + \frac{13}{5} f_1^{*2} f_2^* h + \frac{343}{165} f_1^* f_2^{*2} h^2 + \frac{7}{15} f_2^{*3} h^3 \right). \quad (3.4.83)$$

We see that the smallest power of h in these corrections is nine. Since in the first-order corrections from Δf^* the smallest power is seven, CPM(2) is a *sixth* order method. Clearly when h tends to zero the corrections from Δf^* tend to numerically dominate. Yet, when h which satisfies the condition that $|Z|$ is small is not extremely small in itself (this is the case when F is small), corrections (3.4.80—83) may be equally important.

To illustrate numerically the dependence of the error vs. Z we take the case $g(x) = V(x) - E$, $V(x) = x^2 + x^4$, $X = 3$, $h = 1/2$. Z is then

$$Z = Fh^2 = h^2 \left[\frac{1}{h} \int_X^{X+h} (V(x) - E) dx \right] = \frac{1}{2} \int_3^{3\cdot 5} V(x) dx - \frac{1}{4} E,$$ and it varies with E.

The experimental value of the error in $u(h)$ for CPM(2) is computed, that is

$$\Delta u(h) = u^{\text{exact}}(h) - (\tilde{u}(h) + u_1(h) + u_2(h))$$

where the "exact" $u(h)$ is generated by taking the same CPM(2) with four steps of $h_1 = 1/8$. The results are displayed on Fig. 3 where $\Delta u(h)$ is plotted for $Z \leqslant 0$ and the normalized $\Delta u(h)/\bar{\xi}(Z)$ for $Z > 0$. We see that, as expected, the largest error occurs at $Z = 0$. Note however that the maximal error, big as it looks, is actually much smaller (by an order of magnitude or so) than the one obtained at the same $h = 1/2$ with typical classical methods, for instance by means of the fourth order Runge-Kutta methods discussed in section 2.2. At large $|Z|$, specifically when $Z < -150$ or when $Z > 250$, the error envelope decreases according to $|Z|^{-1}$ rule. At these values the experimental error closely follows the shape of the dominating component $u_1^*(h)$.

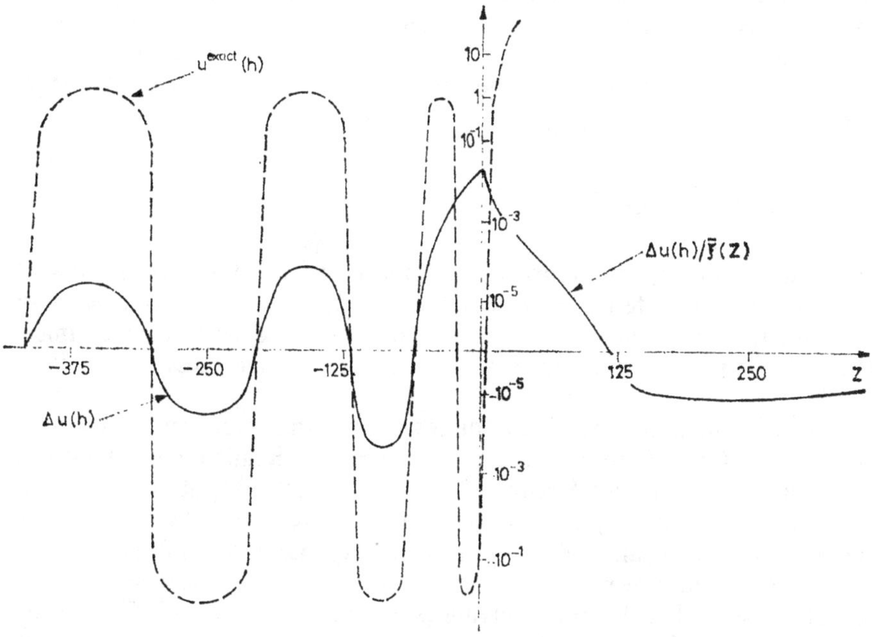

Fig. 3.— Behaviour of the error of CPM(2) against Z.

Practical efficiency

As seen, the order of CPM(Q) is $m = 2Q + 2$ for the relevant $Q = 0, 1$ and 2. As for the computational effort, extensive numerical tests have shown that if τ is the computation time/step for the CPM(0) (to give a scale we note that $\tau \simeq 1.7$ ms when simple precision computation is carried out on an IBM 370/135 computer) then, with careful programming, it is about $1.2\,\tau$ for the CPM(1) and $2.2\,\tau$ for the CPM(2). The reason of such a mode of increase with Q becomes obvious if one examines the structure of the formulas (3.4.7—8), (3.4.46—47, 50—51) (3.4.48—49, 52—53). Indeed, the formulas (3.4.7—8) corresponding to Q=0 imply the computation of trigonometric (or hyperbolic) functions ξ and η_0. This is significantly slower than the time to compute the additional few arithmetic operations to generate the first-order corrections, but approximately equal to that needed to compute the markedly larger number of arithmetic operations occurring in the second-order corrections. We also tested CPM(3). The result was that the number of additional arithmetic operations is substantially larger (they take approximately 5τ) without producing a spectacular gain in accuracy. (The last statement is not surprising at all because for any $Q > 2$, no matter how large, the accuracy remains limited at the level dictated by the residual perturbation.) Altogether it can be concluded that CPM(2) is the optimal version.

Survey of the literature

The idea of solving differential equations of form (3.3.1) by means of appropriately adjusted piecewise reference functions is simple enough to admit that it might have been used accidentaly even in earlier times. Yet, the first systematic approaches occurred as recently as in 1969 [2—4].

From the methods based on piecewise constant reference functions the version CPM(0) enjoyed by far the most detailed consideration. Ixaru [4] and Ixaru and Kreuzer [7] applied it to compute the penetration factor for thick potential barrriers, a problem which is hardly manageable by means of classical methods. Leighton [8] used it to obtain upper and lower bounds for eigenvalues and to localize the focal points associated to Sturm-Liouville problems. Canosa and Gomes de Oliveira [9] and Canosa [10] wrote an algorithm to generate the eigenvalues of a boundary value problem. Its error analysis is given in [11]. Also related is the investigation of Pruess [12]. The influence of round-off errors and the extension of the algorithm when $g(x)$ is a complex function is considered by Adam [13]. More recent investigations are due to Smooke [14] and to Andrew [15].

CPM(1) and CPM(2) are treated in [16] and [17] respectively. The procedure formulated in this book to evaluate the corrections is taken from [18] and the connection between $\eta_0, \eta_1, \eta_2, \eta_3$ and η_4 and the functions $\eta, \zeta, \rho, \varphi$ and τ originally used in [16, 17] is $\eta_0 = \eta$, $\eta_1 = \zeta$, $\eta_2 = -3\rho$, $\eta_3 = -3\varphi$ and $\eta_4 = 30\tau$.

A rigorous mathematical investigation upon the role of successive perturbative corrections in diminishing the error was undertaken by Smooke [19]. However the bounds obtained in this paper substantially overestimate the real size of the error. The problem of choosing step sizes is treated in [20—22] at various degrees of completeness. A recent review is [23].

3.5. LINE PERTURBATION METHODS

We consider again the one-step initial value problem (3.4.1) to describe a piecewise perturbation method based on linear reference functions. Since, as before, the perturbative corrections can be evaluated analytically in closed form only when the perturbation is a polynomial we make use of the same strategy as in section 3.4, that is first introduce a pilot polynomial reference function $f^*(\delta)$ to finally consider only corrections from the pilot perturbation.

To be specific we assume that the pilot reference function is the same as in the previous case, eq. (3.4.31), where again the parameters are computed through the least squares fit, eqs (3.3.30—31) with $t = 2$. The parameters of the working reference function, which is a straight line here viz.

$$\tilde{f}(\delta) = F(\delta) \equiv F_0 + F_1\delta \tag{3.5.1}$$

are calculated by fitting the pilot function $f^*(\delta)$ via the same least square condition. The resultant values are

$$F_0 = f_0^* - \frac{1}{6} f_2^* h^2, \quad F_1 = f_1^* + f_2^* h. \tag{3.5.2}$$

In short the original $f(\delta)$ reads as

$$f(\delta) = F(\delta) + \Delta f^*(\delta) + \Delta\tilde{f}(\delta), \quad \delta \in [0, h] \tag{3.5.3}$$

where the residual perturbation

$$\Delta f^*(\delta) \equiv f(\delta) - f^*(\delta) \tag{3.5.4}$$

is given by eq. (3.4.38) while the pilot perturbation is

$$\Delta\tilde{f}(\delta) \equiv f^*(\delta) - F(\delta) = f_2^* \left(\frac{1}{6} h^2 - h\delta + \delta^2\right). \tag{3.5.5}$$

The reference equation thus is

$$\tilde{y}''(X + \delta) = (F_0 + F_1\delta)\tilde{y}(X + \delta), \quad \delta \in [0, h] \tag{3.5.6}$$

and its propagators are readily expressed in terms of the Airy functions. To see this we introduce a new variable $Z = \alpha(\delta + \beta)$ where α and β are some constants and denote $\tilde{Y}(Z) = \tilde{y}(X + \delta)$. Eq. (3.5.6) becomes

$$\tilde{Y}''(Z) = \frac{1}{\alpha^2}\left[F_0 + \frac{F_1}{\alpha}Z - F_1\beta\right]\tilde{Y}(Z),\ Z \in [\alpha\beta, \alpha(h+\beta)]. \tag{3.5.7}$$

If we choose $\alpha = F_1^{1/3}$ and $\beta = F_0/F_1$ this becomes

$$\tilde{Y}'' = Z\tilde{Y}, \tag{3.5.8}$$

an equation which can be solved analytically in terms of the Bessel functions [24]. Two linear independent solutions of eq. (3.5.8), the so-called Airy functions denoted as $Ai(Z)$ and $Bi(Z)$, are well investigated from numerical point of view [3, 20, 25, 26]; their wronskian is

$$W[Ai, Bi] = Ai(Z)Bi'(Z) - Ai'(Z)Bi(Z) = 1/\pi. \tag{3.5.9}$$

The general solution of the reference equation is a linear combination of the Airy functions, viz.

$$\tilde{y}(X + \delta) = A_1Ai(\alpha(\delta + \beta)) + A_2Bi(\alpha(\delta + \beta)) \tag{3.5.10}$$

while its derivative with respect to δ is

$$\tilde{y}'(X + \delta) = \alpha A_1Ai'(\alpha(\delta + \beta)) + \alpha A_2Bi'(\alpha(\delta + \beta)) \tag{3.5.11}$$

where the primes over Ai and Bi denote the first derivatives with respect to the whole argument $Z = \alpha(\delta + \beta)$.

The propagators $\tilde{u}(\delta)$ and $\tilde{v}(\delta)$ are the two solutions of the reference equation which satisfy the initial conditions $\tilde{u}(0) = 1$, $\tilde{u}'(0) = 0$ and $\tilde{v}(0) = 0$, $\tilde{v}'(0) = 1$ respectively. On introducing these conditions in eqs (3.5.10) and (3.5.11) we find the suitable values of the constants A_1 and A_2. In case of $\tilde{u}(\delta)$, for instance, we have ($\alpha \neq 0$)

$$A_1Ai(\alpha\beta) + A_2Bi(\alpha\beta) = 1$$

$$A_1Ai'(\alpha\beta) + A_2Bi'(\alpha\beta) = 0$$

whence

$$A_1 = \pi Bi'(\alpha\beta),\ A_2 = -\pi Ai'(\alpha\beta). \tag{3.5.12}$$

Analogously, for $\tilde{v}(\delta)$ it results

$$A_1 = -\pi Bi(\alpha\beta)/\alpha, \; A_2 = \pi Ai(\alpha\beta)/\alpha. \tag{3.5.13}$$

To summarize, the propagators of the reference equation (3.5.6) and their first derivatives are in order

$$\tilde{u}(\delta) = \pi(Ai(Z)Bi'(Z_0) - Bi(Z)Ai'(Z_0)) \tag{3.5.14}$$

$$\tilde{v}(\delta) = \pi(Bi(Z)Ai(Z_0) - Ai(Z)Bi(Z_0))/\alpha \tag{3.5.15}$$

$$\tilde{u}'(\delta) = \alpha\pi(Ai'(Z)Bi'(Z_0) - Bi'(Z)Ai'(Z_0)) \tag{3.5.16}$$

$$\tilde{v}'(\delta) = \pi(Bi'(Z)Ai(Z_0) - Ai'(Z)Bi(Z_0)) \tag{3.5.17}$$

with $\alpha = F_1^{1/3} \neq 0$, $\beta = F_0/F_1$, $Z = \alpha(\delta + \beta)$ and $Z_0 = \alpha\beta$.

The method in which the exact propagators and their derivative are replaced by these $\tilde{u}(\delta)$, $\tilde{v}(\delta)$, $\tilde{u}'(\delta)$, and $\tilde{v}'(\delta)$ will be called a zeroth order *line* perturbation method, LPM(0), and it is to be applied only when $\alpha \neq 0$. When $\alpha = 0$ the reference function is a constant and thus the formulas of CPM(0) are to be used.

The key of the whole method is to accurately compute the Airy functions and their first derivative.

Computation of Airy functions

To construct the Airy functions three regimes for Z are usually delimited: small, large negative and large positive. For small values of Z the convergent series expansions given in [25] are readily applied. More difficult is the case of large negative or positive values of Z. A procedure often used in such case is that proposed in [26]. It is based on generalized Gaussian techniques to integral representations of the Airy functions and at large *positive* arguments it yields the following approximation formulas

$$Ai(Z) \simeq A_0(Z)\exp(-\zeta), \; Bi(Z) \simeq B_0(Z)\exp(\zeta) \tag{3.5.18}$$

with $\zeta = \dfrac{2}{3}Z^{3/2}$ and

$$\begin{aligned} A_0(Z) &= \frac{1}{2}\pi^{-1/2}Z^{-1/4}\sum_{i=1}^{n}\frac{w_i}{1+(p_i/\zeta)}, \\ B_0(Z) &= \pi^{-1/2}Z^{-1/4}\sum_{i=1}^{n}\frac{w_i}{1-(p_i/\zeta)}, \end{aligned} \tag{3.5.19}$$

while at large *negative* arguments the resultant approximation formulas are

$$\begin{aligned} Ai(Z) &\simeq A_c(Z)\cos\xi + A_s(Z)\sin\xi, \\ Bi(Z) &\simeq A_s(Z)\cos\xi - A_c(Z)\sin\xi, \end{aligned} \tag{3.5.20}$$

with $\zeta = \frac{2}{3}(-Z)^{3/2}$, $\xi = \zeta - \frac{1}{4}\pi$ and

$$\begin{aligned} A_c(Z) &= \pi^{-1/2}(-Z)^{-1/4} \sum_{i=1}^{n} \frac{w_i}{1 + (p_i/\zeta)^2}, \\ A_s(Z) &= \zeta^{-1}\pi^{-1/2}(-Z)^{-1/4} \sum_{i=1}^{n} \frac{w_i p_i}{1 + (p_i/\zeta)^2}. \end{aligned} \tag{3.5.21}$$

The quadrature weights w_i and the positions p_i were first tabulated in [3] and [20] for $n = 1, 2, 4$ and 10. They are collected in table 11 while tables 12 and 13 give suitable ranges to achieve 6-digit and 14-digit accuracy, respectively

Table 11

Positions p_i and weights w_i for the n-term generalized Gaussian integration for Airy functions

i	p_i	w_i
$n = 1$		
1	0.0694444...	1.0
$n = 2$		
1	3.6800601866153044 D−02	9.6807280595773604 D−01
2	1.0592469382112378 D+00	3.1927194042263958 D−02
$n = 4$		
1	3.9198329554455091 D+00	4.7763903057577263 D−05
2	1.6915619004823504 D+00	4.9914306432910959 D−03
3	5.0275532467263018 D−01	8.6169846993840312 D−02
4	1.9247060562015692 D−02	9.0879095845981102 D−01
$n = 10$		
1	1.4083081072180964 D+01	3.1542515762964787 D−14
2	1.0214885479197331 D+01	6.6394210819584921 D−11
3	7.4416018450450930 D+00	1.7583889061345669 D−08
4	5.3070943061781927 D+00	1.3712392370435815 D−06
5	3.6340135029132462 D+00	4.4350966639284350 D−05
6	2.3310652303052450 D+00	7.1555010917718255 D−04
7	1.3447970824609268 D+00	6.4889566103335381 D−03
8	6.4188858369567296 D−01	3.6440415875773282 D−02
9	2.0100345998121046 D−01	1.4399792418590999 D−01
10	8.0594359172052833 D−03	8.1231141336261486 D−01

Table 12

Ranges of methods for evaluating Airy functions to 6-digit accuracy

Z	Ai(Z)	Bi(Z)
$Z<-40.0$	1-term integration	1-term integration
$-40.0<Z<-6.0$	2-term integration	2-term integration
$-6.0<Z<-3.0$	4-term integration	4-term integration
$-3.0<Z<1.9$	power series	power series
$1.9<Z<5.0$	4-term integration	power series
$5.0<Z<8.0$	4-term integration	4-term integration
$8.0<Z$	2-term integration	2-term integration

Table 13

Ranges of methods for evaluating Airy functions to 14-digit accuracy

Z	Ai(Z)	Bi(Z)
$Z<-200.0$	2-term integration	2-term integration
$-200.0<Z<-15.0$	4-term integration	4-term integration
$-15.0<Z<-5.0$	10-term integration	10-term integration
$-5.0<Z<3.7$	power series	power series
$3.7<Z<8.0$	10-term integration	power series
$8.0<Z<15.0$	10-term integration	10-term integration
$15.0<Z$	4-term integration	4-term integration

To compute the first *derivative* of the Airy functions straightforward differentiation of the above representations is recommended in [20].

It is important to emphasize here that some precaution is necessary when computing the propagators in terms of the Airy functions. As a matter of fact the most direct way, that is when the values of Ai, Bi, Ai' and Bi' at Z and Z_0, calculated separately, are finally introduced in eqs (3.5.14—17), is perhaps the worst one. To see the reason we first focus on the case of large *positive* arguments. The exponential factors in eqs (3.5.18) will cause that $Ai(Z)$ eventually underflows and $Bi(Z)$ overflows and thus some extra rearrangements in eqs (3.5.14—17) are necessary. Specifically, since the first derivative of Ai and of Bi are of the same form as the original functions, viz.

$$Ai'(Z) = A_1(Z)\exp(-\zeta),\ Bi'(Z) = B_1(Z)\exp(\zeta) \tag{3.5.22}$$

$\tilde{u}(\delta)$ can be written as

$$\tilde{u}(\delta) = \pi[A_0(Z)B_1(Z_0)\exp(\zeta_0 - \zeta) - B_0(Z)A_1(Z_0)\exp(\zeta - \zeta_0)] \quad (3.5.23)$$

(similar expressions result for $\tilde{v}(\delta), \tilde{u}'(\delta)$ and $\tilde{v}'(\delta)$) and the point is that now the argument of the exponential functions is the *difference* $\zeta - \zeta_0$ so that the computation becomes free of scaling difficulties.

A similar rearrangement is also applicable in case of large *negative* arguments and the necessity to do so stems from the fact that the trigonometric functions cos and sin of large arguments would be numerically unaccurate.

We thus take the derivative of eq. (3.5.20),

$$Ai'(Z) \simeq D_c(Z)\cos\xi + D_s(Z)\sin\xi \quad (3.5.24)$$

$$Bi'(Z) \simeq D_s(Z)\cos\xi - D_c(Z)\sin\xi \quad (3.5.25)$$

where

$$D_c(Z) = A_c'(Z) - (-Z)^{1/2}A_s(Z) \quad (3.5.26\text{ a})$$

and

$$D_s(Z) = A_s'(Z) + (-Z)^{1/2}A_c(Z), \quad (3.5.26\text{ b})$$

introduce these and also eq. (3.5.20) into eq. (3.5.14) and use a little trigonometry to finally obtain

$$\begin{aligned}\tilde{u}(\delta) = {} & \pi[A_c(Z)D_s(Z_0) - A_s(Z)D_c(Z_0)]\cos(\xi_2 - \xi_1) + \\ & + \pi[A_c(Z)D_c(Z_0) - A_s(Z)D_s(Z_0)]\sin(\xi_2 - \xi_1).\end{aligned} \quad (3.5.27)$$

Again the scaling problems disappear since the argument of the trigonometric functions is now the difference $\xi_2 - \xi_1$.

Although the elimination of the scaling problems significantly improves the numerical accuracy of the reference propagators, the LPM(0) still exhibits some special sort of instability. This instability, which has nothing in common with the instability of the multistep methods discussed in chapter 1, is due to the inherently "subtractive" component of the linear reference function.

To clarify the point we note that coefficient F_1 in eq. (3.5.1) represents an average *slope* of $f(\delta)$ on $[0, h]$ and thus, as any slope, it is expressed in terms of some *differences* of the values of $f(\delta)$ at certain points. If, for instance, eqs (3.4.36 b—c) are applied we find out that

$$F_1 = f_1^* + f_2^* h = \frac{1}{336h}[111(f_6 - f_1) + 425(f_5 - f_2) - 150(f_4 - f_3)]. \tag{3.5.28}$$

For this reason the computation of F_1 is inherently exposed to the danger of near cancellation of like terms and this fact may cause severe decrease in accuracy, particularly when h is small and/or when f is slowly varying over $[0, h]$. In such cases the accuracy loss in F_1 is propagated in α and β and thus in the arguments Z_0 and Z of the Airy functions which define the propagators.

Altogether, even in the case when the accuracy of the computation of the Airy functions is beyond any doubt, the whole accuracy of the LPM(0) is limited at the level imposed by the subtractive nature of its reference function and, to restore the accuracy, extended precision is recommended to generate the arguments Z_0 and Z.

The above remark also suggests that such instabilities remain present and eventually grow up if polynomials of higher degree are tried as reference functions. It is also interesting to note that only the constant reference function is free of such effects and the reason is that F is of *additive* nature; it is computed by quadrature, eq. (3.4.5). Extensive numerical tests have actually confirmed the remarkable stability of the CPM(0) against round-off errors.

The above remark equally seems to imply that versions CPM(1) and CPM(2) will become automatically exposed to the same sort of instability as the LPM(0) whereas the perturbation corrections they use contain the weights f_1^* and f_2^* which are both of subtractive nature. This is however a false impression and the reason is that the presumably distorted f_1^* and f_2^* occur in *corrections*, not in the reference propagators. To be specific we focus on the first-order correction $u_1(h)$ given by eq. (3.4.57) and assume that, due to the inherent near cancellation of like terms in eqs (3.4.36 b—c), f_1^* and f_2^* were falsified by amounts proportional to h^{-1} and to h^{-2} respectively, that is $f_1^* = f_1^{*\,\text{exact}} + R_1 h^{-1}$, $f_2^* = f_2^{*\,\text{exact}} + R_2 h^{-2}$. The point is that the presence of the factor h^3 in eq. (3.4.57) not only does not magnify the influence of such distortions, it simply damps them out. Altogether CPM(1) and CPM(2) will continue to enjoy the same remarkable stability as CPM(0).

Perturbation corrections

We explicitly evaluate only the first-order corrections from the pilot perturbation and consider in detail $u_1(\delta)$. We follow the procedure given in section 3.3 and thus assume that $u_1(\delta)$ is of the form (3.3.57), viz.

$$u_1(\delta) = a_1(\delta)\tilde{u}(\delta) + b_1(\delta)\tilde{v}(\delta) + c_1(\delta)\tilde{u}'(\delta) + d_1(\delta)\tilde{v}'(\delta) \tag{3.5.29}$$

where $a_1(\delta)$, $b_1(\delta)$, $c_1(\delta)$ and $d_1(\delta)$ satisfy system (3.3.61),

$$a_1'' + 2c_1'(F_0 + F_1\delta) + c_1F_1 = f_2^*\left(\delta^2 - h\delta + \frac{1}{6}h^2\right) \tag{3.5.30 a}$$

$$b_1'' + 2d_1'(F_0 + F_1\delta) + d_1F_1 = 0 \tag{3.5.30 b}$$

$$c_1'' + 2a_1' = 0 \tag{3.5.30 c}$$

$$d_1'' + 2b_1' = 0 \tag{3.5.30 d}$$

with the initial conditions

$$a_1(0) + d_1(0) = 0, \quad a_1'(0) + c_1(0)F_0 + b_1(0) + d_1'(0) = 0. \tag{3.5.31}$$

To solve this system we first differentiate eq. (3.5.30 c) and introduce it in eq. (3.5.30 a). We get

$$-\frac{1}{2}c''' + 2c_1'(F_0 + F_1\delta) + c_1F_1 = f_2^*\left(\delta^2 - h\delta + \frac{1}{6}h^2\right). \tag{3.5.32}$$

A particular solution of this differential equation is the polynomial

$$c_1(\delta) = \frac{f_2^*}{F_1}\left[\frac{1}{5}\delta^2 - \frac{1}{3}\left(h + \frac{4}{5}Q\right)\delta + \frac{1}{6}h^2 + \frac{2}{3}Q\left(h + \frac{4}{5}Q\right)\right] \tag{3.5.33}$$

with $Q = F_0/F_1$.

We also differentiate eq. (3.5.30 d) and introduce it in eq. (3.5.30 b). The resultant equation

$$-\frac{1}{2}d_1''' + 2d_1'(F_0 + F_1\delta) + d_1F_1 = 0 \tag{3.5.34}$$

has the particular solution

$$d_1(\delta) = 0. \tag{3.5.35}$$

These particular solutions are further introduced in eq. (3.5.30 c) and eq. (3.5.30 d) respectively and the following *general* solutions $a(\delta)$ and $b(\delta)$ result

$$a(\delta) = \frac{f_2^*}{F_1}\left(-\frac{1}{5}\delta + A_0\right), \quad b(\delta) = B_0 \tag{3.5.36}$$

where A_0 and B_0 are constants which are finally fixed by the two initial conditions (3.5.31); the result is

$$A_0 = 0, \quad B_0 = \frac{1}{5} - F_0\left[\frac{1}{6}h^2 + \frac{2}{3}Q\left(h + \frac{4}{5}Q\right)\right]. \tag{3.5.37}$$

Altogether the expression of $u_1(\delta)$ is

$$u_1(\delta) = \frac{f_2^*}{F_1}\left\{-\frac{1}{5}\delta\bar{u}(\delta) + \left\{\frac{1}{5} - F_0\left[\frac{1}{6}h^2 + \frac{2}{3}Q\left(h + \frac{4}{5}Q\right)\right]\right\}\bar{v}(\delta) + \right.$$
$$\left. + \left[\frac{1}{5}\delta^2 - \frac{1}{3}\left(h + \frac{4}{5}Q\right)\delta + \frac{1}{6}h^2 + \frac{2}{3}Q\left(h + \frac{4}{5}Q\right)\right]\bar{u}'(\delta)\right\}. \tag{3.5.38}$$

The expression of $\bar{u}_1'(\delta)$ results by direct differentiation, see eq. (3.3.58).

The procedure is entirely similar for $v_1(\delta)$ and the result is

$$v_1(\delta) = \frac{f_2^*}{F_1}\left\{-\left[\frac{1}{6}h^2 + \frac{2}{3}Q\left(h + \frac{4}{5}Q\right)\right]\bar{u}(\delta) + \right.$$
$$+ \left[-\frac{1}{5}\delta + \frac{1}{3}\left(h + \frac{4}{5}Q\right)\right]\bar{v}(\delta) +$$
$$\left. + \left[\frac{1}{5}\delta^2 - \frac{1}{3}\left(h + \frac{4}{5}Q\right)\delta + \frac{1}{6}h^2 + \frac{2}{3}Q\left(h + \frac{4}{5}Q\right)\right]\bar{v}'(\delta)\right\}. \tag{3.5.39}$$

Again eq. (3.3.58) must be used to calculate $v_1'(\delta)$.

The algorithm in which these corrections are added to the reference propagators and to their first-order derivative will be called a first-order line perturbation method, LPM(1). This version is used less often that LPM(0) and the expressions of the first-order corrections are merely used to measure the error of LPM(0); see also the comment at the end of this section.

Error analysis

To our knowledge a general systematic error analysis for the LP methods does not exist. However, general arguments suggest an error behaviour similar with that for CP methods namely that if $g(x) = V(x) - E$ where $V(x)$ is a given function and E is a parameter (this means that on the local interval we have $f(\delta) = V(X+\delta) - E$) the error is maximal when E is such that $|f(\delta)|$ is small (or, in a geometrical language, when E intersects V on $[X, X+h]$) and decreases with increasing $|f(\delta)|$.

We thus restrict the error analysis to the case when h is so small that (i) the propagators $\tilde{u}(\delta)$ and $\tilde{v}(\delta)$ are expressed through power series and (ii) only the first term of the perturbation expansion which is neglected in the algorithm is numerically sufficient to measure the error.

In practice the first restriction is stronger than the second and thus "small h" delimited above is roughy the same as "small h" in the sense of classical methods. We do also restrict the analysis to the case of LPM(0) only.

The error is given by the four quantities

$$\Delta u(h) = u_1(h) + u_1^*(h), \quad \Delta v(h) = v_1(h) + v_1^*(h),$$
$$\Delta u'(h) = u_1'(h) + u_1^{*\prime}(h), \quad \Delta v'(h) = v_1'(h) + v_1^{*\prime}(h), \tag{3.5.40}$$

where u_1^*, v_1^*, $u_1^{*\prime}$ and $v_1^{*\prime}$ are the first-order corrections from the residual perturbation.

By virtue of restriction (i) the reference propagators are searched for as power series

$$\tilde{p}(\delta) = \sum_{r=0}^{\infty} \tilde{p}_r \delta^r \tag{3.5.41}$$

We replace it into the reference equation $\tilde{p}'' = (F_0 + F_1\delta)\tilde{p}$ and obtain a recurrence formula for the $\tilde{p}_r$

$$\tilde{p}_2 = \frac{1}{2} F_0\tilde{p}_0,\ \tilde{p}_{r+2} = \frac{1}{(r+2)(r+1)} [F_0\tilde{p}_r + F_1\tilde{p}_{r-1}], \quad r = 1, 2, \ldots \tag{3.5.42}$$

If $p = u$ we have $\tilde{p}_0 = 1$ and $\tilde{p}_1 = 0$ while if $p = v$ we have $\tilde{p}_0 = 0$ and $\tilde{p}_1 = 1$.

Using these we form the product $\Delta\tilde{f}(\delta)\tilde{p}(\delta)$

$$f_2^* S(\delta) \equiv \Delta\tilde{f}(\delta)\tilde{p}(\delta) = f_2^*\left(\delta^2 - \delta h + \frac{1}{6}h^2\right)\tilde{p}(\delta) = f_2^* \sum_{r=0}^{\infty} S_r\delta^r \tag{3.5.43}$$

with

$$S_0 = \frac{1}{6} h^2\tilde{p}_0,\quad S_1 = \frac{1}{6} h^2 p_1 - h\tilde{p}_0,$$
$$S_r = \frac{1}{6} h^2\tilde{p}_r - h\tilde{p}_{r-1} + \tilde{p}_{r-2} \text{ for } r \geqslant 2. \tag{3.5.44}$$

The first-order correction is the solution of the problem

$$p_1'' = (F_0 + F_1\delta)p_1 + f_2^* S(\delta),\ p_1(0) = p_1'(0) = 0, \tag{3.5.45}$$

and it is of the form

$$p_1(\delta) = f_2^* \sum_{r=0}^{\infty} a_r\delta^r \tag{3.5.46}$$

with

$$a_0 = 0,\ a_1 = 0,\ a_2 = \frac{1}{2} S_0,$$
$$a_{r+2} = \frac{1}{(r+2)(r+1)}[F_0 a_r + F_1 a_{r-1} + S_r] \text{ for } r \geqslant 1. \tag{3.5.47}$$

We now particularize this approach for the cases $p = u$ and $p = v$. When $p = u$ we have $\tilde{p}_2 = \frac{1}{2} F_0$, $\tilde{p}_3 = \frac{1}{6} F_1$, $\tilde{p}_4 = \frac{1}{24} F_0^2$, $\tilde{p}_5 = \frac{1}{30} F_0 F_1$. These are introduced in eqs (3.5.44) and (3.5.47) to finally calculate the value of $u_1(\delta)$ at $\delta = h$. After a little algebra it is found that

$$u_1(h) = f_2^* \sum_{r=0}^{\infty} a_r h^r = \frac{1}{2520} f_2^* F_1 h^7 + 0(h^8). \tag{3.5.48}$$

On differentiating eq. (3.5.46) we obtain

$$u_1'(h) = f_2^* \sum_{r=1}^{\infty} r a_r h^{r-1} = \frac{1}{180} f_2^* F_0 h^5 + 0(h^6). \tag{3.5.49}$$

Likewise, in case $p = v$ we have $\tilde{p}_2 = 0$, $\tilde{p}_3 = \frac{1}{6} F_0$, $\tilde{p}_4 = \frac{1}{12} F_1$, $\tilde{p}_5 = \frac{1}{120} F_0^2$, so that

$$v_1(h) = -\frac{1}{180} f_2^* F_0 h^5 + 0(h^6),\quad v_1'(h) = -\frac{1}{2520} f_2^* F_1 h^7 + 0(h^8). \tag{3.5.50}$$

We see that the smallest power of h in these corrections is five. The same procedure can be applied to compute the starred corrections and the result will be that the smallest power of h is there seven. The conclusion thus is that LPM(0) is a *fourth*-order method.

Practical efficiency

The practical efficiency is usually measured in terms of computation time and of the storage volume required to reach some preset accuracy.

The perturbation methods discussed in this and in the previous section are mainly suited for repeated solution of problem (3.3.1) in which $g(x) = V(x) - E$ with fixed $V(x)$ and various values of E. An optimal programming then consists of first generating the piecewise pilot reference function of $V(x)$; its parameters are stored into the memory for further re-use. If n is the number of steps of the partition the total number of memory locations is $3n$ (remember that the pilot function is taken as a piecewise parabola) and the total computation time at each E is Tn where T is the computation time/step for the used method.

Naturally the straight-line segments are better piecewise approximations of $V(x)$ than are the constants and thus, for one and the same accuracy in the results, n for CPM(0) is, as a rule, definitely larger than for LPM(0). The discrepancy is reduced when perturbation corrections are added to CPM(0) but it may still remain significant when $V(x)$ exhibits important variations over $[a, b]$. On the other hand, the evaluation of the elements of the propagation matrix is much slower for LPM(0) than for the three CP versions. As an example, the computation time/step for LPM(0) is about ten times slower than for CPM(0).

Altogether LPM(0) may offer definite advantages with respect to the memory requirements but it is less efficient with respect to the computation time. A reasonable compromise consists perhaps of using LPM(0) only in the subintervals of $[a, b]$ where $V(x)$ exhibits large variation while the CP versions otherwise.

Survey of the literature

LPM(0) is first described by Gordon in [3] and further developed in [20]. These papers also contain procedures to compute perturbation corrections, but it is interesting to mention that the way proposed in these papers is different from that adopted in this book.

To make the difference clear we refer to the general approach developed in section 3.3 and recall that in this approach the perturbation theory is used to construct successive approximations to the exact propagation matrix. In Gordon's approach the exact solution of eq. (3.3.5) is assumed of the form

$$\begin{pmatrix} y(X+\delta) \\ y'(X+\delta) \end{pmatrix} = \begin{pmatrix} \bar{u}(\delta) & \bar{v}(\delta) \\ \bar{u}'(\delta) & \bar{v}'(\delta) \end{pmatrix} \begin{pmatrix} A(\delta) \\ B(\delta) \end{pmatrix}, \tag{3.5.51}$$

that is the propagation matrix is retained in its *reference* form but this is applied on a δ dependent column vector which must be determined.

In short, in our approach the perturbation theory is used to successively improve the propagation matrix while in Gordon's approach it is aimed to successively amend the elements placed in the block of the initial conditions.

The realization of Gordon's approach is as follows. One starts with the perturbed equation

$$y''(X+\delta;\lambda) = \mathscr{F}(\delta;\lambda)y(X+\delta;\lambda)$$

with given $y(X)$ and $y'(X)$ and $\mathscr{F}(\delta;\lambda)$ defined by eq. (3.3.13). Its exact solution is of form (3.5.51),

$$\begin{pmatrix} y(X+\delta;\lambda) \\ y'(X+\delta;\lambda) \end{pmatrix} = \begin{pmatrix} \tilde{u}(\delta) & \tilde{v}(\delta) \\ \tilde{u}'(\delta) & \tilde{v}'(\delta) \end{pmatrix} \begin{pmatrix} A(\delta,\lambda) \\ B(\delta,\lambda) \end{pmatrix} \tag{3.5.52}$$

or, equivalently,

$$\begin{pmatrix} A(\delta;\lambda) \\ B(\delta;\lambda) \end{pmatrix} = \begin{pmatrix} \tilde{v}'(\delta) & -\tilde{v}(\delta) \\ -\tilde{u}'(\delta) & \tilde{u}(\delta) \end{pmatrix} \begin{pmatrix} y(X+\delta;\lambda) \\ y'(X+\delta;\lambda) \end{pmatrix}. \tag{3.5.53}$$

(See also derivation of eq. (3.3.56).)
This is further differentiated once with respect to δ (use eqs (3.3.10) and (3.3.11)),

$$A'(\delta;\lambda) = -\lambda\Delta f(\delta)y(X+\delta;\lambda)\tilde{v}(\delta),\; B'(\delta;\lambda) = \lambda\Delta f(\delta)y(X+\delta;\lambda)\tilde{u}(\delta). \tag{3.5.54}$$

A and B are written as perturbation series

$$A(\delta;\lambda) = y(X) + \sum_{q=1}^{\infty} \lambda^q A_q(\delta),\; B(\delta;\lambda) = y'(X) + \sum_{q=1}^{\infty} \lambda^q B_q(\delta) \tag{3.5.55}$$

(note that this implies $A_q(0) = B_q(0) = 0$ for any q).

They are introduced in eq. (3.5.54) where the like powers of λ are identified. The result is

$$A_q(\delta) = -\int_0^{\delta} d\delta_1 \Delta f(\delta_1)[\tilde{u}(\delta_1)A_{q-1}(\delta_1) + \tilde{v}(\delta_1)B_{q-1}(\delta_1)]\tilde{v}(\delta_1) \tag{3.5.56}$$

$$B_q(\delta) = \int_0^\delta d\delta_1 \, \Delta f(\delta_1)[\tilde{u}(\delta_1) \, A_{q-1}(\delta_1) + \tilde{v}(\delta_1) \, B_{q-1}(\delta_1)] \, \tilde{u}(\delta_1) \qquad (3.5.57)$$

for $q = 1, 2, 3, \ldots$, with the constant starting functions $A_0(\delta) = y(X)$ and $B_0(\delta) = y'(X)$.

Theoretically the two approaches are entirely equivalent but it is important to insist that while the first approach generates corrections which are *linear* in propagators and in their derivative (see eq. (3.3.57)) the second approach leads to eqs (3.5.56—57) which are *quadratic* in propagators. Due to that, when practised on computer the first formulas are definitely less exposed to the unpleasant phenomenon of near cancellation of large terms than the latter formulas. As a matter of fact, the first-order corrections from the linear reference function calculated in [20] on the basis of eqs (3.5.56—57) are reported to suffer of such heavy loss in accuracy and this is actually the reason why LPM(1) was seldom used. We think that use of first-order corrections in form (3.5.38—39) will significantly restore confidence in that method.

Other investigation on LP methods are in [27—32].

3.6. PIECEWISE PERTURBATION METHODS FOR SYSTEMS OF DIFFERENTIAL EQUATIONS

We here consider the straightforward vector extension of problem (3.3.1) and closely follow the approach developed in section 3.3.

Our problem thus is

$$\mathbf{y}'' = \mathbf{g}(x)\mathbf{y}, \; x \in [a, b], \; \mathbf{y}(a) = \mathbf{y}_0, \; \mathbf{y}'(a) = \mathbf{y}_0' \qquad (3.6.1)$$

where $\mathbf{g}(x)$ is a N by N matrix of given real functions $g_{ij}(x), i, j = 1, 2, \ldots, N$; $\mathbf{y}(x), \mathbf{y}_0$ and $\mathbf{y}_0'$ are real column vectors with N elements. We also assume that $\mathbf{g}(x)$ is symmetric, $g_{ij}(x) = g_{ji}(x)$.

We focus on the current interval $[X, X + h]$ where the problem to be solved is

$$\mathbf{y}''(X + \delta) = \mathbf{f}(\delta) \, \mathbf{y}(X + \delta), \; \mathbf{f}(\delta) \equiv \mathbf{g}(X + \delta), \; \delta \in [0, h] \qquad (3.6.2)$$

with known initial conditions $\mathbf{y}(X) = \mathbf{A}$ and $\mathbf{y}'(X) = \mathbf{B}$. Its exact solution is written in terms of the $2N$ by $2N$ propagation matrix, viz.

$$\begin{pmatrix} \mathbf{y}(X + \delta) \\ \mathbf{y}'(X + \delta) \end{pmatrix} = \begin{pmatrix} \mathbf{u}(\delta) & \mathbf{v}(\delta) \\ \mathbf{u}'(\delta) & \mathbf{v}'(\delta) \end{pmatrix} \begin{pmatrix} \mathbf{A} \\ \mathbf{B} \end{pmatrix} \qquad (3.6.3)$$

where the propagators $\mathbf{u}(\delta)$ and $\mathbf{v}(\delta)$ are N by N matrices each, that is solutions of problems

$$\mathbf{u}''(\delta) = \mathbf{f}(\delta)\mathbf{u}(\delta), \quad \mathbf{u}(0) = \mathbf{I},\ \mathbf{u}'(0) = \mathbf{0} \tag{3.6.4}$$

and

$$\mathbf{v}''(\delta) = \mathbf{f}(\delta)\mathbf{v}(\delta), \quad \mathbf{v}(0) = \mathbf{0},\ \mathbf{v}'(0) = \mathbf{I}. \tag{3.6.5}$$

$\mathbf{I}$ and $\mathbf{0}$ are the N by N unit and zero matrix, respectively.

A matrix $\tilde{\mathbf{f}}(\delta)$ is assumed to exist for which the system

$$\tilde{\mathbf{y}}''(X+\delta) = \tilde{\mathbf{f}}(\delta)\,\tilde{\mathbf{y}}(X+\delta), \quad \delta \in [0, h] \tag{3.6.6}$$

has *known* general solution, and the problem is to construct the exact, unknown propagators $\mathbf{u}(\delta)$ and $\mathbf{v}(\delta)$ in terms of the known propagators $\tilde{\mathbf{u}}(\delta)$ and $\tilde{\mathbf{v}}(\delta)$ of the reference system (3.6.6). To this aim the perturbation procedure is used and the following analog of the result given on page 186 is obtained.

The propagator $\mathbf{p}(\delta)$ ($\mathbf{p}$ *stands for any of* $\mathbf{u}$ *or* $\mathbf{v}$) *is of the form*

$$\mathbf{p}(\delta) = \mathbf{p}_0(\delta) + \mathbf{p}_1(\delta) + \mathbf{p}_2(\delta) + \dots\ . \tag{3.6.7}$$

where the first term $\mathbf{p}_0(\delta)$ *is exactly the reference propagator* $\tilde{\mathbf{p}}(\delta)$ *while the correction matrix* $\mathbf{p}_q(\delta)$, $q = 1, 2, 3, \dots$ *is the solution of the problem*

$$\mathbf{p}_q'' = \tilde{\mathbf{f}}(\delta)\mathbf{p}_q + \Delta\mathbf{f}(\delta)\mathbf{p}_{q-1}(\delta),\ \mathbf{p}_q(0) = \mathbf{p}_q'(0) = \mathbf{0} \tag{3.6.8}$$

where

$$\Delta\mathbf{f}(\delta) \equiv \mathbf{f}(\delta) - \tilde{\mathbf{f}}(\delta). \tag{3.6.9}$$

Up to this point no significant distinction between single equations and systems appeared and actually many other issues discussed in section 3.3 remain similar in case of systems.

A major difference occurs when effectively constructing the perturbation corrections.

We thus first construct two special solutions of the reference system. These are the *column vectors* $\tilde{\mathbf{U}}(\delta)$ and $\tilde{\mathbf{V}}(\delta)$ which obey the initial conditions $\tilde{U}_i(0) = \tilde{V}_i'(0) = 1$ and $\tilde{U}_i'(0) = \tilde{V}_i(0) = 0$, $i = 1, 2, \dots, N$. Their components are readily expressed in terms of the very propagators, viz.

$$\tilde{U}_i(\delta) = \sum_{s=1}^{N} \tilde{u}_{is}(\delta), \quad \tilde{V}_i(\delta) = \sum_{s=1}^{N} \tilde{v}_{is}(\delta) \cdot \tag{3.6.10}$$

These $\tilde{\mathbf{U}}(\delta)$, $\tilde{\mathbf{V}}(\delta)$ and their first derivative are used as a basis for the evaluation of corrections. Specifically we focus on the jth column of the correction matrix $\mathbf{p}_q(\delta)$ and denote it as $[\mathbf{p}_q(\delta)]_j$. This is searched for in the form

$$[\mathbf{p}_q(\delta)]_j = \mathbf{a}_q(\delta)\tilde{\mathbf{U}}(\delta) + \mathbf{b}_q(\delta)\tilde{\mathbf{V}}(\delta) + \mathbf{c}_q(\delta)\tilde{\mathbf{U}}'(\delta) + \mathbf{d}_q(\delta)\tilde{\mathbf{V}}'(\delta) \tag{3.6.11}$$

where $\mathbf{a}_q$, $\mathbf{b}_q$, $\mathbf{c}_q$ and $\mathbf{d}_q$ are N by N matrices of functions which must be determined. Note in passing that when $N = 1$ eq. (3.6.11) becomes just eq. (3.3.57).

We repeat the procedure exposed on p. 193; some extra attention is however necessary to put the factors (which are now matrices and vectors) in the right positions. The following expressions result

$$[\mathbf{p}_q'(\delta)]_j = (\mathbf{a}_q' + \mathbf{c}_q\tilde{\mathbf{f}})\tilde{\mathbf{U}} + (\mathbf{b}_q' + \mathbf{d}_q\tilde{\mathbf{f}})\tilde{\mathbf{V}} + (\mathbf{a}_q + \mathbf{c}_q')\tilde{\mathbf{U}}' + (\mathbf{b}_q + \mathbf{d}_q')\tilde{\mathbf{V}}', \tag{3.6.12}$$

$$[\mathbf{p}_q''(\delta)]_j = [\mathbf{a}_q'' + (2\mathbf{c}_q' + \mathbf{a}_q)\tilde{\mathbf{f}} + \mathbf{c}_q\tilde{\mathbf{f}}']\tilde{\mathbf{U}} + [\mathbf{b}_q'' + (2\mathbf{d}_q' + \mathbf{b}_q)\tilde{\mathbf{f}} + \mathbf{d}_q\tilde{\mathbf{f}}']\tilde{\mathbf{V}} +$$
$$+ (\mathbf{c}_q'' + 2\mathbf{a}_q' + \mathbf{c}_q\tilde{\mathbf{f}})\tilde{\mathbf{U}}' + (\mathbf{d}_q'' + 2\mathbf{b}_q' + \mathbf{d}_q\tilde{\mathbf{f}})\tilde{\mathbf{V}}', \tag{3.6.13}$$

$$\tilde{\mathbf{f}}(\delta)[\mathbf{p}_q(\delta)]_j + \Delta\mathbf{f}(\delta)[\mathbf{p}_{q-1}(\delta)]_j = (\tilde{\mathbf{f}}\mathbf{a}_q + \Delta\mathbf{f}\mathbf{a}_{q-1})\tilde{\mathbf{U}} +$$
$$+ (\tilde{\mathbf{f}}\mathbf{b}_q + \Delta\mathbf{f}\mathbf{b}_{q-1})\tilde{\mathbf{V}} + (\tilde{\mathbf{f}}\mathbf{c}_q + \Delta\mathbf{f}\mathbf{c}_{q-1})\tilde{\mathbf{U}}' + (\tilde{\mathbf{f}}\mathbf{d}_q + \Delta\mathbf{f}\mathbf{d}_{q-1})\tilde{\mathbf{V}}'. \tag{3.6.14}$$

Coefficients of $\tilde{\mathbf{U}}, \tilde{\mathbf{V}}, \tilde{\mathbf{U}}'$ and $\tilde{\mathbf{V}}'$ in eqs (3.6.13) and (3.6.14) are identified. Thus

$$\begin{aligned}
&\mathbf{a}_q'' + 2\mathbf{c}_q'\tilde{\mathbf{f}} + [\mathbf{a}_q, \tilde{\mathbf{f}}] + \mathbf{c}_q\tilde{\mathbf{f}}' = \Delta\mathbf{f}\mathbf{a}_{q-1} \\
&\mathbf{b}_q'' + 2\mathbf{d}_q'\tilde{\mathbf{f}} + [\mathbf{b}_q, \tilde{\mathbf{f}}] + \mathbf{d}_q\tilde{\mathbf{f}}' = \Delta\mathbf{f}\mathbf{b}_{q-1} \\
&\mathbf{c}_q'' + 2\mathbf{a}_q' + [\mathbf{c}_q, \tilde{\mathbf{f}}] = \Delta\mathbf{f}\mathbf{c}_{q-1} \\
&\mathbf{d}_q'' + 2\mathbf{b}_q' + [\mathbf{d}_q, \tilde{\mathbf{f}}] = \Delta\mathbf{f}\mathbf{d}_{q-1}
\end{aligned} \tag{3.6.15}$$

where $[\mathbf{m}_1, \mathbf{m}_2]$ is the *commutator* of matrices $\mathbf{m}_1$ and $\mathbf{m}_2$,

$$[\mathbf{m}_1, \mathbf{m}_2] \equiv \mathbf{m}_1\mathbf{m}_2 - \mathbf{m}_2\mathbf{m}_1.$$

The initial conditions of eq. (3.6.8), that is $[\mathbf{p}_q(0)]_{ij} = [\mathbf{p}'_q(0)]_{ij} = 0$ for $i = 1, 2, 3, \ldots, N$, are introduced in eqs (3.6.11) and (3.6.12) and thus the following initial conditions for the solutions $\mathbf{a}_q, \mathbf{b}_q, \mathbf{c}_q$ and $\mathbf{d}_q$ result

$$\sum_{s=1}^{N} \{[\mathbf{a}_q(0)]_{is} + [\mathbf{d}_q(0)]_{is}\} = 0$$

$$\sum_{s=1}^{N} \{[\mathbf{a}'_q(0)]_{is} + [\mathbf{c}_q(0)\tilde{\mathbf{f}}(0)]_{is} + [\mathbf{b}_q(0)]_{is} + [\mathbf{d}'_q(0)]_{is}\} = 0. \tag{3.6.16}$$

If these equations are compared with eqs (3.3.61) and (3.3.62) we see that they are similar except that in the case of systems a new term occurs, namely the commutator.

The occurrence of this extra term has two practical consequences. First, the perturbation corrections are by no means straightforward extensions of the formulas derived in the single equation case; they must be evaluated anew.

Second, the class of candidates for being reasonable reference matrices (that is matrices of functions which secure analytic evaluation of both reference solutions and perturbation corrections) is more restricted than in the single equation case. As a matter of fact, the linear functions cease to belong to this class.

To explicitly construct perturbation algorithms for systems we adopt a way which exploits the experience accumulated with single equations. Thus a pilot reference function matrix $\mathbf{g}^*(X + \delta)$ is introduced and we continue to adopt the convention that it consists of parabolic functions. A procedure as in eq. (3.4.31—38) is applied to each matrix element. Specifically $g_{ij}(X+\delta)$ is written as

$$g_{ij}(X + \delta) = G_{ij} + \Delta g_{ij}(X + \delta) \tag{3.6.17}$$

where

$$G_{ij} = \frac{1}{h}\int_0^h g^*_{ij}(X + \delta)d\delta = [\mathbf{g}^*_0]_{ij} + \frac{1}{2}h[\mathbf{g}^*_1]_{ij} + \frac{1}{3}h^2[\mathbf{g}^*_2]_{ij}, \tag{3.6.18}$$

$$\Delta g_{ij}(X + \delta) = \Delta g^*_{ij}(X + \delta) + \Delta \tilde{g}_{ij}(X + \delta), \tag{3.6.19}$$

with

$$\Delta g^*_{ij}(X+\delta) \equiv g_{ij}(X+\delta) - g^*_{ij}(X+\delta) = \frac{1}{120} g^{(3)}_{ij}(X)h^3[20(\delta/h)^3 -$$

$$- 30(\delta/h)^2 + 12\delta/h - 1] + \frac{1}{840} g^{(4)}_{ij}(X)h^4[35(\delta/h)^4 - 60(\delta/h)^2 +$$

$$+ 32\delta/h - 3] + O(h^5) \tag{3.6.20}$$

and

$$\Delta \tilde{g}_{ij}(X+\delta) \equiv g^*_{ij}(X+\delta) - G_{ij} = [\mathbf{g}^*_1]_{ij}\left(\delta - \frac{1}{2}h\right) +$$

$$+ [\mathbf{g}^*_2]_{ij}\left(\delta^2 - \frac{1}{3}h^2\right) \tag{3.6.21}$$

The matrix of constants $\mathbf{G}$ is real and symmetric and thus it can be diagonalized. Let $\mathbf{D}$ be the unitary diagonalization matrix and $\mathbf{D}^T$ its transposed. In this representation the one-step problem becomes

$$\mathbf{z}''(X+\delta) = \mathbf{f}^D(\delta)\mathbf{z}(X+\delta),\ \delta \in [0, h] \tag{3.6.22}$$

where $\mathbf{z}(X+\delta) = \mathbf{D}^T\mathbf{y}(X+\delta)$,

$$\mathbf{f}^D(\delta) = \mathbf{F} + \Delta\mathbf{f}^*(\delta) + \Delta\tilde{\mathbf{f}}(\delta), \tag{3.6.23}$$

with $\mathbf{F} = \mathbf{D}^T\mathbf{G}\mathbf{D}$, $\Delta\mathbf{f}^*(\delta) = \mathbf{D}^T\Delta\mathbf{g}^*(X+\delta)\mathbf{D}$ and

$$\Delta\tilde{\mathbf{f}}(\delta) = \mathbf{f}^*_1\left(\delta - \frac{1}{2}h\right) + \mathbf{f}^*_2\left(\delta^2 - \frac{1}{3}h^2\right) \tag{3.6.24}$$

with $\mathbf{f}^*_i = \mathbf{D}^T\mathbf{g}^*_i\mathbf{D}$, $i = 1, 2$. The initial conditions are $\mathbf{z}(X) = \mathbf{D}^T\mathbf{y}(X)$ and $\mathbf{z}'(X) = \mathbf{D}^T\mathbf{y}'(X)$ and hence the exact solution of eq. (3.6.2) reads

$$\begin{pmatrix}\mathbf{y}(X+\delta)\\ \mathbf{y}'(X+\delta)\end{pmatrix} = \begin{pmatrix}\mathbf{D} & \mathbf{0}\\ \mathbf{0} & \mathbf{D}\end{pmatrix}\begin{pmatrix}\mathbf{u}(\delta) & \mathbf{v}(\delta)\\ \mathbf{u}'(\delta) & \mathbf{v}'(\delta)\end{pmatrix}\begin{pmatrix}\mathbf{D}^T & \mathbf{0}\\ \mathbf{0} & \mathbf{D}^T\end{pmatrix}\begin{pmatrix}\mathbf{y}(X)\\ \mathbf{y}'(X)\end{pmatrix} \tag{3.6.25}$$

where $\mathbf{u}$ and $\mathbf{v}$ are now the propagators of the almost uncoupled system (3.6.22). This diagonalization procedure offers the advantage that the evaluation of the reference propagators and of the perturbation corrections is greatly simplified.

Constant perturbation methods

Zeroth order version

The reference matrix is here simply the diagonal matrix of constants $\mathbf{F}$, $[\mathbf{F}]_{ij} = F_i\delta_{ij}$. The reference system hence is a set of N *separate* equations of form (3.4.2) and thus the reference propagators are diagonal,

$$\tilde{u}_{ij}(\delta) = \delta_{ij}\xi(F_i, \delta), \quad \tilde{v}_{ij}(\delta) = \delta_{ij}\eta_0(F_i, \delta); \tag{3.6.26}$$

their first-order derivations are also diagonal,

$$\tilde{u}'_{ij}(\delta) = \delta_{ij}F_i\eta_0(F_i, \delta), \ \tilde{v}'_{ij}(\delta) = \delta_{ij}\xi(F_i, \delta) = \tilde{u}_{ij}(\delta). \tag{3.6.27}$$

CPM(0) is the version in which the exact propagation matrix of eq. (3.6.25) is replaced by the reference propagation matrix with the components just determined.

Perturbation corrections

The two particular column solutions $\tilde{\mathbf{U}}(\delta)$ and $\tilde{\mathbf{V}}(\delta)$ have components $\tilde{U}_i(\delta) = \xi(F_i, \delta)$ and $\tilde{V}_i(\delta) = \eta_0(F_i, \delta)$, $i = 1, 2, \ldots, N$. Since $\tilde{\mathbf{U}}'(\delta) = \mathbf{F}\tilde{\mathbf{V}}(\delta)$ and $\tilde{\mathbf{V}}'(\delta) = \tilde{\mathbf{U}}(\delta)$ the sum in eq. (3.6.11) can be restrained to only two terms and thus we search for $[\mathbf{p}_q(\delta)]_j$ of the form

$$[\mathbf{p}_q(\delta)]_j = \mathbf{a}_q(\delta)\tilde{\mathbf{U}}(\delta) + \mathbf{b}_q(\delta)\tilde{\mathbf{V}}(\delta). \qquad (3.6.28$$

This is worked out the same way as eq. (3.6.11) and the result is that the N by N matrices $\mathbf{a}_q$ and $\mathbf{b}_q$ are solutions of the system

$$\mathbf{a}''_q + 2\mathbf{b}'_q + [\mathbf{a}_q, \mathbf{F}] = \Delta\mathbf{f}\,\mathbf{a}_{q-1} \tag{3.6.29 a}$$

$$\mathbf{b}''_q + 2\mathbf{a}'_q\mathbf{F} + [\mathbf{b}_q, \mathbf{F}] = \Delta\mathbf{f}\,\mathbf{b}_{q-1} \tag{3.6.29 b}$$

with initial conditions

$$\sum_{s=1}^{N} [\mathbf{a}_q(0)]_{is} = 0, \quad \sum_{s=1}^{N} \{[\mathbf{a}'_q(0)]_{is} + [\mathbf{b}_q(0)]_{is}\} = 0, \; i = 1,2,3,\ldots, N. \tag{3.6.30}$$

As in the case of single equations we disregard the residual perturbation; $\Delta\mathbf{f}$ is thus reduced to the pilot perturbation $\Delta\tilde{\mathbf{f}}$, eq. (3.6.24).

It is worthy to be noticed that since $\mathbf{F}$ is diagonal the product of any matrix by $\mathbf{F}$ does not mix the components of the matrix, in particular

$$[\mathbf{a}'_q\mathbf{F}]_{is} = F_s[\mathbf{a}'_q]_{is}, \quad [[\mathbf{a}_q, \mathbf{F}]]_{is} = [\mathbf{a}_q]_{is}(F_s - F_i),$$

$$[[\mathbf{b}_q, \mathbf{F}]]_{is} = [\mathbf{b}_q]_{is}(F_s - F_i).$$

For this reason the system (3.6.29 a, b) becomes a system of N^2 *separated* pairs of equations

$$[\mathbf{a}''_q]_{is} + 2[\mathbf{b}'_q]_{is} + [\mathbf{a}_q]_{is}(F_s - F_i) = [\Delta\tilde{\mathbf{f}}\mathbf{a}_{q-1}]_{is} \tag{3.6.31 a}$$

$$[\mathbf{b}''_q]_{is} + 2F_s[\mathbf{a}'_q]_{is} + [\mathbf{b}_q]_{is}(F_s - F_i) = [\Delta\tilde{\mathbf{f}}\mathbf{b}_{q-1}]_{is} \tag{3.6.31 b}$$

for $i, s=1, 2, \ldots, N$ and if the right member of each of the two equations is a polynomial in δ (and so it actually is because $\Delta\tilde{\mathbf{f}}$ is a polynomial), then the pair has a particular solution of polynomial form. Hence the typical strategy consists of separate evaluation of the particular solution of each pair (i, s), $s = 1, 2, \ldots, i-1, i+1, \ldots, N$. When $s = i$ the polynomial solution $[\mathbf{a}_q]_{ii}$ will be determined up to an additive arbitrary constant and the same is true for $[\mathbf{b}_q]_{ii}$. These constants are finally fixed in terms of the initial conditions (3.6.30).

This procedure allows explicit evaluation of the perturbation corrections in any order q of the perturbation theory and for illustration we consider the case of *first-order corrections.*

We continue to refer to the jth column of the correction matrix $\mathbf{p}_1(\delta)$ and, to fix the ideas, we focus on the matrix element $[\mathbf{u}_1(\delta)]_{ij}$ with $i \neq j$. Since $\mathbf{u}_0(\delta) = \tilde{\mathbf{u}}(\delta)$ we have $[\mathbf{a}_0]_{ms} = \delta_{mj}\delta_{sj}$ and $\mathbf{b}_0(\delta) = \mathbf{0}$. Hence

$$[\Delta\tilde{\mathbf{f}}\mathbf{a}_0]_{is} = \delta_{js}\left\{[\mathbf{f}_1^*]_{ij}\left(\delta - \frac{1}{2}h\right) + [\mathbf{f}_2^*]_{ij}\left(\delta^2 - \frac{1}{3}h^2\right)\right\},$$

$$[\Delta\tilde{\mathbf{f}}\mathbf{b}_0]_{is} = 0. \tag{3.6.32}$$

The only case when the right terms in the pair (3.6.31) are not both vanishing is $s = j$. When simultaneously $s \neq j$ and $s \neq i$ it is readily seen that $[\mathbf{a}_1]_{is} = 0$ and $[\mathbf{b}_1]_{is} = 0$ are particular solutions so that only two pairs remain to be examined namely $s = j$ and $s = i$. The two remaining pairs of equations read

$$a_j'' + 2b_j' + a_j(F_j - F_i) = [\mathbf{f}_1^*]_{ij}\left(\delta - \frac{1}{2}h\right) + [\mathbf{f}_2^*]_{ij}\left(\delta^2 - \frac{1}{3}h^2\right)$$

$$b_j'' + 2F_j a_j' + b_j(F_j - F_i) = 0$$

$$a_i'' + 2b_i' = 0 \tag{3.6.33}$$

$$b_j'' + 2F_i a_i' = 0$$

and the initial conditions are reduced to

$$a_i(0) + a_j(0) = 0, \ a_i'(0) + a_j'(0) + b_i(0) + b_j(0) = 0; \tag{3.6.34}$$

we used here the short-hand notations $a_s = [\mathbf{a}_1]_{is}$ and $b_s = [\mathbf{b}_1]_{is}$. As announced, for the first pair a particular polynomial solution is to be constructed while for second pair we take the simplest solution $a_i = C_1$, $b_i = C_2$ with constant C_1 and C_2. The solution of the first pair is

$$a_j(\delta) = [\mathbf{f}_1^*]_{ij} D\left(\delta - \frac{1}{2}h\right) + [\mathbf{f}_2^*]_{ij} D\left(\delta^2 - \frac{1}{3}h^2 + 2D + 8F_j D^2\right),$$

$$b_j(\delta) = -2[\mathbf{f}_1^*]_{ij} D^2 F_j - 4[\mathbf{f}_2^*]_{ij} D^3 F_j \delta. \tag{3.6.35}$$

with $D = (F_j - F_i)^{-1}$.

The initial conditions (3.6.34) furnish the values of C_1 and C_2 such as

$$a_i(\delta) = \frac{1}{2} hD[\mathbf{f}_1^*]_{ij} - D[\mathbf{f}_2^*]_{ij}\left(-\frac{h^3}{3} - 2D + 8F_j D^2\right)$$

$$b_i(\delta) = [\mathbf{f}_1^*]_{ij}(F_i + F_j)D^2. \tag{3.6.36}$$

We collect these results into eq. (3.6.28) and put $\delta = h$. The result is

$$[\mathbf{u}_1(h)]_{ij} = h^3[\mathbf{f}_1^*]_{ij}A^1_{ij} + h^4[\mathbf{f}_2^*]_{ij}A^2_{ij} \tag{3.6.37}$$

with

$$A^1_{ij} = \Delta^{-2}\left[\frac{1}{2}\Delta\bar{\xi}(Z_i) + (\Delta + 2Z_i)\bar{\eta}_0(Z_i) + + \frac{1}{2}\Delta\bar{\xi}(Z_j) - 2(\Delta + Z_i)\bar{\eta}_0(Z_j)\right]$$

$$A^2_{ij} = \Delta^{-3}\left[-\left(\frac{1}{3}\Delta^2 + 6\Delta + 8Z_i\right)\bar{\xi}(Z_i) + + \left(\frac{2}{3}\Delta^2 + 6\Delta + 8Z_i\right)\bar{\xi}(Z_j) - 4\Delta(\Delta + Z_i)\bar{\eta}_0(Z_j)\right] \tag{3.6.38}$$

with $Z_i = F_i h^2$, $Z_j = F_j h^2$, $\Delta = Z_j - Z_i$.

When Z_i and Z_j are close together, the accuracy of the computation of A^1_{ij} and A^2_{ij} deteriorates due to the heavy cancellation of like terms. To cure this, series in powers of Δ are taken for $\bar{\xi}(Z_j)$ and for $\bar{\eta}_0(Z_j)$

$$\bar{\xi}(Z_i + \Delta) = \bar{\xi}(Z_i) + \sum_{n=1}^{\infty} \frac{\Delta^n}{n!2^n}\bar{\eta}_{n-1}(Z_i), \tag{3.6.39}$$

$$\bar{\eta}_0(Z_i + \Delta) = \bar{\eta}_0(Z_i) + \sum_{n=1}^{\infty} \frac{\Delta^n}{n!2^n}\bar{\eta}_n(Z_i); \tag{3.6.40}$$

these expansions can be readily derived from eq. (3.4.14). We thus find

$$A^1_{ij} = -\frac{1}{4}\bar{\eta}_1(Z_i) + \sum_{n=1}^{\infty} \frac{\Delta^n}{(n+2)!2^{n+2}}[n\bar{\eta}_n(Z_i) - 2\bar{\eta}_{n+1}(Z_i)], \tag{3.6.41}$$

$$A^2_{ij} = -\frac{1}{4}\bar{\eta}_1(Z_i) + \sum_{n=1}^{\infty} \frac{\Delta^n}{(n+3)!2^n}\left[\frac{1}{3}n(n+2)\bar{\eta}_n(Z_i) - -\frac{3}{2}(n+1)\bar{\eta}_{n+1}(Z_i)\right]. \tag{3.6.42}$$

These series converge rather fast; as a matter of fact, five terms are enough if $|\Delta| \leqslant 2$ to obtain A^1_{ij} and A^2_{ij} with four exact figures for any Z_i. The case $\Delta = 0$ directly furnishes the diagonal correction

$$[\mathbf{u}_1(h)]_{ii} = -\frac{1}{4} h^3\{[\mathbf{f}_1^*]_{ii} + h[\mathbf{f}_2^*]_{ii}\}\bar{\eta}_1(Z_i),$$

a formula which, as expected, coincides with the formula (3.4.57) for single equations.

The expressions of the other first-order correction matrices are

$$[\mathbf{u}_1'(h)]_{ij} = h^2[\mathbf{f}_1^*]_{ij}A^{1'}_{ij} + h^3[\mathbf{f}_2^*]_{ij}A^{2'}_{ij} \tag{3.6.43}$$

$$[\mathbf{v}_1(h)]_{ij} = h^4[\mathbf{f}_1^*]_{ij}B^1_{ij} + h^5[\mathbf{f}_2^*]_{ij}B^2_{ij} \tag{3.6.44}$$

$$[\mathbf{v}_1'(h)]_{ij} = h^3[\mathbf{f}_1^*]_{ij}B^{1'}_{ij} + h^4[\mathbf{f}_2^*]_{ij}B^{2'}_{ij} \tag{3.6.45}$$

with

$$A^{1'}_{ij} = \Delta^{-2}\Big\{(\Delta + 2Z_i)(\bar{\xi}(Z_i) + \bar{\xi}(Z_j)) + \frac{1}{2}\Delta[Z_i\bar{\eta}_0(Z_i) + $$

$$+ (\Delta + Z_i)\bar{\eta}_0(Z_j)]\Big\}, \tag{3.6.46}$$

$$A^{2'}_{ij} = \Delta^{-3}\Big[-Z_i\Big(-\frac{1}{3}\Delta^2 + 6\Delta + 8Z_i\Big)\bar{\eta}_0(Z_i) - 2\Delta(\Delta+2Z_i)\bar{\xi}(Z_j) + $$

$$+ (\Delta + Z_i)\Big(\frac{2}{3}\Delta^2 + 2\Delta + 8Z_i\Big)\bar{\eta}_0(Z_j)\Big], \tag{3.6.47}$$

$$B^1_{ij} = \Delta^{-2}\Big[2(\bar{\xi}(Z_i) - \bar{\xi}(Z_j)) + \frac{1}{2}\Delta(\bar{\eta}_0(Z_i) + \bar{\eta}_0(Z_j))\Big], \tag{3.6.48}$$

$$B^2_{ij} = \Delta^{-3}\Big[-\Big(-\frac{1}{3}\Delta^2 + 2\Delta + 8Z_i\Big)\bar{\eta}_0(Z_i) - 4\Delta\bar{\xi}(Z_j) + $$

$$+ \Big(\frac{2}{3}\Delta^2 + 6\Delta + 8Z_i\Big)\bar{\eta}_0(Z_j)\Big], \tag{3.6.49}$$

$$B_{ij}^{1\prime} = \Delta^{-2}\left[\frac{1}{2}\Delta(\bar{\xi}(Z_i) + \bar{\xi}(Z_j)) + 2Z_i\bar{\eta}_0(Z_i) - (\Delta + 2Z_i)\bar{\eta}_0(Z_j)\right], \tag{3.6.50}$$

$$B_{ij}^{2\prime} = \Delta^{-3}\left[-\left(-\frac{1}{3}\Delta^2 + 2\Delta + 8Z_i\right)\bar{\xi}(Z_i) + \right.$$

$$\left. + \left(\frac{2}{3}\Delta^2 + 2\Delta + 8Z_i\right)\bar{\xi}(Z_j) - 2\Delta(\Delta + 2Z_i)\bar{\eta}_0(Z_j)\right]. \tag{3.6.51}$$

Case $Z_j \simeq Z_i$ can be equally accommodated by means of series expansions for $\bar{\xi}(Z_j)$ and for $\bar{\eta}_0(Z_j)$. The expressions of the diagonal corrections are identical with eqs (3.4.47), (3.4.50) and (3.4.51).

The knowledge of the first-order corrections is exploited in two ways. One is to measure the accuracy of the CPM(0) and for illustration we focus on the case when h is so small that only the leading term in the power series expansions of functions A_{ij}^1, A_{ij}^2, B_{ij}^1, B_{ij}^2, $A_{ij}^{1\prime}$, $A_{ij}^{2\prime}$, B_{1j}' and $B_{ij}^{2\prime}$ is numerically significant. The point is that all these series begin with the Δ^0 term except for $A_{ij}^{1\prime}$,

$$A_{ij}^{1\prime} = -\sum_{n=1}^{\infty} \frac{n}{(n+2)!\,2^{n+2}} \Delta^n[3\bar{\eta}_n(Z_i) + \bar{\eta}_{n-1}(Z_i)] \tag{3.6.52}$$

and for B_{ij}^1 which begin with the Δ^1 term. For this reason $A_{ij}^{1\prime}$ and B_{ij}^1 essentially behave as h^2 as Δ does. Also the leading term of $A_{ij}^{2\prime}$ which is $Z_i\bar{\eta}_2(Z_i)/12$ equally behaves as h^2 due to the explicit presence of factor $Z_i = F_i h^2$. Hence although $[\mathbf{u}_1'(h)]_{ij}$ given by eq. (3.6.43) seems to behave as h^2 it actually behaves as h^4. Two other first-order corrections i.e. $[\mathbf{u}_1(h)]_{ij}$ and $[\mathbf{v}_1'(h)]_{ij}$ behave as h^3 while the fourth, $[\mathbf{v}_1(h)]_{ij}$, behaves as h^5 (since B_{ij}^1 begins with the Δ^1 term). The local truncation error of CPM(0) thus globally behaves as h^3 and therefore CPM(0) is a *second*-order method, exactly as in case of single equations.

Rigorously speaking we should have also considered the behaviour of the corrections from the residual perturbation. Their influence is of minor importance and to see it we only remark that all corresponding expansions in powers of Δ have the single equation expressions (3.4.54—56) as leading terms and thus these corrections collectively behave as h^7.

Another use of first-order corrections is of just adding them in the propagation matrix and thus CPM(1) is obtained. It can be proved that this is a fourth-order method, that is exactly as in the single equation case.

A comment is now in order with respect to the practical efficiency of these methods. The most costly portion of the algorithm consists of the diagonalization of $\mathbf{G}$; the number of operations of this sequence is proportional to N^3. Hence these methods are particularly useful when $\mathbf{g}(x)$ is of the form $\mathbf{g}(x) = \mathbf{V}(x) + E \cdot \mathbf{I}$ where $\mathbf{V}(x)$ is a given matrix and E is a scalar, and the integration of the system is to be repeated at various values of E. In fact, the diagonalization of $\mathbf{G}$ in each interval means essentially only the piecewise diagonalization of $\mathbf{V}(x)$. Thus it is to be performed *once*, at the beginning of the whole calculation, and the corresponding values of the unitary diagonalization matrices $\mathbf{D}$ and of the resultant quantities $\mathbf{F}, \mathbf{f}_1^*$ and $\mathbf{f}_2^*$ on each step of the partition must be stored for direct reuse during the very integration of the system at various values of E.

Quite frequent in practice are the cases when *one and the same system* (that is with the same $\mathbf{V}(x)$ and E) is to be solved under *different* initial conditions. In such a case the simultaneous propagation of all solutions is preferred to the separate computation of each of them and the reason is that the one-step propagation matrices are *common* for all solutions.

It is also important to be noted that the computational effort to generate the reference propagators is proportional to N because the reference matrix $\mathbf{F}$ is diagonal. It is proportional to N^2 for the first-order corrections since these are linear in perturbation weights $\mathbf{f}_1^*$ and $\mathbf{f}_2^*$ but is proportional to N^3 for the second-order corrections since these are quadratic in perturbation weights *. For this reason the inclusion of the second-order corrections should be made with some precaution. As a matter of fact, the optimal CP version [33] consists of taking the zeroth order propagators and the first-order corrections in full while only the leading components from the second-order corrections. Its order is six. The computer program [34] is available on request.

Earlier investigations on the CPM(0) are in [3, 20, 35].

Line perturbation methods

The diagonal components of $\mathbf{f}^D(\delta)$, eq. (3.6.23), are here organized as follows

$$f_{ii}^D(\delta) = F_i^0 + F_i^1\delta + [\Delta\mathbf{f}^*(\delta)]_{ii} + [\mathbf{f}_2^*]_{ii}\left(\delta^2 - h\delta + \frac{1}{6}h^2\right) \qquad (3.6.53)$$

* The computation time to evaluate numerically higher order corrections, $q = 3, 4, \ldots,$ remains proportional to N^3 but the constant of proportionality strongly increases with q.

where $F_i^0 = [\mathbf{f}_0^*]_{ii} - \dfrac{1}{6} h^2[\mathbf{f}_2^*]_{ii}$ and $F_i^1 = [\mathbf{f}_1^*]_{ii} + h[\mathbf{f}_2^*]_{ii}$.

The off-diagonal components remain unchanged.

The *diagonal* matrix $\tilde{\mathbf{f}}(\delta)$ with components

$$\tilde{\mathbf{f}}_{ij}(\delta) = \delta_{ij}(F_i^0 + F_i^1\delta) \tag{3.6.54}$$

is taken as reference function matrix and its propagators are the diagonal matrices with the elements calculated through eqs (3.5.14—17). The resultant algorithm is coined LPM(0) and its order is two, the same as for CPM(0).

Paradoxical at it seems at first sight the last statement is just normal. In fact, although it contradicts the situation in the single equation case where CPM(0) was of order two while LPM(0) was of order four, the very picture is significantly changed in the case of systems. To get the point we recall that in the case of single equations the difference between the orders of the two versions emerged from the fact that the neglected pilot perturbation was given by eq. (3.4.35) for the first method and its size is more substantial than that of the pilot perturbation (3.5.5) which is neglected in the second method. (The residual perturbations were the same.) In contrast, in the case of systems only the diagonal components of the neglected perturbation are different while the off-diagonal components are the same and thus the above statement appears as normal.

Perturbation corrections to algorithm LPM(0) cannot be derived in closed forms except for the particular case when the slopes are equal, $F_1^1 = F_2^1 = \ldots = F_N^1$. The expressions of the first-order corrections for this particular case were reported in [3, 20] in the frame of the vector extension of the formalism given on p. 232. (The same can be obtained as well in the frame of the propagation formalism of this book.) An average slope is suggested in [20] to be used in the evaluation of the first-order corrections. The order of the resultant LPM(1) is unknown. A computer code is [36].

3.7. PERTURBATION SOLUTION OF DIFFERENTIAL EQUATIONS WITH SINGULAR COEFFICIENTS

The piecewise perturbation methods of the previous sections were devised for the solution of the initial value problem (3.3.1) or of its vector extension, eq. (3.6.1), provided that $g(x)$ (in the first case) or the components of matrix $\mathbf{g}(x)$ (in the second) are bounded functions.

However quite often are in practice the cases when these are singular functions and to be specific we assume that the point at which they are singular is $x = 0$ and that the singularities are of the type x^{-2} and x^{-1}. We first take the *single equation* case and assume that $g(x)$ is of the form

$$g(x) = \frac{1}{x^2} g_{-2}(x) + \frac{1}{x} g_{-1}(x) + g_0(x) \tag{3.7.1}$$

where $g_{-2}(x)$, $g_{-1}(x)$ and $g_0(x)$ are well-behaved functions.

To generate numerically the regular solution of the differential equation

$$Y'' = g(x)Y, \quad x > 0 \tag{3.7.2}$$

a point $a > 0$ close to the origin is delimited and the regular solution and its first derivative are calculated on $(0, a]$. Their values at $x = a$ are further used as initial conditions to propagate the solution ahead by any standard procedure. Clearly the accuracy of the whole chain closely depends on the quality of the solution on the initial interval $(0, a]$ and to construct such a solution we develop a suitable perturbation approach.

The introductory step into the approach consists of approximating $g_{-2}(x)$, $g_{-1}(x)$ and $g_0(x)$ by polynomials ; some standard fit procedure can be used as for instance the least squares fit. We choose to explicitly retain only the terms up to x^2, viz.

$$g(x) = \frac{1}{x^2} g_{-2}^* + \frac{1}{x} g_{-1}^* + g_0^* + g_1^* x + g_2^* x^2 + \Delta g(x) \tag{3.7.3}$$

where g_i^*, $i = -2, -1, 0, 1$ and 2 are numbers; $\Delta g(x)$ is the residual deviation and from now on it will be neglected. We also choose to transform the interval $(0, a]$ onto $(0, 1]$ and to this aim we use the variable $\delta = x/a$. Upon denoting $y(\delta) = Y(x)$, $f_p = a^{p+2} g_p^*$ the problem becomes of constructing the regular solution of

$$y'' = f(\delta)y, \quad \delta \in (0, 1] \tag{3.7.4}$$

with

$$f(\delta) = \frac{1}{\delta^2} f_{-2} + \sum_{p=-1}^{2} f_p \delta^p. \tag{3.7.5}$$

The central point is that the dominant term in $f(\delta)$, that is

$$\tilde{f}(\delta) = f_{-2}\delta^{-2} \tag{3.7.6}$$

is appropriate to be taken as reference function in a perturbation scheme. In fact, it meets the two conditions: (i) the reference solution has analytic expression; (ii) any correction to the reference solution can be evaluated in analytic form.

Reference solution. We search for the solution of the reference equation

$$\tilde{y}'' = f_{-2}\delta^{-2}\tilde{y} \tag{3.7.7}$$

in the form $\tilde{y} = \delta^{l+1}$. We introduce this into the equation and find out that l must satisfy the algebraic equation

$$l(l+1) = f_{-2}. \tag{3.7.8}$$

This has two solutions: $l_1 = -\dfrac{1}{2} + A$, $l_2 = -\dfrac{1}{2} - A$, with $A = \dfrac{1}{2}(1 + 4f_{-2})^{1/2}$. Two solutions of the reference equation then are $\tilde{y}_1(\delta) = \delta^{l_1+1}$ and $\tilde{y}_2(\delta) = \delta^{l_2+1} = \delta^{-l_1}$ and they are different when $f_{-2} \neq -1/4$. When $f_2 > -1/4$ $\tilde{y}_1$ is regular and $\tilde{y}_2$ is singular while when $f_{-2} < -1/4$ both are singular. When $f_{-2} = -1/4$ the general regular solution is of the form $y(\delta) = \delta^{1/2}(C_1 + C_2 \ln \delta)$. In short, eq. (3.7.7) has regular solution only if $f_{-2} \geqslant -1/4$ but its form is different in cases $f_{-2} > -1/4$ and $f_{-2} = -1/4$.

In the following we consider only the case $f_{-2} > -1/4$ and thus the regular reference solution is

$$\tilde{y}(\delta) = \delta^{l+1} \tag{3.7.9}$$

where l is the largest solution of the algebraic equation (3.7.8).

Perturbation corrections. We assume that the $(k-1)$th-order correction is of form

$$y_{k-1}(\delta) = \delta^{l+1} \sum_{q=q_0}^{Q} a_q(k-1)\delta^q \tag{3.7.10}$$

and show that y_k results of the same form. $\Delta f y_{k-1}$ is first evaluated for

$$\Delta f = \sum_{p=-1}^{2} f_p \delta^p, \tag{3.7.11}$$

to get

$$\Delta f y_{k-1} = \delta^{l+1} \sum_{q=q_0-1}^{Q+2} A_q \delta^q$$

with

$$A_q = \sum_{p=-1}^{q-q_0} f_p a_{q-p}(k-1)\theta(p-q+Q)\theta(2-p), \tag{3.7.12}$$

where $\theta(s)$ is the usual step function

$$\theta(s) = \begin{cases} 1 & \text{if } s \geqslant 0 \\ 0 & \text{if } s < 0. \end{cases} \tag{3.7.13}$$

The equation of the kth correction,

$$y_k'' = \tilde{f} y_k + \Delta f y_{k-1}, \tag{3.7.14}$$

has the solution

$$y_k(\delta) = \delta^{l+1} \sum_{q=q_0+1}^{Q+4} a_q(k)\delta^q,\ a_q(k) = \frac{A_{q-2}}{(q+l+1)(q+l) - l(l+1)}. \tag{3.7.15}$$

Q.E.D. The perturbation scheme for single equations is such completed in full. The values at $\delta = 1$ of the kth correction and of its first derivative thus are

$$y_k(1) = \sum_{q=q_0+1}^{Q+4} a_q(k),\ y_k'(1) = \sum_{q=q_0+1}^{Q+4} (l+1+q)a_q(k)$$

and they furnish directly the approximate values at $x = a$ of the solution of the original equation (3.7.2) and of its first derivative as follows

$$Y(a) = 1 + y_1(1) + y_2(1) + y_3(1) + \dots$$

$$Y'(a) = \frac{1}{a}(l + 1 + y_1'(1) + y_2'(1) + y_3'(1) + \dots)$$

These $Y(a)$ and $Y'(a)$ will be used as initial conditions to solve the equation (3.7.2) for $x > a$.

In case of *systems of equations* the problem to be solved is the matrix extension of eqs (3.7.4) and (3.7.5), specifically

$$\mathbf{y}'' = \mathbf{f}(\delta)\mathbf{y}, \quad \delta \in (0, 1], \tag{3.7.16}$$

with

$$\mathbf{f}(\delta) = \frac{1}{\delta^2}\mathbf{f}_{-2} + \Delta\mathbf{f}(\delta), \ \Delta\mathbf{f}(\delta) = \sum_{p=-1}^{2} \mathbf{f}_p\delta^p, \tag{3.7.17}$$

where $\mathbf{f}_{-2}$, $\mathbf{f}_{-1}$, $\mathbf{f}_0$, $\mathbf{f}_1$ and $\mathbf{f}_2$ are N by N *numerical* matrices. The only restrictions which we impose are that $\mathbf{f}_{-2}$ is diagonal and that $[\mathbf{f}_{-2}]_{ii} > -1/4$ for all $i = 1, 2, \ldots, N$.

We must here construct the set of the N linear independent regular solutions $\mathbf{y}^{(n)}$, $n = 1, 2, \ldots, N$. Each $\mathbf{y}^{(n)}(\delta)$ is a column vector with N components. In the following we focus of some fixed n and omit the upper label. The matrix $\tilde{\mathbf{f}}(\delta) = \mathbf{f}_{-2}\delta^{-2}$ satisfies the two conditions for being taken as a reference matrix.

Reference solution. Since the reference matrix is diagonal the regular solution of the reference equation $\tilde{\mathbf{y}}'' = \mathbf{f}_{-2}\delta^{-2}\tilde{\mathbf{y}}$ has the components

$$[\tilde{\mathbf{y}}(\delta)]_i = \delta^{l_n+1}\delta_{in}, \quad i = 1, 2, \ldots, N \tag{3.7.18}$$

where δ_{in} is the usual Kronecker symbol and l_n is the largest root of the equation $l_n(l_n + 1) = [\mathbf{f}_{-2}]_{nn}$.

Perturbation corrections. We assign the following form to $\mathbf{y}_{k-1}$

$$\mathbf{y}_{k-1}(\delta) = \sum_{q=q_0}^{Q} \mathbf{z}_{k-1}^q, \ \ \mathbf{z}_{k-1}^q = \delta^{l_n+1} \sum_{r=r_0}^{R} \mathbf{a}_{qr}(k-1)d_{qr} \tag{3.7.19}$$

with

$$d_{qr} \equiv \delta^q(\ln\delta)^r \tag{3.7.20}$$

where q_0, Q, r_0 and R are known positive integers; ln is the natural logarithm function.

Note that an important difference in comparison with the single equation case occurs, namely that now the basic functions are d_{qr} in lieu of the power functions δ^q. (This resembles the situation encountered in the previous section and the ground is similar, that is the explicit occurrence of the commutator.)

We now show that $\mathbf{y}_k$ results of the same form. As before, first construct $\Delta\mathbf{f}\mathbf{y}_{k-1}$

$$\Delta\mathbf{f}\mathbf{y}_{k-1} = \delta^{l_n+1} \sum_{q=q_0-1}^{Q+2} \sum_{r=r_0}^{R} \mathbf{A}_{qr} d_{qr},$$

$$\mathbf{A}_{qr} = \sum_{p=-1}^{q-q_0} \mathbf{f}_p \mathbf{a}_{q-p,r}(k-1)\,\theta(p-q+Q)\theta(2-p). \tag{3.7.21}$$

Thus the kth correction, which is the solution of

$$\mathbf{y}_k'' = \tilde{\mathbf{f}}\mathbf{y}_k + \Delta\mathbf{f}\mathbf{y}_{k-1}, \tag{3.7.22}$$

can be written as

$$\mathbf{y}_k = \sum_{q=q_0+1}^{Q+4} \mathbf{z}_k^q,$$

where $\mathbf{z}_k^q$ is the regular solution of the system

$$\mathbf{z}_k^{q''} = \frac{1}{\delta^2}\mathbf{f}_{-2}\mathbf{z}_k^q + \delta^{l_n+1} \sum_{r=r_0}^{R} \mathbf{A}_{q-2,r} d_{q-2,r} \tag{3.7.23}$$

which on components reads

$$[\mathbf{z}_k^q]_i'' = \frac{l_i(l_i+1)}{\delta^2}[\mathbf{z}_k^q]_i + \delta^{l_n+q-1} \sum_{r=r_0}^{R} [\mathbf{A}_{q-2,r}]_i d_{0r}. \tag{3.7.24}$$

It remains to show that $\mathbf{z}_k^q$ is of a form similar to that of $\mathbf{z}_{k-1}^q$, that is

$$\mathbf{z}_k^q = \delta^{l_n+1} \sum_r \mathbf{a}_{qr}(k) d_{qr}, \tag{3.7.25}$$

and the way we choose consists of effectively constructing the $\mathbf{a}_{q,r}(k)$ and finding the limits of summation over r. We introduce the parameter $\varepsilon = l_i - l_n - q$ and the threshold $\eta > 0$ which, for computational convenience, must be a small number (see below).

Three cases are distinguished

Case (a): $|\varepsilon| > \eta$

From eq. (3.7.24) we obtain after some algebra that the coefficients $[\mathbf{a}_{qr}(k)]_i$ are given by the following recurrence formula

$$[\mathbf{a}_{qr}(k)]_i = \frac{1}{(q+l_n+1)(q+l_n) - l_i(l_i+1)} \{[\mathbf{A}_{q-2,r}]_i - (r+1)(r+2)\theta(R-r-2)[\mathbf{a}_{q,r+2}(k)]_i - (2q+2l_n+1)(r+1)\theta(R-r-1)[\mathbf{a}_{q,r+1}(k)]_i\} \tag{3.7.26}$$

with $r = R, R-1, \ldots, r_0$. The limits of the sum over r are then r_0 and R.

Case (b): $\varepsilon = 0$

In this case eq. (3.7.26) cannot be applied simply because the denominator of its front factor vanishes. Eq. (3.7.24) which now reads

$$[\mathbf{z}_k^q]_i'' = \frac{(l_n+q)(l_n+q+1)}{\delta^2}[\mathbf{z}_k^q]_i + \delta^{l_n+q-1}\sum_{r=0}^{R}[\mathbf{A}_{q-2,r}]_i d_{or} \quad (3.7.27)$$

yields the following recurrence relations for $[\mathbf{a}_{qr}(k)]_i$:

$$[\mathbf{a}_{qr}(k)]_i = \frac{1}{(2l_n+2q+1)r}\{[\mathbf{A}_{q-2,r-1}]_i - r(r+1)\,\theta(R-r)\,[\mathbf{a}_{q,r+1}(k)]_i\}, \quad (3.7.28)$$

with $r = R+1, R, \ldots, r_0+2, r_0+1$. The limits of the sum over r are thus r_0+1 and R_0+1.

Case (c): $0 < |\varepsilon| \leqslant \eta$.

Let us denote $\bar{q} = q + \varepsilon$. Eq. (3.7.24) becomes

$$[\mathbf{z}_k^q]_i'' = \frac{(l_n+\bar{q})(l_n+\bar{q}+1)}{\delta^2}[\mathbf{z}_k^q]_i + \delta^{l_n+\bar{q}-1-\varepsilon}\sum_{r=r_0}^{R}[\mathbf{A}_{q-2,r}]_i\, d_{or}. \quad (3.7.29)$$

It has exactly the same form as eq. (3.7.27) except for the factor $\delta^{-\varepsilon}$ in the inhomogeneous term. If, however, use is made of the series expansion of $\delta^{\pm\varepsilon}$ in terms of d_{om},

$$\delta^{\pm\varepsilon} = \sum_{m=0}^{\infty}\frac{(\pm\varepsilon)^m}{m!}d_{om} \quad (3.7.30)$$

and when ε is sufficiently small that only a few terms are numerically relevant (up to M, say), then eq. (3.7.29) becomes

$$[\mathbf{z}_k^q]_i'' = \frac{(l_n+\bar{q})(l_n+\bar{q}+1)}{\delta^2}[\mathbf{z}_k^q]_i + \delta^{l_n+\bar{q}-1}\sum_{r=r_0}^{R+M}[\mathbf{B}_{q-2,r}]_i d_{or} \quad (3.7.31)$$

with

$$\mathbf{B}_{q-2,r} = \sum_{m=0}^{r-r_0} \frac{(-\varepsilon)^m}{m!} \mathbf{A}_{q-2,r-m} \theta(m - r + R)\theta(M - m). \tag{3.7.32}$$

Eq. (3.7.31) is of form (3.7.27) except for the coefficients **B** in place of **A**; also the summation over r now runs from r_0 to $R + M$. Hence $[\mathbf{z}_k^q]_i$ is of the form

$$[\mathbf{z}_k^q]_i = \delta^{l_n+1+\varepsilon} \sum_{r=r_0+1}^{R+M+1} [\mathbf{b}_{qr}]_i d_{qr} \tag{3.7.33}$$

where $[\mathbf{b}_{qr}]_i$ results by the analogue of eq. (3.7.28),

$$[\mathbf{b}_{qr}]_i = \frac{1}{(2l_n + 2\bar{q} + 1)r} \{[\mathbf{B}_{q-2,r-1}]_i - r(r+1)\theta(R+M-r)[\mathbf{b}_{q,r+1}]_i\},$$

$$r = R + M + 1, \quad R + M, \ldots, r_0 + 1. \tag{3.7.34}$$

Eq. (3.7.33) is finally reconverted into the standard form (3.7.25) by means of the expansion (3.7.30). After some algebra we find that the sum over r in eq. (3.7.25) is now between $r_0 + 1$ and $R + 2M + 1$ and the coefficients $\mathbf{a}_{qr}(k)$ are given by

$$[\mathbf{a}_{qr}(k)]_i = \sum_{m=0}^{r-r_0-1} \frac{\varepsilon^m}{m!} [\mathbf{b}_{q,r-m}]_i \theta(m - r + R + M + 1)\theta(M - m). \tag{3.7.35}$$

The proof is thus completed.

For illustration results at $\delta = 1$ are reported in table 14 from the system (3.7.16) of $N = 2$ equations with

$$\mathbf{f}_{-2} = \begin{pmatrix} l_1(l_1+1) & 0 \\ 0 & l_2(l_2+1) \end{pmatrix}, \quad \mathbf{f}_{-1} = \begin{pmatrix} 3 & 2 \\ 2 & 3 \end{pmatrix}, \quad \mathbf{f}_0 = \begin{pmatrix} 2 & 3 \\ 3 & 2 \end{pmatrix},$$

$$\mathbf{f}_1 = \begin{pmatrix} 2 & 1 \\ 1 & 2 \end{pmatrix}, \quad \mathbf{f}_2 = \begin{pmatrix} 1 & 2 \\ 2 & 1 \end{pmatrix},$$

$l_1 = 1$ and (i) $l_2 = 2$; (ii) $l_2 = 2 + 9 \cdot 10^{-3}$. The computation was carried out on a computer IBM 370/135 in simple precision. The used parameters are $\eta = 10^{-2}$ and $M = 2$ so that the computation meets only the situations (a) and (b) for (i) while all the three situations for (ii).

Table 14

Numerical values at $\delta = 1$ of the two regular independent solutions $y_i^{(1)}$ and $y_i^{(2)}$ of the test system, their first derivatives and the corresponding errors, for various values of k, the number of perturbative corrections included. (Upper entry $i = 1$, lower entry $i = 2$)

k	Solutions		Derivatives		Errors *	
	$y^{(1)}$	$y^{(2)}$	$y^{(1)\prime}$	$y^{(2)\prime}$	$\Delta^{(1)}$	$\Delta^{(2)}$
$l_1 = 1,\ l_2 = 2$						
0	1.00000	0.00000	2.00000	0.00000	−8.0(0)	−8.0(0)
	0.00000	1.00000	0.00000	3.00000	−8.0(0)	−8.0(0)
3	2.78000	0.83768	10.08875	4.77992	−2.4(0)	−1.6(0)
	1.31096	2.21169	7.75833	9.40011	−2.4(0)	−1.6(0)
6	2.80088	0.86142	10.35040	5.00781	6.7(−3)	−5.7(−3)
	1.33077	2.23694	8.02139	9.64111	6.7(−3)	−5.7(−3)
9	2.80078	0.86145	10.34939	5.00828	1.7(−5)	−2.2(−6)
	1.33066	2.23698	8.02033	9.64160	1.7(−5)	−2.2(−6)
$l_1 = 1,\ l_2 = 2.009$						
0	1.00000	0.00000	2.00000	0.00000	−8.0(0)	−8.0(0)
	0.00000	1.00000	0.00000	3.00900	−8.0(0)	−8.0(0)
3	2.78321	0.83241	10.10579	4.75704	−2.4(0)	−1.6(0)
	1.32087	2.20659	7.80484	9.39101	−2.4(0)	−1.6(0)
6	2.80484	0.85586	10.37322	4.98252	6.3(−3)	−5.6(−3)
	1.34152	2.23156	8.07433	9.62957	6.3(−3)	−5.6(−3)
9	2.80474	0.85590	10.37226	4.98299	1.6(−5)	−2.1(−6)
	1.34142	2.23160	8.07331	9.63005	1.6(−5)	−2.1(−6)

* Notation $a(n)$ means $a \cdot 10^n$

For each k the numerical values at $\delta = 1$ of the two regular, independent solutions $\mathbf{y}^{(1)}$ and $\mathbf{y}^{(2)}$ are given,

$$y_i^{(n)}(1) = [\tilde{\mathbf{y}}^{(n)}(1)]_i + [\mathbf{y}_1^{(n)}(1)]_i + \ldots + [\mathbf{y}_k^{n}(1)]_i, \quad i = 1, 2, \quad n = 1, 2,$$

of their derivatives, and also of the deviations

$$\Delta_i^{(n)} \equiv y_i^{(n)\prime\prime}(1) - l_i(l_i + 1)y_i^{(n)}(1) - \sum_{p=-1}^{2} \sum_{j=1}^{2} [\mathbf{f}_p(1)]_{ij} y_j^{(n)}(1).$$

The upper and lower entries correspond to $i = 1$ and to $i = 2$, respectively. It is seen that the first three perturbations do not decrease significantly the deviations Δ, which remain very large. When the perturbation order is further increased the deviations decrease faster and faster with each new order. As a matter of fact, for $k > 6$ the decrease is of about one order of magnitude at each new step of the perturbation approach.

This behaviour is natural because the range (0, 1] is so large in this example that the size of the perturbation is also substantial; in particular, when $\delta = 1$ it becomes larger than all the components of the very reference matrix weight $\mathbf{f}_{-2}$. However, the user can easily avoid such situations if smaller value is assigned to a in the original equation. In fact, the perturbation series will converge faster when a is decreased because the reference matrix weight $\mathbf{f}_{-2}$ is *invariant* to the transformation $\delta = x/a$ while the perturbation *decreases* with a.

The method presented in this section was developed in [37]. Its program is part of package [34].

3.8. A WAY TO IMPROVE THE ACCURACY OF CLASSICAL ALGORITHMS

It was mentioned in section 3.1 that the derivation of the classical algorithms is based on the general belief that the solution of any differential equation is well approximated by piecewise polynomials. It was also mentioned there that in many cases such a belief is justified only if the corresponding interval partition is sufficiently fine; in particular the more oscillating the solution, the narrower the steps of the partition.

It follows that the efficiency of the classical methods will be certainly improved if their algorithms are devised on a modified, more relaxed basis, namely that the piecewise representation of the solution is not a polynomial but some other function appropriately chosen. The expression of such function obviously depends on the shape of the solution and hence some a priori qualitative knowledge on the solution is of major importance.

In most cases such knowledge cannot be attained. It becomes attainable only for some special differential equations and the second-order equation

$$y'' = g(x)y, \tag{3.8.1}$$

discussed in several sections of this chapter, is one of them. This permits to illustrate the above ideas by showing how the knowledge acquired by means of constant perturbation technique of section 3.4 can be used to improve the efficiency of a widely used classical method, the two-step method of Numerov.

In section 3.4 the exact propagators $u(\delta)$ and $v(\delta)$ were first approximated by the reference propagators $\tilde{u}(\delta) = \xi(F, \delta)$ and $\tilde{v}(\delta) = \eta_0(F, \delta)$ and each of these is a linear combination of the functions $\exp(\pm F^{1/2}\delta)$. Successive perturbation corrections add terms containing the functions $\eta_1(F, \delta)$, $\eta_2(F, \delta)$, $\eta_3(F, \delta)$, ..., and it is easily seen that $\eta_1(F, \delta)$ is a linear combination of $\exp(\pm F^{1/2}\delta)$, and $\delta \exp(\pm F^{1/2}\delta)$, $\eta_2(F, \delta)$ is constructed from $\exp(\pm F^{1/2}\delta)$, $\delta \exp(\pm F^{1/2}\delta)$ and $\delta^2 \exp(\pm F^{1/2}\delta)$, etc. It can be equivalently said that the exact propagators (and hence the piecewise representation of the exact solution) are actually linear combinations of terms of the form $\delta^p \exp(\pm F^{1/2}\delta)$, $p = 0, 1, 2, \ldots$

The standard Numerov algorithm is the two-step case of the Cowell algorithm.

It reads

$$\bar{y}_{k+1} - 2\bar{y}_k + \bar{y}_{k-1} = h^2[\beta_0(f(x_{k+1}, \bar{y}_{k+1}) + f(x_{k-1}, \bar{y}_{k-1})) + \\ + \beta_1 f(x_k, \bar{y}_k)] \tag{3.8.2}$$

with $\beta_0 = \dfrac{1}{12}$ and $\beta_1 = \dfrac{5}{6}$, see eqs (1.11.16) or (1.13.24). The values of these constants resulted from the condition that the associated two-step differential-difference equation

$$z(x + h) - 2z(x) + z(x - h) = h^2[b_2 z''(x + h) + b_1 z''(x) + \\ + b_0 z''(x - h)] \tag{3.8.3}$$

is identically satisfied by the power functions $z(x) = x^p, p = 0, 1, 2, \ldots$ up to the highest possible p. This is a particular form of problem (3.1.16)–(3.1.17) and its solution is that the maximal p is $p = 5$ and this is achieved with $b_2 = b_0 = \beta_0$ and $b_1 = \beta_1$.

We continue to assume that $b_2 = b_0 \equiv \beta_0$ and $b_1 \equiv \beta_1$ but now think of β_0 and β_1 as being functions of h which must be determined. We thus write eq. (3.8.3) as

$$z(x + h) - 2z(x) + z(x - h) = h^2[\beta_0(h)(z''(x + h) + z''(x - h)) + \\ + \beta_1(h) z''(x)] \tag{3.8.4}$$

and search for the expressions of $\beta_0(h)$ and of $\beta_1(h)$ such as eq. (3.8.4) is identically satisfied by functions of the form $z(x) = x^q \exp(\pm F^{1/2}x)$, $q = 0, 1, \ldots$ up to the highest possible q, [23, 38]. (General problems of this type were investigated by Lyche [39].)

The result is that the maximal q is $q = 1$ and this is attained with the functions

$$\beta_0(h) = \begin{cases} \dfrac{1}{Z}\left(1 - \dfrac{1}{p}\tanh p\right) & \text{for } Z > \varepsilon \\ \dfrac{1}{12}\left(1 - \dfrac{1}{10}Z + \dfrac{17}{1680}Z^2 - \dfrac{31}{30240}Z^3 + \dfrac{691}{6652800}Z^4\right) & \text{for } -\varepsilon \leqslant Z \leqslant \varepsilon \\ \dfrac{1}{Z}\left(1 - \dfrac{1}{p}\tan p\right) & \text{for } Z < -\varepsilon \end{cases} \tag{3.8.5}$$

and

$$\beta_1(h) = \begin{cases} \dfrac{2}{Z}\left(\dfrac{1}{p}\tanh p \cdot \cosh(2p) - 1\right) & \text{for } Z > \varepsilon \\ \dfrac{5}{6}\left(1 + \dfrac{1}{50}Z + \dfrac{1}{336}Z^2 - \dfrac{29}{151200}Z^3 + \dfrac{139}{6652800}Z^4\right) & \text{for } -\varepsilon \leqslant Z \leqslant \varepsilon \\ \dfrac{2}{Z}\left(\dfrac{1}{p}\tan p \cdot \cos(2p) - 1\right) & \text{for } Z < -\varepsilon \end{cases} \tag{3.8.6}$$

where $Z = Fh^2$ and $p = \dfrac{1}{2}|Z|^{1/2}$. ε is a threshold value which depends on the word length used in the computation. As a matter of fact, if an IBM 370/135 computer is used in double precision arithmetic (this means about 16 exact decimal figures in the mantissa) then $\varepsilon = 10^{-2}$ is recommended.

We note in passing that these expressions of $\beta_0(h)$ and $\beta_1(h)$ also ensure that the power functions 1 and x identically satisfy eq. (3.8.4).

The algorithm (3.8.2) in which these $\beta_0(h)$ and $\beta_1(h)$ are used will be called a *modified* Numerov algorithm. By its very construction it will be particularly useful to solve second-order equations of the form

$y'' = f(x, y)$ where solutions are well approximated by piecewise linear combinations of the functions $\exp(\pm F^{1/2}\delta)$ and $\delta \cdot \exp(\pm F^{1/2}\delta)$ and, as we have seen before, this is the case when $f(x, y) = g(x)y$ and F is defined as the average value of g on the subinterval of interest. As a matter of fact, F defined in eq. (3.4.3) referred to each step of the partition. For the modified Numerov algorithm the average F will refer to at least two adjacent steps.

The advantages of the modified over the standard Numerov method are readily seen on the corresponding local error expressions.

The local error for the standard Numerov method (N) is given by eq.(1.11.24) with the coefficients given in Table 8 for Cowell method with $r = 2$

$$e_{k+1}^{N} \simeq -\frac{1}{240} h^6 y^{(6)}(x_k); \tag{3.8.7}$$

$y^{(6)}(x_k)$ is explicited in terms of the very equation (3.8.1), thus

$$e_{k+1}^{N} \simeq -\frac{1}{240} h^6[(g^{(4)} + 7gg'' + 4g'^2 + g^3)y + (4g^{(3)} + 6gg')y'] \tag{3.8.8}$$

where y, g and their derivatives are taken at $x = x_k$.

For the modified Numerov algorithm (MN) the local error is

$$\begin{aligned} e_{k+1}^{MN} &\simeq -\frac{1}{240} h^6(y^{(6)}(x_k) - 2Fy^{(4)}(x_k) + F^2 y''(x_k)) \\ &= -\frac{1}{240} h^6\{[g^{(4)} + 7g''\Delta + 4g'^2 + \Delta^3 + F(5g'' + \Delta^2)]y + \\ &\quad + (4g^{(3)} + 6g'\Delta + 2Fg')y'\} \end{aligned} \tag{3.8.9}$$

with $\Delta = g(x_k) - F$.

We see that the front factor $-h^6/240$ is common in the two formulas but the remaining factors differ. The magnitude of the difference depends very much on the magnitude of $g(x_k)$. For illustration we think of the case $g(x) = V(x) - E$ where $V(x)$ is a given function and E a numerical parameter. Since $F = \tilde{V} - E$ where $\tilde{V}$ is some average of $V(x)$ around x_k we can admit that $F \simeq g(x_k)$, hence $\Delta \simeq 0$. We distinguish two cases. If $E \simeq V(x_k)$ we have $g(x_k) \simeq 0$ and the two factors are numerically close to each other. The two versions are then expected to furnish results of comparable accuracy. If E is such that $|V(x_k) - E| \gg 0$ i.e. $|g(x_k)| \simeq |F| \gg 0$ the asymptotic expressions of the errors should be used, viz.

$$e_{k+1}^{N} \simeq -\frac{1}{240} h^6 [g(x_k)]^3 y(x_k), \tag{3.8.10}$$

$$e_{k+1}^{MN} \simeq -\frac{1}{240} h^6 g(x_k)[5g''(x_k)y(x_k) + 2g'(x_k)y'(x_k)] \tag{3.8.11}$$

and thus the error of N increases as the *cube* of $g(x_k)$ while the error of MN is only *linear* in $g(x_k)$. For this reason the modified Numerov algorithm will be definitely more accurate that the standard algorithm in the subintervals of the equation domain on which $|g(x)|$ is large.

Such behaviour is illustrated on the following numerical example:

$$V(x) = u_0/(1 + q) + u_1 q/(1 + q)^2, \; q = \exp((x - X_0)/a), \tag{3.8.12}$$

$u_0 = -50$, $a = 0.6$, $X_0 = 7$ and $u_1 = -u_0/a$; E is in the range $[1, 1000]$. Eq. (3.8.1) is solved on $[x_{\min} = 0, x_{\max} = 15]$ with the boundary conditions

$$y(x_{\min}) = 0, \; E^{1/2}\sin(E^{1/2}x_{\max})y(x_{\max}) + \cos(E^{1/2}x_{\max})y'(x_{\max}) = 0. \tag{3.8.13}$$

This is an eigenvalue problem (but notice that it is not of orthodox Sturm-Liouville form because the second boundary condition explicitly depends on the parameter E) and it has eleven eigenvalues in the quoted range of E. It is solved numerically by means of three methods, N, MN and CPM(2) of section 3.4. For MN the reference average constant F is taken as

$$F = \begin{cases} -50-E & \text{for } x \in [0, 6.5] \\ -E & \text{for } x > 6.5. \end{cases} \tag{3.8.14}$$

Since each of the three methods are methods for initial value problems a shooting procedure is used to solve for eigenvalues, see section 4.4.

The accuracy in eigenvalues directly reflects the accuracy of the numerical methods and detailed results for three eigenvalues are collected in table 15.

Table 15

Deviations of the computed eigenvalues of problem $y'' = (V(x) - E)y$, with $V(x)$ given by (3.8.12) and the boundary conditions by (3.8.13), from the exact ones, in 10^{-6} units, for the two versions of the Numerov scheme (N and MN) and for CPM(2). Empty areas in N indicate that the corresponding deviations are larger than the format allowed in the table. An IBM 370/135 computer was used in double precision

h	N	MN	h	CPM(2)
$E_{exact} = 53.5888719$				
1/16	−230727	−1205	1/2	−5615
1/32	−14110	−75	1/4	0.4
1/64	−879	−4.5		
1/128	−56	−0.3		
$E_{exact} = 341.4958743$				
1/16		−35267	1/2	5470
1/32	−7536068	−1979	1/4	−0.6
1/64	−436825	−121		
1/128	−27127	−7		
$E_{exact} = 989.7019159$				
1/16		−394003	1/2	4786
1/32		−16064	1/4	−0.8
1/64		−927		
1/128	−1637376	−53		

It is seen that CPM(2) gives the best results and it is followed in order by MN and N. It is also seen that the dependence of the experimental errors against E is different for the three methods; the errors remain in the same limits for CPM(2) (and thus they are consistent with a E^0 dependence) but asymptotically increase as E^2 in case of MN and as E^4 in the case of N.

Quite strange may seem to the reader the fact that powers of E resulted from this experiment are larger by one unit than the ones which would be normally expected on the basis of the error analysis for CPM(2) made on p. 215 (which shows that, due to the neglect of the residual perturbation, the local error decreases as Z^{-1} or, equivalently, as E^{-1}) and on eqs (3.8.10) and (3.8.11). Careful examination shows however that this discrepancy has no connection with the numerics; it is entirely due to the peculiarity of the very mathematical problem, more exactly to the fact that the second boundary condition explicitly depends on E. The discrepancy will be absent if both boundary conditions are E independent.

Finally note an earlier modification of the Numerov method which is due to Raptis and Allison [40]. In their approach eq. (3.8.4) is required to be identically satisfied by the functions $\exp(\pm \omega_1 x)$ and $\exp(\pm \omega_2 x)$. ω_1 and ω_2 are finally identified as $\omega_1 = F^{1/2}$ and $\omega_2 = 0$ respectively and the resultant values of $\beta_0(h)$ and $\beta_1(h)$ are

$$\beta_0(h) = \begin{cases} \dfrac{1}{Z}\left(1 - \dfrac{Z}{4\sinh^2 p}\right) & \text{for } Z > \varepsilon \\ \dfrac{1}{12}\left(1 - \dfrac{1}{20}Z + \dfrac{1}{504}Z^2 + \dfrac{1}{43200}Z^3\right) & \text{for } -\varepsilon \leqslant Z \leqslant \varepsilon \\ \dfrac{1}{Z}\left(1 + \dfrac{Z}{4\sin^2 p}\right) & \text{for } Z < -\varepsilon \end{cases} \qquad (3.8.15)$$

$$\beta_1(h) = 1 - 2\beta_0(h) \qquad (3.8.16)$$

Algorithm (3.8.2) with coefficients (3.8.15) and (3.8.16) integrates exactly the functions $1, x, x^2, x^3$ and $\exp(\pm F^{1/2}x)$. When applied to eq. (3.8.1) its results are better than of N but worse than of MN. When $|g(x)|$ is large the local error is *quadratic* in $g(x_k)$.

REFERENCES

1. R. Bellman, *Perturbation Techniques in Mathematics, Physics and Engineering*, Dover Publications, Inc., New York, 1966.
2. L. Gr. Ixaru, *An algebraic solution of the Schrödinger equation*, Internal Report IC/69/6, International Centre for Theoretical Physics, Trieste 1969; *The algebraic approach to the scattering problem*, Internal Report IC/69/7, International Centre for Theoretical Physics, Trieste, 1969.
3. R. G. Gordon, *New method for constructing wavefunctions for bound states and scattering*, J. Chem. Phys., **51**, 14 (1969).
4. L. Gr. Ixaru, *On the exact solution of the external problem in the theory of alpha decay*, MPIH-1969-V14, Max-Planck Institut für Kernphysik, Heidelberg, 1969.
5. L. Gr. Ixaru, *A new method for solving linear inhomogeneous second order differential equations*, Rev. Roum. Math. Pures et Appl., **19**, 199 (1974).

6. E. Kamke, *Differentialgleichungen Lösungsmethoden und Lösungen*, vol. 1. Akademische Verlaggeselschaft Geest und Portig K.-G, Leipzig, 1956.
7. L. Gr. Ixaru and H.-J. Kreuzer, *On the accuracy of JWKB method in alpha-decay*, Energia Nucl., **17**, 62 (1970).
8. W. Leighton, *Upper and lower bounds for eigenvalues*, J. Math. Anal. Appl., **35**, 381 (1971); *On approximating conjugate, focal and σ-points for linear differential equations of second order*, Ann. Math. Pura. Appl., **107**, 373 (1976).
9. J. Canosa and R. Gomes de Oliveira, *A new method for the solutions of the Schrödinger equation*, J. Comput. Phys., **5**, 188 (1970).
10. J. Canossa, *Numerical solution of Mathieu's equation*, J. Comput. Phys., **7**, 255 (1971).
11. L. Gr. Ixaru, *The error analysis of the algebraic method for solving the Schrödinger equation*, J. Comput. Phys., **9**, 159 (1972).
12. S. A. Pruess, *Solving linear boundary value problems by approximating the coefficients*, Math. Comp., **27**, 551 (1973); *Estimating the eigenvalues of Sturm-Liouville problems by approximating the differential equation*, SIAM J. Numer. Anal., **10**, 55 (1973).
13. Gh. Adam, *Piecewise analytical perturbation series methods in quantum mechanics*, Ph. D. Thesis, Bucharest, 1979 (unpublished).
14. M. D. Smooke, *Piecewise analytical perturbation series solutions of the radial Schrödinger equation — one dimensional case*, preprint SAND 80-8687, Sandia Laboratories, Albuquerque, New Mexico, 1980; SIAM J. Sci. Statist. Comput., **3**, 195 (1982).
15. A. L. Andrew, *Computation of higher Sturm-Liouville eigenvalues*, an invited lecture at the 11th Manitoba Conference on Numerical Mathematics and Computing, Winnipeg, 1981; Congressus Numerantium, **34**, 3 (1982).
16. Gh. Adam, L. Gr. Ixaru, and A. Corciovei, *A first order perturbative numerical method for the solution of the radial Schrödinger equation*, J. Comput. Phys., **22**, 1 (1976).
17. L. Gr. Ixaru and Gh. Adam, *A second order perturbative numerical method for the solution of the radial Schrödinger equation*, Rev. Roum. Phys., **24**, 723 (1979).
18. L. Gr. Ixaru, *Simple procedure to compute accurate energy levels of an anharmonic oscillator*, Phys. Rev. D, **25**, 1557 (1982).
19. M. D. Smooke, *Error estimates for piecewise perturbation series solutions of the radial Schrödinger equation: I. one-dimensional case*, preprint SAND 80-8611, Sandia Laboratories, Albuquerque, New Mexico, 1980; SIAM J. Numer. Anal., **20**, 279 (1983).
20. R. G. Gordon, *Quantum scattering using piecewise analytic solutions*, in Methods in Computational Physics, vol. 10, Academic Press, 1971, p. 81.
21. J. P. Riehl, D. J. Diestler and A. F. Wagner, *Comparison of perturbative and direct-numerical-integration techniques for the calculation of phase shifts for elastic scattering*, J. Comput. Phys., **15**, 212 (1974).

22. L. Gr. Ixaru, M. I. Cristu and M. S. Popa, *Choosing step sizes for perturbative methods of solving the Schrödinger equation*, J. Comput. Phys., **36**, 170 (1980).
23. L. Gr. Ixaru, *Perturbative numerical methods to solve the Schrödinger equation*, Computer Phys. Commun., **20**, 97 (1980).
24. G. N. Watson, *A treatise on the theory of Bessel functions*, Cambridge Univ. Press., Cambridge, England, 1962.
25. M. Abramowitz and I. A. Stegun, eds., *Handbook of Mathematical Functions*, Appl. Math. Ser., **55**, Nat. Bur. Stand. U.S. Govt. Print. Off. Washington, D.C. 1965.
26. R. G. Gordon, *Error bounds in equilibrium statistical mechanics*, J. Math. Phys., **9**, 655 (1968).
27. Z. A. Luthey, *Piecewise analytical solution method for the radial Schrödinger equation*, Thesis, Harvard University, 1974.
28. L. D. Tolsma, *A fast method for nuclear coupled-channel calculations including Coulomb excitation*, J. Comput. Phys., **17**, 384 (1975).
29. A. Rosenthal and R. G. Gordon, *Piecewise analytic wavefunctions for bound-states and scattering*, J. Chem. Phys., **69**, 1621 (1976).
30. M. H. Alexander, *Stable solution of close-coupled equations for ro-vibrational inelastic collisions*, J. Comput. Phys., **20**, 248 (1976).
31. Z. Schulten, D. G. M. Anderson and R. G. Gordon, *An algorithm for the evaluation of the complex Airy functions*, J. Comput. Phys., **31**, 60 (1979).
32. M. H. Alexander, *Close-coupling studies of rotationally inelastic HF-HF collisions at hyperthermal energies*, J. Chem. Phys., **73**, 5153 (1980).
33. L. Gr. Ixaru, *A highly convergent perturbative method for the solution of systems of coupled equations arising from the Schrödinger equation*, J. Comput. Phys., **36**, 182 (1980).
34. * * * *Numerical solution of systems of coupled equations arising from the Schrödinger equations. Subroutines: RANGE, PERTH, LI and LIX; User's guide*, Central Institute of Physics, Bucharest, 1980.
35. M. Beiner and P. Gara, *Direct numerical integration of coupled one-dimensional Schrödinger equation*, Comput. Phys. Commun., **4**, 1 (1972).
36. R. G. Gordon, *Program 187, Quantum Chemistry Program Exchange (QCPE)*, Indiana University, Bloomington, In.
37. L. Gr. Ixaru and M. Micu, *On the perturbative numerical approach to the Schrödinger equation with singular potential*, in Topics in Theoretical Physics (I. A. Dorobanţu, Ed.) p. 259, Central Institute of Physics, Bucharest, 1978.
38. L. Gr. Ixaru and M. Rizea, *A Numerov-like scheme for the numerical solution of the Schrödinger equation in the deep continuum spectrum of energies*, Comput. Phys. Commun., **19**, 23 (1980).
39. T. Lyche, *Chebyshevian multistep methods for ordinary differential equations* Numerische Math., **19**, 65 (1972).
40. A. Raptis and A. C. Allison, *Exponential-fitting methods for the numerical solution of the Schrödinger equation*, Comput. Phys. Commun., **14**, 1 (1978).

Chapter 4

Boundary value problems

In the previous chapters methods have been devised for the numerical solution of the initial value problem for differential equations, a problem also referred to as the Cauchy problem.

In many practical cases problems occur in which the subsidiary conditions are imposed at several *distinct* points. Naturally, such problems may occur only in connection with *systems* of first-order equations or with differential equations of higher order whereas for a single equation of the first order only *one* condition is necessary and sufficient to fix the solution.

The subsidiary conditions are typically imposed at the boundaries a and b of the domain of the equation and for this reason such problems are called *boundary value problems.*

4.1. GENERALITIES

The simplest boundary value problem is perhaps

$$y'' = f(x, y), \; x \in [a, b], \; y(a) = A, \; y(b) = B \tag{4.1.1}$$

where A and B are given constants.

No matter how simple this problem looks like its mathematical theory is much more complicated than the theory of the initial value problem associated to the same equation $y'' = f(x, y)$. In particular the *existence* of a solution of problem (4.1.1) depends not only on the properties of $f(x, y)$ but also on the values of the constants A and B. We consider for illustration the simple case $f(x, y) = -\pi^2 y$, $a = 0$, $b = 1$. If $A = B = 0$, a solution is $y(x) = C\sin(\pi x)$ where C is an arbitrary constant (thus the solution exists but it is not unique);

if $A = 0$ and $B = 1$ problem (4.1.1) has *no* solution. For contrast we recall that the *only* condition that the initial value problem associated to equation $y'' = f(x, y)$ admits a unique solution refers to the function $f(x, y)$ and this is that $f(x, y)$ must satisfy the Lipschitz condition.

A particular case in which problem (4.1.1) admits unique solution for *arbitrary* A and B is when $f_y(x, y) \geqslant 0$. In this case problem (4.1.1) is called a boundary value problem of class M, see [1, p. 347].

One possible strategy to solve numerically a boundary value problem is to reduce it to a sequence of initial value problems and as an illustration we refer to problem (4.1.1) to which the following initial value problem

$$y'' = f(x, y),\ x \in [a, b],\ y(a) = A,\ y'(a) = \alpha \tag{4.1.2}$$

is associated. $f(x, y)$ and A are the same as in eq. (4.1.1) while α is an arbitrary parameter. The solution $y(x; \alpha)$ of this problem can be calculated without difficulty by any of the methods developed in the previous chapters. This $y(x; \alpha)$ will be also a solution of the original boundary value problem only if it happens that the value of α is some $\bar{\alpha}$ such as the value of $y(x; \bar{\alpha})$ at $x = b$ equals B. In other words $y(x; \alpha)$ is a solution of problem (4.1.1) when $\alpha = \bar{\alpha}$ where $\bar{\alpha}$ is the root of the algebraic (generally nonlinear) equation $G(\alpha) \equiv y(b; \alpha) - B = 0$.

In practice successive values $\alpha_0, \alpha_1, \alpha_2, \ldots$ are tried and $G(\alpha_0), G(\alpha_1), G(\alpha_2), \ldots$ are correspondingly constructed until it is observed that $G(\alpha_i)$ and $G(\alpha_{i+1})$ have different signs. $\bar{\alpha}$ therefore lies somewhere between α_i and α_{i+1} and it is further located with the desired accuracy by some standard rootfinder technique as, for instance, by bisection or by *regula falsi* method [2]. On the opposite, if sufficient computational evidence is accumulated to indicate that $G(\alpha)$ has no root the conclusion will be that problem (4.1.1) has no solution.

The above procedure which uses numerical methods for initial value problems plus some rootfinder technique is known as a *shooting procedure* and it is described in many books, see for instance [3].

Although the shooting procedure is quite general and easily applicable, some other, more direct procedures may be preferable in certain particular cases.

This will be illustrated in the two forthcoming sections in which we refer to a boundary value problem which is more relaxed than eq. (4.1.1) with respect to both the form of the boundary conditions and the number of arguments in f. Specifically f is allowed to depend also on y' while the two boundary conditions are taken as linear combinations of the solution and of its first derivative. The only restriction is that f must be linear in y and in y'. Altogether the problem is

$$y'' + a(x)y' + b(x)y + c(x) = 0, \quad x \in [a, b] \tag{4.1.3}$$

with the boundary conditions

$$p_0y'(a) + q_0y(a) = u_0, \; p_1y'(b) + q_1y(b) = u_1 \tag{4.1.4}$$

where $a(x)$, $b(x)$ and $c(x)$ are well-behaved functions and p_0, p_1, q_0, q_1, u_0 and u_1 are given numbers so that $|p_0| + |q_0| \neq 0$ and $|p_1| + |q_1| \neq 0$.

Problems of this type are much more frequent in physics and engineering than problem (4.1.1) and they also received more attention from mathematicians.

4.2. SHOOTING METHODS

As stated, the shooting procedure is a way to solve a boundary value problem by reducing it to a sequence of initial value problems.

In case of problem (4.1.3—4) the mise en scène of such a procedure is greatly facilitated by the special form of eq. (4.1.3). In fact, since the latter is a linear, nonhomogeneous second-order equation, its general solution is the sum of the general solution of the associated homogeneous equation

$$z'' + a(x)z' + b(x)z = 0 \tag{4.2.1}$$

and a particular solution $y^p(x)$ of the original equation.

As for the general solution of eq. (4.2.1), this is a linear combination of two linear independent solutions $z_1(x)$ and $z_2(x)$,

$$z(x) = Az_1(x) + Bz_2(x) \tag{4.2.2}$$

so that altogether

$$y(x) = y^p(x) + Az_1(x) + Bz_2(x) \tag{4.2.3}$$

with arbitrary constants A and B. This $y(x)$ becomes the solution of the quoted boundary value problem if it also satisfies the two boundary conditions in eq. (4.1.4),

$$p_0(y^{p\prime}(a) + Az_1'(a) + Bz_2'(a)) + q_0(y^p(a) + Az_1(a) + Bz_2(a)) = u_0, \tag{4.2.4 a}$$

$$p_1(y^{p\prime}(b) + Az_1'(b) + Bz_2'(b)) + q_1(y^p(b) + Az_1(b) + Bz_2(b)) = u_1 \tag{4.2.4 b}$$

which form an algebraic system of two linear equations for A and B. As a matter of fact, this system can be written in standard form

$$\begin{aligned} \alpha_{11}A + \alpha_{12}B &= \beta_1 \\ \alpha_{21}A + \alpha_{22}B &= \beta_2 \end{aligned} \tag{4.2.5}$$

and, as known from the general theory of linear systems, see for instance [2, chapter 8], it has unique solution if the matrix $\boldsymbol{\alpha}$ is nonsingular that is $\alpha_{11}\alpha_{22} - \alpha_{21}\alpha_{12} \neq 0$. A solution still exists but it is *not unique* when $\alpha_{11}\alpha_{22} - \alpha_{21}\alpha_{12} = 0$ provided that $\beta_2\alpha_{11} - \beta_1\alpha_{21} = 0$ and $\beta_1\alpha_{22} - \beta_2\alpha_{12} = 0$; otherwise there is *no solution.* We can thus state that problem (4.1.3—4) has unique, not unique or no solution at all as it is dictated by system (4.2.4 a, b).

In practice the construction of the solution of the quoted boundary value problem depends on the ability of generating the components $y^p(x)$, $z_1(x)$ and $z_2(x)$. As a rule, some point $X \in [a, b]$ is chosen and six numbers $y_0^p, y_0^{p\prime}, z_{10}, z_{10}', z_{20}, z_{20}'$ are taken to fix the values of the components and of their first derivatives at $x = X$. The functions $y^p(x)$, $z_1(x)$ and $z_2(x)$ are effectively computed by applying any numerical method for initial value problems in each of the two directions $X \to a$ and $X \to b$. The resultant values at a and at b are introduced in eqs (4.2.4 a) and (4.2.4 b) to determine A and B. Only one precaution is required, and this consists of taking the values of z_{10}, z_{10}', z_{20} and z_{20}' such that $z_{10}z_{20}' - z_{10}'z_{20} \neq 0$ since otherwise the resultant $z_1(x)$ and $z_2(x)$ will be linear dependent. [We recall that two solutions of eq. (4.2.1) are linear independent when their wronskian is nonvanishing. Since the values of the wronskian of z_1 and z_2 at any two points x_1 and x_2 are connected by $W(x_2) = W(x_1) \exp\left[-\int_{x_1}^{x_2} a(x)\,dx\right]$ we see that if the wronskian is nonvanishing at the initial point it remains nonvanishing at any other point.]

Two versions are particularly popular in practice namely $X = a$ and $X = b$ since they lead to pure forward integration in the first case, and to pure backward integration in the second.

A third version which is a hybrid of the two previous ones is also frequent. In this version a set of initial conditions is imposed at $X = a$ and the solution components are propagated forward up to some $x_m \in$ $\in [a, b]$; also initial conditions are imposed at $X = b$ and the components are now propagated backward to the same x_m. The solution is then

$$y^F(x) = y^{pF}(x) + A^F z_1^F(x) + B^F z_2^F(x) \quad \text{on } [a, x_m] \tag{4.2.6}$$

and

$$y^B(x) = y^{pB}(x) + A^B z_1^B(x) + B^B z_2^B(x) \quad \text{on } [x_m, b]. \tag{4.2.7}$$

F and B stand for "forward" and "backward", respectively. A^F and B^F satisfy eq. (4.2.4 a) while A^B and B^B satisfy eq. (4.2.4 b).

At $x = x_m$ we require that $y^F = y^B$ and $y^{F'} = y^{B'}$; these conditions furnish two additional linear equations for the unknowns A^F, B^F, A^B and B^B. We thus obtain a system of *four* linear equations which, when solved, yields the value of the four constants. These are introduced in eqs (4.2.6) and (4.2.7) to furnish the solution of the original boundary value problem. This version is usually referred to as "matching in the middle".

In all three versions a direct possibility exists to reduce the amount of computation from *three* initial value problems to only *two*. In fact, what we are merely interested in is not to compute the general solution of eq. (4.1.3) but a family of solutions which automatically satisfy either the first boundary condition when $X = a$, or the second when $X = b$. Such a family is clearly of the form

$$y(x) = y^*(x) + A^* z^*(x) \tag{4.2.8}$$

where y^* and z^* are solutions of the original equation (4.1.3) and of its associated homogeneous equation (4.2.1) with initial conditions $y^*(a)$, $y^{*\prime}(a)$, $z^*(a)$, $z^{*\prime}(a)$ such as

$$p_0 y^{*\prime}(a) + q_0 y^*(a) = u_0 \tag{4.2.9 a}$$

and

$$p_0 z^{*\prime}(a) + q_0 z^*(a) = 0, \ |z^{*\prime}(a)| + |z^*(a)| \neq 0 \tag{4.2.9 b}$$

if $X = a$, or with $y^*(b)$, $y^{*\prime}(b)$, $z^*(b)$ and $z^{*\prime}(b)$ such as

$$p_1 y^{*\prime}(b) + q_1 y^*(b) = u_1 \tag{4.2.10 a}$$

and

$$p_1 z^{*\prime}(b) + q_1 z^*(b) = 0, \quad |z^{*\prime}(b)| + |z^*(b)| \neq 0 \tag{4.2.10 b}$$

if $X = b$. In both cases A^* is a constant which is finally fixed in terms of the other boundary condition on $y(x)$. (Note in passing that now we have only one parameter to be fixed instead of two, A and B, in the previous case.) In the third version we will have *two* constants A^{*F} and A^{*B} and these will be finally fixed by matching in the middle.

All three shooting versions, either in general form, eq. (4.2.3), or in compressed form, eq. (4.2.8), are entirely equivalent from theoretical point of view. Yet, the numerical efficiency may sometimes differ from one version to the other.

Consider, for example, the problem given in [4, p. 60], that is

$$y'' - 11y' - 12y + 12x + 11 = 0, \quad x \in [0, 1] \tag{4.2.11}$$

with the boundary conditions

$$y(0) = 1, \; y(1) = 1 + \exp(-1) = 1.3679\ldots \tag{4.2.12}$$

The general solution of eq. (4.2.11) is

$$y(x) = x + A\exp(-x) + B\exp(12x) \tag{4.2.13}$$

Note in passing that eq. (4.2.11) is a *stiff* equation.*

---

* In general, a system of N first-order equations

$$\mathbf{y}' = \mathbf{f}(x, \mathbf{y})$$

will be called stiff if the eigenvalues of the N by N matrix $\mathbf{F}$ with components $F_{ij} \equiv \dfrac{\partial f_i}{\partial y_j}$ are not all of the same order of magnitude. If eq. (4.1.3) is written as a system, it will consist of two first-order equations with $f_1(x, y_1, y_2) = y_2$ and $f_2(x, y_1, y_2) = -a(x)y_2 - b(x)y_1 - c(x)$, so that

$$\mathbf{F}(x) = \begin{pmatrix} 0 & 1 \\ -b(x) & -a(x) \end{pmatrix}.$$

The eigenvalues of $\mathbf{F}$, that is the solutions of the algebraic equation $\det(\mathbf{F}(x) -$

We take the first version in compressed form (4.2.8) and specifically assume that $X = 0$, $y^*(0) = 1$, $y^{*\prime}(0) = s$, $z^*(0) = 0$, $z^{*\prime}(0) = t$, where s and t are arbitrary numbers and $t \neq 0$. These initial conditions are consistent with the initial conditions (4.2.9 a, b) since $p_0 = 0$, $q_0 = 1$ and $u_0 = 1$. $y(x)$ is of form (4.2.8) with

$$y^*(x) = x + \frac{13 - s}{13} \exp(-x) + \frac{s}{13} \exp(12x) \tag{4.2.14}$$

and

$$z^*(x) = \frac{t}{13} (- \exp(- x) + \exp(12x)). \tag{4.2.15}$$

On using the second boundary condition we find out directly that $A^* = s/t$ and therefore the exact solution of the given boundary value problem is simply $y(x) = x + \exp(-x)$.

It is seen that the family (4.2.8) of candidates for constructing the solution of the given problem consists of three partial solutions x, $\exp(- x)$ and $\exp(12x)$ while in the final result the last partial solution is absent. Moreover, since the last partial solution is numerically dominant in each $y^*(x)$ and $z^*(x)$, the true moderately sized solution of the given boundary value problem appears as a result of the near cancellation of the two components which are very large indeed, in particular at $x = 1$ where the second boundary condition will be imposed to fix A^*. Specifically, to find an accurate solution of problem (4.2.11—12), say to four decimals, we have to accumulate the partial components $y^*(x)$ and $z^*(x)$ to four decimals which means a large number of significant figures.

This phenomenon is illustrated on the left in Table 16, where $y^*(x)$ and $z^*(x)$ were computed numerically by the Fox-Goodwin algorithm

---

$- \lambda(x) \cdot \mathbf{I}) = 0$ ($\mathbf{I}$ is the 2×2 unit matrix), are obviously

$$\lambda^{\pm}(x) = \frac{1}{2} [- a(x) \pm (a^2(x) - 4b(x))^{1/2}]$$

and eq. (4.1.3) will be stiff in the regions of x where $|\lambda^+(x)| \gg |\lambda^-(x)|$ or $|\lambda^-(x)| \gg |\lambda^+(x)|$. In particular, eq. (4.2.11) gives $\lambda^+ = 12$ and $\lambda^- = -1$ and thus it is stiff everywhere on the quoted interval [0, 1].

The numerical solution of stiff equations generally raises some special problems, see e.g. [5]. As a matter of fact, the difficulties to be seen further on in the main text can be interpreted as being just a direct reflection of the stiff character of eq. (4.2.11).

Table 16

Collection of data for the components $y^*(x)$ and $z^*(x)$ and for the solution $y(x)$, eq. (4.2.8), of the boundary value problem (4.2.11). The components were calculated by means of algorithm (4.2.16) in forward and backward direction.

	forward			backward			
x	$y^*(x)$	$z^*(x)$	$y(x)$	x	$y^*(x)$	$z^*(x)$	$y(x)$
0.0	1.0000	0.0000	1.0000	1.0	1.3679	0.0000	1.3679
0.1	0.0000	1.0000	1.0050	0.9	0.0000	1.0000	1.3072
0.2	−3.7156	4.7111	1.0190	0.8	−0.5378	1.3677	1.2501
0.3	−17.8024	18.7501	1.0412	0.7	−0.8685	1.5803	1.1973
0.4	−71.3954	72.1067	1.0707	0.6	−1.1569	1.7644	1.1495
0.5	−275.3834	275.1190	1.1070	0.5	−1.4476	1.9544	1.1072
0.6	−1051.8221	1047.7487	1.1493	0.4	−1.7537	2.1609	1.0710
0.7	−4007.1124	3988.4284	1.1971	0.3	−2.0803	2.3881	1.0414
0.8	−15255.4401	15181.0172	1.2498	0.2	−2.4304	2.6389	1.0192
0.9	−58068.2551	57781.5388	1.3070	0.1	−2.8067	2.9160	1.0051
1.0	−221019.9703	219925.0791	1.3678	0.0	−3.2120	3.2222	1.0000
$A^* = \dfrac{221021.3382}{219925.0791} = 1.004484693$				$A^* = \dfrac{4.2120}{3.2222} = 1.3072$			

(1.13.21) which for eq. (4.2.11) and $h = 0.1$ is

$$0.45\bar{y}_{k+1} - 2.12\bar{y}_k + 1.55\bar{y}_{k-1} + 0.12x_k + 0.11 = 0. \tag{4.2.16}$$

Values of $y^*(x)$ are generated by this algorithm with starting data $y_0^* = y^*(0) = 1$, $y_1^* = y^*(0.1) = 0$ while to obtain the values of $z^*(x)$ the same algorithm with last two terms suppressed is used and with $z_0^* = z^*(0) = 0$, $z_1^* = z^*(0.1) = 1$. As a matter of fact, the knowledge of the values at h is necessary to start algorithm (4.2.16) but for the discussed problem these values are as arbitrary as s and t were above.

The value of A^* is finally fixed from the condition that $y^*(1) + A^*z^*(1) = 1.3679$.

It is important to understand that *all* the figures given in columns y^* and z^* are necessary to provide four decimal accuracy in y although the number of figures needed in A^* for this purpose varies with the size of the components. For instance, four decimals in A^* are sufficient for $y(0.1)$ whereas nine exact decimals are necessary for $y(0.9)$.

We now refer to the second version, that is $X = 1$ and apply the same algorithm backwards; the data are collected on the right in table 16. Although the expressions of $y^*(x)$ and of $z^*(x)$ will be different from those in eqs (4.2.14) and (4.2.15) the structure remains the same, i.e. the three partial solutions x, $\exp(-x)$ and $\exp(12x)$ remain present. The point is that when x runs backwards from $X = 1$ the "unwanted" term $\exp(12x)$ is a decreasing function and the same is valid for x. Only $\exp(-x)$ increases but slowly. Altogether the size of $y^*(x)$ and of $z^*(x)$ slowly varies on $[0, 1]$ and the whole computation becomes practicable even on a five-figure machine. It follows that the second version is to be preferred for the quoted problem.

The third version is particularly suited when the variation of the size of the components is comparable in forward and backward directions. Such a case may appear, for example, in the case of the problem

$$y'' - 4y + x = 0, \; x \in [0, 1], \; y(0) = y(1) = 1 \tag{4.2.17}$$

since the general solution of the differential equation is

$$y(x) = \frac{1}{4} x + A \exp(2x) + B \exp(-2x) \tag{4.2.18}$$

and thus the forward solution essentially increases as $\exp(2x)$ that is as fast as does $\exp(-2x)$ in the backward solution. Naturally, an optimal x_m exists for such cases and its position is to be evaluated in terms of the relative rates of the increase in the forward and backward solutions. In the case of eq. (4.2.17) $x_m = 1/2$ is perhaps optimal.

In principle *any* of the initial value methods presented in the previous chapters may be used to numerically advance the components but some distinction must be made in terms of the explicit expressions of the boundary conditions. The reason is that some of the methods compute simultaneously the solution *and* its first derivative whereas the others compute *only* the solution. The one-step and piecewise perturbation methods belong to the first class while the multistep methods (except for the method of de Vogelaere) to the second. Moreover, for the multistep methods some additional starting values are necessary to advance the solution.

It follows that when none of the two boundary conditions explicitly contains the derivative of y, that is when $p_0 = p_1 = 0$, any initial value method is directly applicable while when it happens that p_0 and p_1 are not both vanishing the application of the methods of the first class remains as direct as before but some extra work is necessary to prepare the additional starting values for the multistep methods and also to reconstruct numerically the value of the derivative at the other boundary point in terms of the very values of the solutions at that point and at some other neighboring points.

As for the matching in the middle this raises no special problem since the conditions $y^F(x_m) = y^B(x_m)$ and $y^{F'}(x_m) = y^{B'}(x_m)$ can be replaced by $y^F(x_m) = y^B(x_m)$ and $y^F(x_{m'}) = y^B(x_{m'})$ where $x_{m'}$ is some mesh point close to x_m; the latter conditions involve only values of y and thus the methods of the second class can be applied without difficulty.

The errors of the results furnished by shooting procedures are readily expressed in terms of the accumulated errors of the initial value methods used to construct the components.

Consider for illustration the first version. The exact A^* is the solution of the linear equation

$$p_1(y^{*\prime}(b) + A^*z^{*\prime}(b)) + q_1(y^*(b) + A^*z^*(b)) = u_1 \tag{4.2.19}$$

while what is actually computed is a value $\tilde{A}^*$ which is the solution of the same equation in which the exact values at b of the components are replaced by the numerical values resulted from the initial value scheme used, hence

$$p_1(\tilde{y}^{*\prime}(b) + \tilde{A}^*\tilde{z}^{*\prime}(b)) + q_1(\tilde{y}^*(b) + \tilde{A}^*\tilde{z}^*(b)) = u_1. \tag{4.2.20}$$

On using the accumulated errors at b, $\Delta y = y^*(b) - \tilde{y}^*(b)$ etc., eq. (4.2.20) is subtracted from eq. (4.2.19),

$$p_1(\Delta y' + A^*z^{*\prime}(b) - \tilde{A}^*\tilde{z}^{*\prime}(b)) + q_1(\Delta y + A^*z^*(b) - \tilde{A}^*\tilde{z}^*(b)) = 0. \tag{4.2.21}$$

The error in A^*, $\Delta A = A^* - \tilde{A}^*$ is explicited at once (only linear terms in errors are retained)

$$\Delta A \simeq -\frac{1}{p_1z^{*\prime}(b) + q_1z^*(b)}[A(p_1\Delta z' + q_1\Delta z) + p_1\Delta y' + q_1\Delta y] \tag{4.2.22}$$

If the initial value method is of order $m \geqslant 1$, that is each of Δy, $\Delta y'$, Δz and $\Delta z'$ behaves as h^m, then ΔA is of the same order. It follows that the numerical values at mesh points,

$$\tilde{y}(x_k) = \tilde{y}^*(x_k) + \tilde{A}^* \tilde{z}^*(x_k) \tag{4.2.23}$$

are affected by errors of the *same order* m.

Note finally that in the general case when the differential equations are non-linear we cannot superpose solutions and the shooting approach is necessarily more tentative. Such problems are treated in [4].

4.3. DIRECT METHODS

The example discussed in the previous section has shown that, although the shooting methods are fully justified from a theoretical point of view, their practical application may sometimes raise special problems. The reason was seen to lie in the fact that the solution components contain terms which may be absent in the true solution of the boundary problem which is solved. Direct methods, that is methods which avoid the use of the components, are desired and they can be generated in a very simple way.

More exactly, a direct method will be a method which seeks directly the solution of the given boundary value problem by taking *both* boundary conditions from the very beginning and on equal footing.

It must be understood that a direct method does not mean at all a new algorithm; it consists of just the same algorithms as for the initial value problem. The novelty actually regards only the way in which such algorithms are implemented, and to illustrate the point we refer to the Fox-Goodwin algorithm for eq. (4.1.3). It is given by formula (1.13.21) which we write here again as

$$A_k \bar{y}_{k+1} + B_k \bar{y}_k + C_k \bar{y}_{k-1} = D_k, \qquad k = 1, 2, \ldots, n-1 \tag{4.3.1}$$

where

$$A_k = 1 + \frac{1}{2} ha_k,\ B_k = -(2 - h^2 b_k),\ C_k = 1 - \frac{1}{2} ha_k,\ D_k = - h^2 c_k \tag{4.3.2}$$

are known numbers.

Let us look at the set (4.3.1) as at a selfstanding system of $n-1$ linear equations for $n+1$ unknowns $\bar{y}_0, \bar{y}_1, \bar{y}_2, \ldots, \bar{y}_n$. Such a system is clearly undetermined because the number of unknowns exceeds the number of equations. It becomes determined if the values of any *two* unknowns are fixed.

One possibility is to fix just $\bar{y}_0$ and $\bar{y}_1$. If so, the number of effective unknowns is reduced to $n-1$ and these are $\bar{y}_2, \bar{y}_3, \bar{y}_4, \ldots, \bar{y}_n$. The system thus becomes fully determined and its unknowns can be calculated directly by recurrence

$$\bar{y}_{k+1} = \frac{1}{A_k}[-B_k\bar{y}_k - C_k\bar{y}_{k-1} + D_k], \quad k = 1, 2, 3, \ldots, n-1 \quad (4.3.3)$$

Another possibility is to fix the values of $\bar{y}_0$ and of $\bar{y}_n$, say $\bar{y}_0 = u_0$, $\bar{y}_n = u_1$. The system (4.3.1) is again determined but now its $n-1$ unknowns are $\bar{y}_1, \bar{y}_2, \ldots, \bar{y}_{n-1}$. In contrast to the previous case its solution can be no longer constructed by recurrence. Some appropriate method for solving systems of linear equations must be used instead.

The two possibilities considered above represent two alternative uses of the Fox-Goodwin formula associated to the differential equation (4.1.3). The first choice is nothing else than the standard use when *initial conditions* are imposed. The second is what was called before a direct method to solve the same differential equation when the *boundary conditions* $y(a) = u_0$ and $y(b) = u_1$ are imposed. The direct method associated to the Fox-Goodwin algorithm thus consists of effectively solving the following system of $n-1$ linear equations

$$\mathbf{M}\bar{\mathbf{y}} = \bar{\mathbf{f}} \quad (4.3.4)$$

where $\mathbf{M}$ is the $(n-1)$ by $(n-1)$ matrix

$$\mathbf{M} = \begin{pmatrix} B_1 & A_1 & 0 & 0 & \ldots 0 & 0 & 0 & 0 \\ C_2 & B_2 & A_2 & 0 & \ldots 0 & 0 & 0 & 0 \\ 0 & C_3 & B_3 & A_3 & \ldots 0 & 0 & 0 & 0 \\ \cdot & \cdot & \cdot & \cdot & \ldots \cdot & \cdot & \cdot & \cdot \\ \cdot & \cdot & \cdot & \cdot & \ldots \cdot & \cdot & \cdot & \cdot \\ \cdot & \cdot & \cdot & \cdot & \ldots \cdot & \cdot & \cdot & \cdot \\ 0 & 0 & 0 & 0 & \ldots C_{n-3} & B_{n-3} & A_{n-3} & 0 \\ 0 & 0 & 0 & 0 & \ldots 0 & C_{n-2} & B_{n-2} & A_{n-2} \\ 0 & 0 & 0 & 0 & \ldots 0 & 0 & C_{n-1} & B_{n-1} \end{pmatrix} \quad (4.3.5)$$

while $\bar{\mathbf{y}}$ and $\bar{\mathbf{f}}$ are the following vectors with $n-1$ components

$$\bar{\mathbf{y}} = \begin{pmatrix} \bar{y}_1 \\ \bar{y}_2 \\ \vdots \\ \bar{y}_{n-2} \\ \bar{y}_{n-1} \end{pmatrix}, \qquad \bar{\mathbf{f}} = \begin{pmatrix} D_1 - C_1 u_0 \\ D_2 \\ \vdots \\ D_{n-2} \\ D_{n-1} - C_{n-1} u_1 \end{pmatrix}. \tag{4.3.6}$$

A suitable direct method based on the same Fox-Goodwin formula can also be devised for the case when $p_0 \neq 0$ and $p_1 \neq 0$ that is when the derivatives explicitly occur in the boundary conditions (4.1.4). The only difficulty is that the derivatives are not accepted in the Fox-Goodwin formula and for this reason we have to introduce an extra step in the procedure, namely to convert the original boundary conditions in forms that involve the values of the very solution at two neighboring mesh points. We thus look for the constants P_i, Q_i, U_i, $i = 0, 1$ such as equations

$$P_0 y(h) + Q_0 y(a+h) = U_0, \; P_1 \bar{y}(b-h) + Q_1 y(b) = U_1 \tag{4.3.7}$$

are equivalent with the two original boundary conditions. Indeed, if such constants can be constructed, then we simply add two new equations

$$P_0 \bar{y}_0 + Q_0 \bar{y}_1 = U_0, \quad P_1 \bar{y}_{n-1} + Q_1 \bar{y}_n = U_1 \tag{4.3.8}$$

to the previous $n-1$ linear equations (4.3.1) and thus form a linear algebraic system of $n+1$ equations for the $n+1$ unknowns $\bar{y}_0, \bar{y}_1, \ldots, \bar{y}_n$.

To construct the above constants we assume that $a(x)$, $b(x)$ and $c(x)$ are differentiable sufficiently many times and make use of the Taylor series for the solution,

$$y(x+h) = y(x) + \frac{h}{1!} y'(x) + \frac{h^2}{2!} y''(x) + \frac{h^3}{3!} y'''(x) + \ldots \tag{4.3.9}$$

We remark that each higher order derivative of y can be expressed as a linear combination of y and y' plus a free term, for instance

$$y''(x) = -ay' - by - c \tag{4.3.10}$$

$$y'''(x) = (y''(x))' = -a'y' - ay'' - b'y - by' - c'$$

$$= -a'y' - a(-ay' - by - c) - b'y - by' - c'$$

$$= (-a' + a^2 - b)y' + (ab - b')y + ac - c' \tag{4.3.11}$$

and so on. We thus conclude that $y(x + h)$ is of the form

$$y(x + h) = \alpha(x; h)y(x) + \beta(x; h)y'(x) + \gamma(x; h) \tag{4.3.12}$$

where $\alpha(x; h)$, $\beta(x; h)$ and $\gamma(x; h)$ are readily constructed in terms of the values $a(x)$, $b(x)$, $c(x)$ and of their derivatives. We now put $x = a$,

$$y(a + h) = \alpha(a; h)y(a) + \beta(a; h)y'(a) + \gamma(a; h) \tag{4.3.13}$$

and also recall that $y(a)$ and $y'(a)$ must satisfy the first boundary condition

$$p_0 y'(a) + q_0 y(a) = u_0. \tag{4.3.14}$$

Eqs (4.3.13) and (4.3.14) are combined to eliminate $y'(a)$ and the result is

$$[p_0\alpha(a; h) - q_0\beta(a; h)]y(a) - p_0 y(a + h) = -\beta(a; h)u_0 - p_0\gamma(a; h) \tag{4.3.15}$$

that is exactly the first of eqs (4.3.7) with $P_0 = p_0\alpha(a; h) - q_0\beta(a; h)$, $Q_0 = -p_0$ and $U_0 = -\beta(a; h)u_0 - p_0\gamma(a; h)$; P_1, Q_1 and U_1 are constructed in the same way. (In practice only four terms in the right-hand member of eq. (4.3.9) must be actually taken since this is enough to saturate the accuracy of the Fox-Goodwin method which is a second order method.)

Altogether the direct method to solve the boundary value problem (4.1.3—4) by means of the Fox-Goodwin formula consists of solving a linear system of $n + 1$ equations, viz.

$$\mathbf{N}\bar{\mathbf{z}} = \bar{\mathbf{g}} \tag{4.3.16}$$

where **N** is the $(n+1)$ by $(n+1)$ matrix

$$\mathbf{N} = \begin{pmatrix} P_0 & Q_0 & 0 & 0 & \dots 0 & 0 & 0 & 0 \\ C_1 & B_1 & A_1 & 0 & \dots 0 & 0 & 0 & 0 \\ 0 & C_2 & B_2 & A_2 & \dots 0 & 0 & 0 & 0 \\ \cdot & \cdot & \cdot & \cdot & \dots \cdot & \cdot & \cdot & \cdot \\ \cdot & \cdot & \cdot & \cdot & \dots \cdot & \cdot & \cdot & \cdot \\ \cdot & \cdot & \cdot & \cdot & \dots \cdot & \cdot & \cdot & \cdot \\ 0 & 0 & 0 & 0 & \dots C_{n-2} & B_{n-2} & A_{n-2} & 0 \\ 0 & 0 & 0 & 0 & \dots 0 & C_{n-1} & B_{n-1} & A_{n-1} \\ 0 & 0 & 0 & 0 & \dots 0 & 0 & P_1 & Q_1 \end{pmatrix}; \tag{4.3.17}$$

$\bar{\mathbf{z}}$ and $\bar{\mathbf{g}}$ are vectors with $n+1$ components,

$$\bar{\mathbf{z}} = \begin{pmatrix} \bar{y}_0 \\ \bar{y}_1 \\ \bar{y}_2 \\ \vdots \\ \bar{y}_{n-1} \\ \bar{y}_n \end{pmatrix}, \qquad \bar{\mathbf{g}} = \begin{pmatrix} U_0 \\ D_1 \\ D_2 \\ \vdots \\ D_{n-1} \\ U_1 \end{pmatrix}. \tag{4.3.18}$$

Systems (4.3.4) and (4.3.16) are particular cases of the general system of linear equations

$$\mathbf{Ax} = \mathbf{d} \tag{4.3.19}$$

where **A** is a given N by N nonsingular matrix, and **d** is a given vector with N components. Problem (4.3.19) is one of the standard problems in the numerical analysis and many computer codes are available for its solution. It is however important to note that such codes are typically designated for the general case of *full* matrix **A** and due to that they require a computation effort proportional to N^2. Use of such codes should be therefore avoided in favor of other, more direct and simpler procedures which explicitly account for the fact that in the case of eqs (4.3.4) and (4.3.16) the matrix **A** is tridiagonal.

Such a technique, described in many books, as for instance in [2, sec. 8.12], is based on the elimination method.

Specifically, let us assume that **A** is tridiagonal,

$$\mathbf{A} = \begin{pmatrix} b_1 & c_1 & 0 & 0 & \dots 0 & 0 & 0 \\ a_2 & b_2 & c_2 & 0 & \dots 0 & 0 & 0 \\ 0 & a_3 & b_3 & c_3 & \dots 0 & 0 & 0 \\ \cdot & \cdot & \cdot & \cdot & \dots \cdot & \cdot & \cdot \\ \cdot & \cdot & \cdot & \cdot & \dots \cdot & \cdot & \cdot \\ \cdot & \cdot & \cdot & \cdot & \dots \cdot & \cdot & \cdot \\ 0 & 0 & 0 & 0 & \dots a_{N-1} & b_{N-1} & c_{N-1} \\ 0 & 0 & 0 & 0 & \dots 0 & a_N & b_N \end{pmatrix} \tag{4.3.20}$$

and that $b_i \neq 0$, $i = 1, 2, \dots, N$.

We start with the first two equations in system (4.3.19)

$$b_1x_1 + c_1x_2 = d_1$$

$$a_2x_1 + b_2x_2 + c_2x_3 = d_2$$

and eliminate x_1 from the second; to this end the first equation is multiplied by a_2/b_1 and then subtracted from the second. The result is

$$\beta_2x_2 + c_2x_3 = \delta_2, \tag{4.3.21}$$

where $\beta_2 = b_2 - a_2c_1/b_1$ and $\delta_2 = d_2 - a_2d_1/b_1$.
Next eq. (4.3.21) and the third equation in the system (4.3.20)

$$a_3x_2 + b_3x_3 + c_3x_4 = d_3 \tag{4.3.22}$$

are considered and the same procedure is applied to eliminate x_3. The result is

$$\beta_3x_3 + c_3x_4 = \delta_3 \tag{4.3.23}$$

with $\beta_3 = b_3 - a_3c_2/b_2$ and $\delta_3 = d_3 - a_3\delta_2/b_2$.

After N such steps the system (4.3.19) takes the form

$$\bar{\mathbf{A}}\mathbf{x} = \boldsymbol{\delta} \tag{4.3.24}$$

where

$$\bar{\mathbf{A}} = \begin{pmatrix} \beta_1 & c_1 & 0 & 0 & \dots 0 & 0 & 0 \\ 0 & \beta_2 & c_2 & 0 & \dots 0 & 0 & 0 \\ 0 & 0 & \beta_3 & c_3 & \dots 0 & 0 & 0 \\ \cdot & \cdot & \cdot & \cdot & \dots \cdot & \cdot & \cdot \\ \cdot & \cdot & \cdot & \cdot & \dots \cdot & \cdot & \cdot \\ 0 & 0 & 0 & 0 & \dots 0 & \beta_{N-1} & c_{N-1} \\ 0 & 0 & 0 & 0 & \dots 0 & 0 & \beta_N \end{pmatrix}, \qquad \boldsymbol{\delta} = \begin{pmatrix} \delta_1 \\ \delta_2 \\ \delta_3 \\ \cdot \\ \cdot \\ \delta_{N-1} \\ \delta_N \end{pmatrix} \tag{4.3.25}$$

with $\delta_1 = d_1$.

Matrix $\bar{\mathbf{A}}$ is a double-band matrix and thus the unknown vector $\mathbf{x}$ is obtained directly by backward iteration: the last component is simply $x_N = \delta_N/\beta_N$, the last by first component is $x_{N-1} = (\delta_{N-1} - c_{N-1}x_N)/\beta_{N-1}$, etc.

In short, the solution of eq. (4.3.19) when $\mathbf{A}$ is tridiagonal is achieved by *two* sets of iterations. In the first set starting values, $\beta_1 \equiv b_1$ and $\delta_1 \equiv d_1$ are assigned and successively compute

$$m_i \equiv a_{i+1}/b_i, \; \beta_{i+1} = b_{i+1} - m_i c_i, \; \delta_{i+1} = d_{i+1} - m_i \delta_i;$$

$$i = 1, 2, 3, \dots, N-1. \tag{4.3.26}$$

In the second step the starting value $x_N = \delta_N/\beta_N$ is taken and the backward iteration is used to generate $x_{N-1}, x_{N-2}, \dots$, viz.:

$$x_i = (\delta_i - c_i x_{i+1})\beta_i, \; i = N-1, N-2, \dots, 2, 1. \tag{4.3.27}$$

The number of operations of this double iteration solver of eq. (4.3.19) is obviously proportional to N and thus such a solver is clearly faster than any general solver.

The quality of the numerical results generated by the discussed direct method for the boundary value problem (4.1.3—4) is governed by the quality of the Fox-Goodwin formula on which it is based. We mentioned in section 1.13 that, when this formula is applied to an initial value problem, its order is *two*. The same remains valid also for the boundary value problems in the sense that the deviation $\Delta_k \equiv y(x_k) - \bar{y}_k$ is proportional to h^2 for any $k = 0, 1, 2, \ldots, n$; a proof is given in [2, chapter 12].

As a matter of fact, such a relationship is general. It is actually valid for any other numerical method of integration, in the sense that a method which was found to be of order m for initial value problems will remain of the same order also when it is formulated as a direct method for a boundary value problem. For this reason, whenever the structure of the differential equation (4.1.3) allows to, making use of higher order algorithms is highly recommended. If $a(x) = 0$, for instance, the use of the Numerov algorithm, eq. (1.13.26), will be more efficient than (4.3.1) because it is of the fourth order; such option is also considered in [2].

Another way to improve the accuracy of the results is the so-called "deferred" correction and to illustrate it we refer to the system (4.3.16) based on the Fox-Goodwin scheme. The idea consists of writing the *exact* solution of the problem (4.1.3—4) in form (4.3.16) where, of course, the unhomogeneous term is appropriately modified.
We thus assume that the vector $\mathbf{z}$ of exact values

$$\mathbf{z} = \begin{pmatrix} y(x_0) \\ y(x_1) \\ y(x_2) \\ \vdots \\ y(x_{n-1}) \\ y(x_n) \end{pmatrix} \tag{4.3.28}$$

satisfies the algebraic system of $n + 1$ linear equations

$$\mathbf{N}\mathbf{z} = \mathbf{g} \tag{4.3.29}$$

where $\mathbf{N}$ is given by eq. (4.3.17).

On subtracting eq. (4.3.16) from the exact equation (4.3.29) we obtain the algebraic system for the correction vector $\Delta\mathbf{z} = \mathbf{z} - \bar{\mathbf{z}}$,

$$\mathbf{N}\Delta\mathbf{z} = \Delta\mathbf{g}, \tag{4.3.30}$$

with $\Delta\mathbf{g} = \mathbf{g} - \bar{\mathbf{g}}$.

To construct Δg we distinguish two situations. On one hand, the first and last equation in systems (4.3.16) and (4.3.29) account for the boundary conditions only and thus $\bar{g}_0 = g_0 = U_0$ and $\bar{g}_n = g_n = U_1$ so that $\Delta g_0 = \Delta g_n = 0$. (Two more terms should be used in eq. (4.3.9) to generate P_0, Q_0, P_1 and Q_1.) On the other hand, inner equations in the two systems differ on the account that the Fox-Goodwin formula is an approximation to the exact formula (1.13.19) and thus

$$\Delta g_i = -\left[-\left(\frac{\delta^4}{12}+\frac{\delta^6}{90}\right)+ha_i\mu\left(-\frac{\delta^3}{6}+\frac{\delta^5}{30}\cdots\right)\right]y_i \tag{4.3.31}$$

for $i = 1, 2, \ldots, n-1$. This is an exact representation of Δg_i but usually only an approximation Δg_i^* to Δg_i can be effectively constructed in practice and, as a rule, this consists of retaining only the leading term in the right member of eq. (4.3.31). On explicitly constructing the third and fourth central differences and the product $\mu\delta^3$ (use eqs (1.13.2—5)) we get the approximation

$$\begin{aligned}\Delta g_i^* &= \frac{1}{12}(\delta^4 + 2ha_i\mu\delta^3)y(x_i)\\ &= \frac{1}{12}\Big\{y(x_i+2h) - 4y(x_i+h) + 6y(x_i) - 4(x_i-h) + y(x_i-2h) +\\ &\quad + a_ih[y(x_i+2h) - 2y(x_i+h) + 2y(x_i-h) - y(x_i-2h)]\Big\}.\end{aligned} \tag{4.3.32}$$

In practice the solution of the basic system (4.3.16) must be first evaluated and the results $\bar{y}_0, \bar{y}_1, \bar{y}_2, \ldots, \bar{y}_{n-1}$ and $\bar{y}_n$ are identified as $y(x_0), y(x_1), \ldots, y(x_{n-1}), y(x_n)$ in eq. (4.3.32). Note however that these data can furnish only Δg_i^* for $i = 2, 3, \ldots, n-2$. To generate g_1^* and g_{n-1}^* some $\bar{y}_{-1}$ and $\bar{y}_{n+1}$ must be extra evaluated and this is done by using eq. (4.3.1) with $k = 0$ and $k = n$, respectively. (It is tacitly assumed that $a(x)$, $b(x)$ and $c(x)$ are also defined in some small regions on the left side of x_0 and on the right side of x_n.)

Once the vector $\Delta\mathbf{g}^*$ is generated, the system $\mathbf{N}\Delta\mathbf{z}^* = \Delta\mathbf{g}^*$ is solved and the sum $\mathbf{z}^* = \bar{\mathbf{z}} + \Delta\mathbf{z}^*$ will be a better approximation of the exact $\mathbf{z}$ than is the original $\bar{\mathbf{z}}$. The correction $\Delta\mathbf{z}^*$ is clearly "deferred" since it can be evaluated only after the basic approximation $\bar{\mathbf{z}}$ has been calculated.

As stated before, any initial value algorithm can be implemented as a direct method to solve boundary value problems and such a specification seems to be important because, surprisingly enough, only the implement of two-step methods is usually examined in the literature.

Consider, for illustration, the algorithm developed in section 3.2. It is a one-step algorithm which can be conveniently written as

$$\bar{y}_{k+1} = S_k\bar{y}_k + T_k\bar{y}'_k + V_k \tag{4.3.33 a}$$

$$\bar{y}'_{k+1} = P_k\bar{y}_k + Q_k\bar{y}'_k + W_k, \quad k = 0, 1, 2, \ldots, n-1 \tag{4.3.33 b}$$

where S_k, T_k, V_k, P_k, Q_k and W_k can be readily identified from eqs (3.2.10), (3.2.11) and (3.2.12).

If we look at it as at a system of $2n$ linear equations for $2n + 2$ unknowns $\bar{y}_0, \bar{y}'_0, \bar{y}_1, \bar{y}'_1, \ldots, \bar{y}_n, \bar{y}'_n$, its application as an initial value method corresponds to the case when the values of $\bar{y}_0$ and of $\bar{y}'_0$ are fixed, while if we want to apply it as a direct method for the boundary value problem (4.1.3—4) we have to supplement the previous $2n$ equations by the boundary conditions

$$p_0\bar{y}'_0 + q_0\bar{y}_0 = u_0, \quad p_0\bar{y}'_n + q_1\bar{y}_n = u_1. \tag{4.3.34}$$

A linear system of $2n + 2$ equations for $2n + 2$ unknowns results, viz.

$$\mathbf{R}\bar{\mathbf{w}} = \mathbf{d} \tag{4.3.35}$$

where $\mathbf{R}$ is a $(2n + 2)$ by $(2n + 2)$ matrix

$$\mathbf{R} = \begin{pmatrix}
q_0 & p_0 & 0 & 0 & 0 & \ldots & 0 & 0 & 0 & 0 & 0 \\
S_0 & T_0 & -1 & 0 & 0 & \ldots & 0 & 0 & 0 & 0 & 0 \\
P_0 & Q_0 & 0 & -1 & 0 & \ldots & 0 & 0 & 0 & 0 & 0 \\
0 & 0 & S_1 & T_1 & -1 & \ldots & 0 & 0 & 0 & 0 & 0 \\
0 & 0 & P_1 & Q_1 & 0 & \ldots & 0 & 0 & 0 & 0 & 0 \\
\cdot & \cdot & \cdot & \cdot & \cdot & \ldots & 0 & 0 & 0 & 0 & 0 \\
\cdot & \cdot & \cdot & \cdot & \cdot & \ldots & \cdot & \cdot & \cdot & \cdot & \cdot \\
\cdot & \cdot & \cdot & \cdot & \cdot & \ldots & \cdot & \cdot & \cdot & \cdot & \cdot \\
0 & 0 & 0 & 0 & 0 & \ldots & Q_{n-2} & 0 & -1 & 0 & 0 \\
0 & 0 & 0 & 0 & 0 & \ldots & 0 & S_{n-1} & T_{n-1} & -1 & 0 \\
0 & 0 & 0 & 0 & 0 & \ldots & 0 & P_{n-1} & Q_{n-1} & 0 & -1 \\
0 & 0 & 0 & 0 & 0 & \ldots & 0 & 0 & 0 & q_1 & p_1
\end{pmatrix} \tag{4.3.36}$$

while $\bar{\mathbf{w}}$ and $\mathbf{d}$ are vectors with $2n + 2$ components

$$\bar{\mathbf{w}} = \begin{pmatrix} \bar{y}_0 \\ \bar{y}'_0 \\ \bar{y}_1 \\ \bar{y}'_1 \\ \cdot \\ \cdot \\ \cdot \\ \bar{y}'_{n-1} \\ \bar{y}_n \\ \bar{y}'_n \end{pmatrix}, \qquad \mathbf{d} = \begin{pmatrix} u_0 \\ -V_0 \\ -W_0 \\ -V_1 \\ \cdot \\ \cdot \\ \cdot \\ -V_{n-1} \\ -W_{n-1} \\ u_1 \end{pmatrix}. \tag{4.3.37}$$

The advantage is that the presence of the derivative in the boundary conditions does not involve extra troubles as it were for two-step algorithms. On the other hand, the number of equations in the resultant system is now twice as large as in the previous case and the **R** matrix, although still sparse, is a quadruple band matrix. For these reasons the effective computation of system (4.3.36) will be significantly slower for one and the same n. However, in many cases the step sizes associated to algorithm (4.3.33 a, b) are larger than for algorithm (4.3.1) hence n in eq. (4.3.35) is smaller than in eq. (4.3.16). As such, the method of choice depends very much on the form of functions $a(x)$, $b(x)$ and $c(x)$ in eq. (4.1.3).

4.4. EIGENVALUE PROBLEMS

In section 4.1 an example was given to illustrate that the existence of a solution for a boundary value problem is strikingly more accidental than for an initial value problem. Indeed, a solution of the first problem exists only when the differential equation and the boundary conditions associated to it happen to be "cooperating" some way. It follows that, if the differential equation is given, solution will exist only if the boundary conditions are of suitable form. Conversely, if the form of the boundary conditions is specified, a solution will exist only for differential equations whose form is "cooperating" with the preset boundary conditions.

Particularly important in practical applications are the cases which involve second-order linear *homogeneous* differential equations and linear *homogeneous* boundary conditions and it is the form of the boundary conditions that is normally fixed. The problem thus is to construct the coefficients of the second-order equation such as the problem admits a solution.

The most popular problem of this type refers to the following second-order equation,

$$(d(x)y')' - g(x)y + \lambda r(x)y = 0, \quad x \in [a, b] \tag{4.4.1}$$

where $d(x)$, $g(x)$ and $r(x)$ are given bounded functions of x such that $d(x)$ and $r(x)$ are positive while λ is an arbitrary parameter. The boundary conditions are assumed of the form

$$q_0 y'(a) + p_0 y(a) = 0, \quad q_1 y'(b) + p_1 y(b) = 0 \tag{4.4.2}$$

where q_0, p_0, q_1 and p_1 are given numbers such that $|q_0| + |p_0| \neq 0$ and $|q_1| + |p_1| \neq 0$ and the problem actually consists of evaluating the values of λ such that the boundary value problem (4.4.1—2) admits solution.

This problem is known as the *Sturm-Liouville* problem, and it is examined at large in standard books of mathematical physics, for instance in [6]. The basic results of such analysis are as follows

(i) there is an infinite set of values $\lambda_0 < \lambda_1 < \lambda_2 < \lambda_3 \ldots$ for which the quoted problem has solutions; each λ_i is called an *eigenvalue* and the solution $y_i(x)$ corresponding to λ_i, is called an *eigenfunction;*

(ii) the eigenfunction $y_i(x)$ is unique except for an arbitrary factor; the latter statement is just normal whereas both eq. (4.4.1) and (4.4.2) are homogeneous;

(iii) the eigenfunctions $y_0(x), y_1(x), y_2(x), \ldots$ are orthogonal functions, so that

$$\int_a^b y_i(x) y_j(x)\, r(x) dx = 0 \quad \text{if } i \neq j \tag{4.4.3}$$

and they form a complete set;

(iv) the number of points in (a, b) at which $y_i(x)$ vanishes is just i.

The numerical solution of the eigenvalue problem (4.4.1—2) is achieved by adapting the two kinds of methods for boundary value problems.

Shooting methods. Second-order equation (4.4.1) can be readily written as a system of two first-order equations, viz.

$$\mathbf{y}' = \mathbf{f}(x, \mathbf{y}) \tag{4.4.4}$$

where

$$\mathbf{y} \equiv \begin{pmatrix} y_1 \\ y_2 \end{pmatrix} = \begin{pmatrix} y \\ d(x)y' \end{pmatrix}; \quad \mathbf{f}(x, \mathbf{y}) = \begin{pmatrix} \dfrac{1}{d(x)} y_2 \\ (g(x) - \lambda r(x))y_1 \end{pmatrix}. \tag{4.4.5}$$

It follows that any of the general initial value methods for first-order equations can be readily applied.

If we additionally assume that $d(x)$ is differentiable, eq. (4.4.1) reads

$$y'' + a(x)y' + (b^0(x) + \lambda b^1(x))y = 0 \tag{4.4.6}$$

with $a(x) = d'(x)/d(x)$, $b^0(x) = -g(x)/d(x)$, $b^1(x) = r(x)/d(x)$, and thus it can be equally solved numerically by the special methods for second-order linear differential equations.

Once the method of integration is chosen a forward shooting version is as follows. First some values for initial conditions are taken which are consistent with the first boundary condition. Three situations occur: when $p_0 \neq 0$ and $q_0 \neq 0$ we take some nonvanishing value for $y(a)$, say $y(a) = y_0 \neq 0$ and thus $y'(a) = -q_0 y_0/p_0$; when $p_0 \neq 0$ but $q_0 = 0$ we can take $y(a) = y_0 \neq 0$ and $y'(a) = 0$; when $p_0 = 0$ but $q_0 \neq 0$ a possible choice is $y(a) = 0$, $y'(a) = y_0' \neq 0$. (In case when the chosen method of integration is a two-step method some extra work is necessary to convert these initial values to forms which imply $y(a)$ and $y(a + h)$, as in the previous section.)

Some successive values $\lambda^0, \lambda^1, \lambda^2, \ldots$ are next tried and the initial value problem for eq. (4.4.1) is solved at each λ^s. The values of the resultant $\bar{y}(b; \lambda^s)$ and $\bar{y}'(b; \lambda^s)$ are used to form the function $P(\lambda^s) = p_1\bar{y}'(b; \lambda^s) + q_1\bar{y}(b; \lambda^s)$ and the procedure is repeated until it is observed that $P(\lambda^s)$ and $P(\lambda^{s+1})$ have different signs. The value of $\bar{\lambda}$ such as $P(\bar{\lambda}) = 0$ is further located on $[\lambda^s, \lambda^{s+1}]$ to the desired accuracy. By the very construction, this $\bar{\lambda}$ will be an eigenvalue of the problem (4.4.1—2). To identify its label i the computation of the initial value problem corresponding to this $\bar{\lambda}$ is repeated and the number of oscillations of $\bar{y}(x; \bar{\lambda})$ on (a, b) is counted; see property (iv).

Frequent in application is the situation when the user wants to compute only the eigenvalue and the eigenfunction which correspond to some *preset* i. In such a case, if the user has no qualitative knowledge on region where such λ_i is expected to be placed he is to start with some λ^0 and simply count the number of zeros of $y(x, \lambda^0)$. If this is greater than i then the next value to be tried must be smaller than λ^0; in the opposite case some $\lambda^1 > \lambda^0$ will be tried next.

As for the errors in eigenvalues and in eigenfunctions these are obviously determined by the error of the chosen method of integration. In particular if the method is a mth order method the errors

$$\Delta\lambda_i = \lambda_i - \bar{\lambda}_i \tag{4.4.7}$$

$$\Delta y_{i;k} = y_i(x_k) - \bar{y}_{i;k} \tag{4.4.8}$$

will tend to zero as h^m when h is decreased.

The practical adaptation of backward shooting method or of the "matching in the middle" raises no special problem. The behaviour of the errors remains as before.

Direct methods. Since the eigenvalue problem (4.4.1—2) is a special case of the boundary problem (4.1.3—4) a direct method for the former means suitable adaptation of the system of algebraic equations derived for the latter. If the Fox-Goodwin formula is chosen to generate the direct method, the solution of problem (4.4.1—2) means the solution of system (4.3.16).

To write this system explicitly we use form (4.4.6) for the original equation (4.4.1). If so, the following identifications result: A_k and B_k remain given as in eq. (4.3.2) but $B_k = -2 + h^2(b_k^0 + \lambda b_k^1)$ and $D_k = 0$. As for the boundary conditions, use is made of the fact that both differential equation (4.4.1) and its boundary conditions (4.4.2) are homogeneous. The first circumstance implies that in the representation (4.3.12) the last term is absent, $\gamma(a; h) = 0$. In turn both conditions in form (4.3.8) will be also homogeneous that is $U_0 = U_1 = 0$. Moreover, since as it was specified in the previous section, only four terms are sufficient in eq. (4.3.9), $\alpha(x; h)$ and $\beta(x; h)$ will be linear in λ so that P_0 and Q_1 in eqs (4.3.8) will be *linear* in λ, while Q_0 and P_1 are independent of λ.

On collecting these results we find that all elements of the $(n + 1)$ by $(n + 1)$ matrix $\mathbf{N}$ are λ independent, except for the diagonal elements which are *linear* in λ. We also find that all components of the free term $\bar{g}$ are zero. Altogether the system (4.3.16) is actually homogeneous

$$\mathbf{N}(\lambda)\bar{\mathbf{z}} = \mathbf{0} \tag{4.4.9}$$

and thus it has nontrivial solution only if the matrix $\mathbf{N}(\lambda)$ is singular, that is when the λ are solutions of the algebraic equation

$$\det \mathbf{N}(\lambda) = 0. \tag{4.4.10}$$

What is important to notice is that, due to the quoted λ dependence of the matrix elements of $\mathbf{N}(\lambda)$, $\det \mathbf{N}(\lambda)$ is a $(n+1)$th-degree polynomial in λ, hence eq. (4.4.10) will have just $n+1$ roots. If these roots (supposed all real) are labelled in increasing order, $\bar{\lambda}_0 < \bar{\lambda}_1 < \bar{\lambda}_2 < \ldots$ $\ldots < \bar{\lambda}_n$, they are directly interpreted as approximations of the eigenvalues of the problem (4.4.1—2)*. Once these eigenvalue approximations were found, the computation of the associated eigenfunctions is made at once at any $0 \leqslant i \leqslant n$ by effectively solving eq. (4.4.9) in which one of the unknowns, $\bar{y}_0$ for instance, is fixed. All other values $\bar{y}_1, \bar{y}_2, \ldots, \bar{y}_n$ are uniquely determined in terms of this $\bar{y}_0$. Of course, different values of $\bar{y}_0$ will generate different values for the other $\bar{y}_j$ but this situation is just normal because it is in the spirit of the property (ii).

It is equally important to insist that, if any other classical algorithm is chosen to generate a direct method, the λ dependence of the matrix elements of $\mathbf{N}(\lambda)$ remains of *polynomial* form so that the algebraic equation (4.4.10) is a polynomial equation. In general, the degree of the polynomial $\det \mathbf{N}(\lambda)$ is $n+r$ where r increases with the order of the used algorithm. (For illustration we assume that $a(x) = 0$ in eq. (4.4.6) to examine the Numerov algorithm which is of higher order ($m = 4$) than the Fox-Goodwin algorithm ($m = 2$). Most of the elements of the associated matrix $\mathbf{N}(\lambda)$ are now linear in λ, namely $A_1, B_1, C_1, A_2, B_2, C_2, \ldots, A_{n-1}, B_{n-1}$ and C_{n-1} while Q_0 and P_1 remain constants. The difference is that now more terms, namely six, must be taken in series (4.3.9) to saturate the accuracy of this algorithm and for this reason P_0 and Q_1 become now second-degree polynomials in λ and thus $r = 3$.) For this reason eq. (4.4.10) will generally have $n+r$ roots and therefore if we want to obtain approximations to λ_i for $i \geqslant n+r$ we must use larger and larger n, that is finer and finer partitions. The

* The complex roots of eq. (4.4.10), if any, would be ignored since only the real roots can be interpreted as approximations to the eigenvalues. It follows that, in general, $n+1$ represents only the *maximal* number of eigenvalues which may be approximated by the discussed method.

situation is changed in case of piecewise perturbation algorithms. In fact the particularization of eq. (4.3.35) for problem (4.4.1—2) is

$$\mathbf{R}(\lambda)\overline{\mathbf{w}} = \mathbf{0} \tag{4.4.11}$$

so that the eigenvalues are the roots of the algebraic equation

$$\det \mathbf{R}(\lambda) = 0. \tag{4.4.12}$$

The point is that the matrix elements S_k, T_k, P_k and Q_k are now some transcendental functions of λ, *not* polynomials as before, hence eq. (4.4.12) has infinite number of roots. It follows that such an algorithm will be able to furnish approximations to *all* eigenvalues, not only to the first $n + r$ of them.

This difference is not surprising at all. It just reflects in some specific form the observation made in section 3.1 namely that the classical algorithms are based on the belief that the *solution* is well approximated through piecewise polynomials whereas the perturbation algorithms are constructed on the basis of the approximation of the equation *coefficients*.

Thus, if some partition happens to be sufficiently narrow to admit that the first eigenfunction $y_0(x)$ is well approximated by piecewise polynomials (remember that $y_0(x)$ has no zero on (a, b), property (iv)) it will become less and less suited for eigenfunctions $y_i(x)$ with higher i, simply because these are more and more oscillating. It follows that finer partitions are required to restore the accuracy for higher i when classical algorithms are chosen. In contrast, the accuracy is almost uniform for any i in case of perturbation algorithms.

Consider for illustration the simple eigenvalue problem

$$y'' + \lambda y = 0, \; y(0) = y(1) = 0 \tag{4.4.13}$$

with eigenvalues $\lambda_i = (i + 1)^2\pi^2$ and eigenfunctions $y_i(x) = K\sin(\lambda_i^{1/2}x)$, $i = 0, 1, 2, \ldots$ where K is a nonvanishing, arbitrary constant.

Fox-Goodwin formula. Since the derivatives do not occur in the boundary conditions, eq. (4.3.4) is used. We take $n = 4$ so that eq. (4.3.4) reads as

$$\begin{pmatrix} w & 1 & 0 \\ 1 & w & 1 \\ 0 & 1 & w \end{pmatrix} \begin{pmatrix} \bar{y}_1 \\ \bar{y}_2 \\ \bar{y}_3 \end{pmatrix} = \begin{pmatrix} 0 \\ 0 \\ 0 \end{pmatrix} \tag{4.4.14}$$

with $w = -2 + h^2\lambda$

Eq. Det$\mathbf{M}(\lambda) = 0$ reads

$$w^3 - 2w = 0, \tag{4.4.15}$$

with the solutions $\bar{w}_0 = -\sqrt{2}$, $\bar{w}_1 = 0$, $\bar{w}_2 = \sqrt{2}$. Since $\lambda = \frac{1}{h^2}(w+2)$ and $h = 1/4$ the resultant approximate eigenvalues are

$$\bar{\lambda}_0 = 9.44\ldots,\ \bar{\lambda}_1 = 32,\ \text{and}\ \bar{\lambda}_2 = 54.56\ldots \tag{4.4.16}$$

(For comparison here are the values of the exact eigenvalues $\lambda_0 = 9.86\ldots$, $\lambda_1 = 35.46\ldots$, $\lambda_2 = 88.80\ldots$)

Constant perturbation formula. We put $n = 1$. S_0, T_0, P_0 and Q_0 are just propagators (3.4.7) and their derivatives (3.4.8). System (4.3.35) becomes

$$\begin{pmatrix} 1 & 0 & 0 & 0 \\ \xi(-\lambda, 1) & \eta_0(-\lambda, 1) & -1 & 0 \\ -\lambda\eta_0(-\lambda, 1) & \xi(-\lambda, 1) & 0 & -1 \\ 0 & 0 & 1 & 0 \end{pmatrix} \begin{pmatrix} \bar{y}_0 \\ \bar{y}'_0 \\ \bar{y}_1 \\ \bar{y}'_1 \end{pmatrix} = \begin{pmatrix} 0 \\ 0 \\ 0 \\ 0 \end{pmatrix}. \tag{4.4.17}$$

Equation Det$\mathbf{R}(\lambda) = 0$ is simply

$$\eta_0(-\lambda, 1) = 0. \tag{4.4.18}$$

It has nontrivial solutions only when $\lambda > 0$. The upper representation in eq. (3.4.6) is used, hence $\sin(|-\lambda|^{1/2}) = 0$ with the roots $\bar{\lambda}_i = (i+1)^2\pi^2$, $i = 0, 1, 2, 3, \ldots$.

It is seen that the first method furnishes approximations to only the *first three* eigenvalues and the quality of the approximation quickly deteriorates with increasing i while the second produces *all* eigenvalues exactly.

Other numerical procedures for eigenvalue problems are examined in [4].

4.5. THE SCHRÖDINGER EQUATION

The Schrödinger equation is the fundamental equation of nonrelativistic quantum mechanics. Its importance for physicists is thus obvious but at the same time the Schrödinger equation is almost equally interesting as a selfstanding mathematical object for computation.

The interest basically emerges from the fact that this equation, or, better saying, its radial form, is perhaps the simplest differential equation which still preserves the capability of compactly raising many of the problems that may be imagined in connection to the numerical solution of differential equations. This is actually the reason for which the Schrödinger equation is often taken as an eloquent test equation to compare various numerical approaches for quality. As a matter of fact, repeated references to this equation were made also in this book, particularly in chapter 3, although we avoided mentioning its name.

We first briefly review some physical aspects concerning the Schrödinger equation.

The behaviour of a spinless quantum particle of mass m in a potential $v(\vec{x})$ is governed by the three-dimensional Schrödinger equation

$$-\frac{\hbar^2}{2m}\Delta\psi(\vec{x})+(v(\vec{x})-\varepsilon)\psi(\vec{x})=0 \tag{4.5.1}$$

where Δ is the Laplace operator

$$\Delta=\frac{\partial^2}{\partial x_1^2}+\frac{\partial^2}{\partial x_2^2}+\frac{\partial^2}{\partial x_2^2}, \tag{4.5.2}$$

$\hbar$ is the reduced Planck's constant and ε is the particle energy. This is clearly a partial differential equation and the only physical condition is that the solution $\psi(\vec{x})$ must be finite everywhere. Solution $\psi(\vec{x})$ is called the particle wavefunction and it characterizes the localization of the particle in the sense that the quantity $|\psi(\vec{x})|^2 dV$ describes the probability of the presence of the particle inside the infinitesimal volume $\mathrm{d}V$ centred at $\vec{x}$. (Of course, some more specifications should be added here as, for example, that eq. (4.5.1) is a particular, time-independent version of the general Schrödinger equation, that we have to define a norm of $\psi(\vec{x})$ and so on, but we omit such details, whereas they can be found in any standard book of quantum mechanics, e.g. [7].)

For simplicity it will be further assumed that the potential $v(\vec{x})$ admits spherical symmetry and for this reason it is convenient to use spherical coordinates $x \in [0, \infty)$, $\theta \in [0, \pi]$ and $\psi \in [0, 2\pi]$ in place of the original cartesian coordinates x_1, x_2, x_3. The connection between the two types of coordinates is

$$x_1 = x \sin\theta \cdot \cos\varphi,\ x_2 = x \sin\theta \cdot \sin\varphi,\ x_3 = x\cos\theta$$

and the fact that $v(\vec{x})$ admits spherical symmetry means that v depends only on x. We also assume that the function $v(x)$ is real and bounded at any $x \in [0, \infty)$ and that $\lim_{x \to \infty} v(x) = 0$.

The solution $\psi(\vec{x})$ is expanded on the complete set of the spherical functions $Y_{lm}(\theta, \varphi)$,

$$\psi(\vec{x}) = \psi(x, \theta, \varphi) = x^{-1} \sum_{l=0}^{\infty} \sum_{m=-l}^{l} y_l(x) Y_{lm}(\theta, \varphi). \tag{4.5.3}$$

We recall here that the spherical functions satisfy the equation

$$\left[\frac{1}{\sin\theta}\frac{\partial}{\partial\theta}\left(\sin\theta\frac{\partial}{\partial\theta}\right)+\frac{1}{\sin^2\theta}\frac{\partial^2}{\partial\varphi^2}+l(l+1)\right]Y_{lm}(\theta, \varphi) = 0, \tag{4.5.4}$$

$\theta \in [0, \pi]$, $\varphi \in [0, 2\pi]$, with the orthonormalization condition

$$\int_0^{\pi}\int_0^{2\pi} Y_{lm}^*(\theta, \varphi) Y_{l'm'}(\theta, \varphi) \sin\theta \, d\theta \, d\varphi = \delta_{ll'}\delta_{mm'}, \tag{4.5.5}$$

and that the Laplace operator in spherical coordinates reads

$$\Delta = \frac{l}{x^2}\frac{\partial}{\partial x}\left(x^2\frac{\partial}{\partial x}\right)+\frac{1}{x^2}\left[\frac{1}{\sin\theta}\frac{\partial}{\partial\theta}\left(\sin\theta\frac{\partial}{\partial\theta}\right)+\frac{1}{\sin^2\theta}\frac{\partial^2}{\partial\varphi^2}\right], \tag{4.5.6}$$

see [7]. On introducing expansion (4.5.3) into original equation (4.5.1), multiplying the result by $Y_{lm}^*(\theta, \varphi)$ and integrating over θ and φ (properties (4.5.4—6) are appropriately used in this chain), we find that $y_l(x)$ satisfy

$$y_l'' + (E - V_l(x))y_l = 0, \; x \in (0, \infty), \; l = 0, 1, 2, \ldots \tag{4.5.7}$$

where

$$E = \frac{2m}{\hbar^2}\varepsilon, \; V_l(x) = \frac{l(l+1)}{x^2} + \frac{2m}{\hbar^2} v(x) \tag{4.5.8}$$

The original problem (4.5.1) is thus reduced to the infinite set (4.5.7) of separate equations; each equation of this set is known among physicists as a *radial* Schrödinger equation. The important particularities of the radial equations are that $V_l(x)$ is singular at the origin when $l \neq 0$ and that, due to the presence of factor x^{-1} in eq. (4.5.3) and to the physical requirement that the original $\psi(\vec{x})$ must be finite everywhere, the solution of eq. (4.5.7) must vanish at the origin, i.e. $y_l(0) = 0$ for any l.

To fix the ideas we will focus on one radial equation, that with $l = L \neq 0$ say, and omit the subscript.

The problem to be considered numerically hence is

$$y'' + (E - V(x))y = 0, \quad x \in (0, \infty) \tag{4.5.9}$$

with the conditions

$$y(0) = 0, \tag{4.5.10}$$

$$y(x) \text{ is finite at any } x > 0. \tag{4.5.11}$$

We distinguish two regimes in E.

$E < 0$. Our first aim is to convert the problem (4.5.9—11) into a standard mathematical form and this essentially raises the problem of finding an equivalent of standard form for the somewhat unusual condition (4.5.11). We start with the remark that, since $\lim_{x\to\infty} V(x) = 0$, we can always choose some $x_{\max}$ in the asymptotic range of x in such a way that $|E| \gg |V(x)|$ for $x \geqslant x_{\max}$. If so, approximation $E - V(x) \simeq$ $\simeq E$ becomes justified and eq. (4.5.9) assumes the form

$$y'' + Ey = 0, \quad x \geqslant x_{\max}, \tag{4.5.12}$$

with the general solution

$$y(x) = A\exp((-E)^{1/2}x) + B\exp(-(-E)^{1/2}x) \tag{4.5.13}$$

where A and B are arbitrary constants.

We see that the first component exponentially increases with x and therefore the only point where the condition (4.5.11) may be violated is at infinity. The condition will be automatically satisfied only if $A = 0$ or, in other words, condition (4.5.11) is equivalent to the behaviour

$$y(x) = B\exp(-(-E)^{1/2}x), \quad x \geqslant x_{\max}. \tag{4.5.14}$$

At its turn, this $y(x)$ is the solution of the following differential equation

$$y'(x) + (-E)^{1/2}y(x) = 0, \quad x \geqslant x_{\max} \tag{4.5.15}$$

so that, altogether, it can be said that the unusual condition (4.5.11) is equivalent to the condition (4.5.15) which resembles the second boundary condition of a standard Sturm-Liouville problem, see eq. (4.4.2).

In conclusion, the original problem (4.5.9—11) is found equivalent to a Sturm-Liouville-like problem, eqs (4.5.9, 10, 15).

Three striking differences actually exist between the latter and the standard Sturm-Liouville problem (4.4.1—2). First, the domain is finite in the standard Sturm-Liouville problem, while infinite in problem (4.5.9, 10, 15). Second, only the boundary conditions on the left are similar while the second condition which refers to the very point b in eq. (4.4.2) is now replaced by an asymptotic condition. Also one of the coefficients of the latter boundary condition is E-dependent while in the orthodox Sturm-Liouville problem the coefficients were λ-independent. Third, d, g and r of eq. (4.4.1) were bounded functions while $V(x)$ is singular.

For these reasons, although problem (4.5.9, 10, 15) is really an eigenvalue problem as it is the original Sturm-Liouville problem, some of the properties (i)—(iv) associated to the former will be now modified. Specifically, property (i) is modified in the sense that the set of eigenvalues $E_0, E_1, E_2, \ldots$ (these are called eigenenergies) is in general *finite*; the corresponding eigenfunctions $y_i(x)$, $i = 0, 1, 2, \ldots, N$ remain orthogonal as in eq. (4.4.3) (the integration limits are here $a=0$ and $b=\infty$) but they do no longer form a *complete* set as was indicated under (iii). Properties (ii) and (iv) remain unchanged.

The number of eigenenergies, N, depends on the shape of $v(x)$. Eigenenergies exist only if $v(x)$ assumes *negative* values in some subranges of $(0, \infty)$ and their number actually increases with the length of such subranges and also with the depth of $v(x)$ on each subrange. (Note for completeness that N becomes infinite for the singular function $v(x) = -\alpha/x$, $\alpha > 0$, a function known as the attractive Coulomb potential.)

To solve numerically problem (4.5.9, 10, 15) two main issues must be accounted for adequately, namely the singularity of V at $x = 0$ and the fact that the range of the independent variable is infinite. First issue is usually treated by separating a small subrange $(0, a]$ on which the solution is constructed by some ad-hoc numerical procedure as, for instance, the procedure given in section 3.7. As for the second, it must be noted that any attempt to make the range finite by a suitable change of variables is nonoperational since it will produce a more complicated equation, possibly with singular coefficients. It is therefore preferred to operate with the original equation and to approximate to the infinite range.

Two recipes are often taken. First is simply the replacement of condition (4.5.14) or (4.5.15) (both imply that $y(x) \to 0$ when $x \to \infty$) by

$$y(b) = 0 \tag{4.5.16}$$

where b is some value in the asymptotic zone, $b > x_{\max}$. The second consists of taking just (4.5.15) at $x = b$,

$$y'(b) + (-E)^{1/2}\, y(b) = 0. \tag{4.5.17}$$

Both are used in practical application. Surprisingly enough, the first is even more popular than the second, although the latter is definitely more realistic in the sense that it simulates better the original condition (4.5.11) which means $A = 0$ in eq. (4.5.13). In fact, the first condition will only guarantee that the coefficient A is a small number, not exactly zero as it must be. For this reason, the first condition will typically require values of b significantly larger than the second, therefore longer time to compute the solution.

To summarize, when $E < 0$ the original problem (4.5.9, 10, 11) is reduced to two problems which are treated sequentially. First of them refers to the small interval $(0, a]$ in which the values at a of the regular solution and of its derivative are constructed. These $\bar{y}(a)$ and $\bar{y}'(a)$ are obviously defined up to some arbitrary factor. However, the factor is the same in each of them and thus the ratio $C(E) = \bar{y}'(a)/\bar{y}(a)$ is uniquely determined. The second problem is that on $[a, b]$ and it consists of solving eq. (4.5.9) with the boundary conditions

$$y'(a) - C(E)y(a) = 0 \tag{4.5.18}$$

and (4.5.16), when the first recipe is taken, or (4.5.17) in the case of the second recipe.

The computation of this eigenvalue problem can be made by any formula in either shooting or direct form. The shooting is perhaps more appropriate since the analytic expression of $C(E)$ is not known; $C(E)$ is to be computed numerically at each test value E. Direct methods may be used without difficulty only when a is so small that the zeroth-order approximation $y(a) = a^{l+1}$ (see eq. (3.7.9)) is sufficiently accurate; in this case C becomes independent of E, namely $C = (L + 1)/a$.

If a two-step formula is used for the eigenvalue problem, the condition (4.5.18) can be easily converted in a more convenient form, viz.

$$y(a) - S(E)y(a+h) = 0. \tag{4.5.19}$$

In fact a regular solution is generated on $(0, a+h)$ and its values at a and $a+h$ furnish $S(E)$ as $S(E) = \bar{y}(a)/\bar{y}(a+h)$. The condition (4.5.17) can be converted much simpler whereas eq. (4.5.14) equivalently reads as $y(x-h) = y(x)\exp[(-E)^{1/2}h]$ so that

$$y(b-h) - \exp[(-E)^{1/2}h]y(b) = 0 \tag{4.5.20}$$

which is the desired form.

$E > 0$. In this case the general solution of the asymptotic equation (4.5.12) is

$$y = A\sin(E^{1/2}x) + B\cos(E^{1/2}x) \tag{4.5.21}$$

and therefore condition (4.5.11) is automatically satisfied. Its numerical solution consists of first solving on $(0, a]$ to generate $\bar{y}(a)$ and $\bar{y}'(a)$, then on $[a, b]$ by any initial value method. The resultant $\bar{y}(b)$ and $\bar{y}'(b)$ are equated with the values at b of the asymptotic solution (4.5.21) and of its first derivative, respectively; an algebraic linear system for A and B results which, when solved, completely characterizes $y(x)$ for $x \geqslant b$.

Note for completness that when $E = 0$ the general solution of eq. (4.5.12) is $y(x) = Ax + B$, so that $|y(x)|$ increases with x.

A problem which may be posed in such a case is to construct eigen-*potentials*, but this is out of our concern here.

REFERENCES

1. P. Henrici, *Discrete Variable Methods in Ordinary Differential Equations*, John Wiley and Sons, New York, 1964.
2. G. M. Phillips and P. J. Taylor, *Theory and Applications of Numerical Analysis*, Academic Press, London, 1973.
3. S. M. Roberts and J. S. Shipman, *Two point Value Problems: Shooting Methods*, Elsevier, New York, 1972.

4. L. Fox, *Numerical Solution of Ordinary and Partial Differential Equations*, Pergamon Press, New York, 1962.
5. J. D. Lambert, *Computational Methods in Ordinary Differential Equations*, John Wiley and Sons, London, 1973.
6. H. Sagan, *Boundary and Eigenvalue Problems in Mathematical Physics*, John Wiley and Sons, New York, 1961.
7. L. D. Landau and E. M. Lifchitz, *Mécanique Quantique*, Editions Mir, Moscow, 1966; English translation also exists.

Chapter 5

Practical applications

The numerical solution of differential equations is of considerable interest in current applications since such equations govern a large variety of phenomena in science and engineering.

While many such equations are simple enough from the numerical point of view to be solved even by beginners, the number of difficult cases is not negligible at all and then the solution requires some culture in the field of numerical methods, computational experience and, last but not least, inventivity.

As a matter of fact, the literature is rich in papers in which the efficiency of a certain numerical method is analyzed for some difficult equation and appropriate programming tricks are proposed to make the computation faster. A selection of examples of this type to be included in this book is however difficult since the very act of selection may favour some readers in prejudice of the others.

More instructive seems the way in which a *single* example is investigated by means of *several* distinct methods, with programs included. This is the way we adopt with the hope that so the relative merits and limits of various methods are better understood. It also enables to see how the theoretical predictions are reflected and eventually exploited in current practice on computer and equally offers a good introduction into the nice world of programming tricks.

5.1. FORMULATION OF THE TEST PROBLEM

The test problem we choose is the eigenvalue problem for the radial Schrödinger equation corresponding to the so-called S partial wave, that is eq. (4.5.7) with $l = 0$.

We put down here the problem again

$$y'' + (E - V(x))y = 0, \; x \in [0, +\infty), \; E < 0, \tag{5.1.1}$$

with the conditions

$$y(0) = 0, \tag{5.1.2}$$

$$y(x) \text{ is finite everywhere on } (0, +\infty). \tag{5.1.3}$$

The potential $V(x)$ is taken in the so-called Woods-Saxon form, that is

$$V(x) = PP/B(x) + QQ \exp\left[\frac{x - x_0}{a_0}\right]/B^2(x) \tag{5.1.4}$$

with

$$B(x) = 1 + \exp\left[\frac{x - x_0}{a_0}\right] \tag{5.1.5}$$

where PP, QQ, a_0 and x_0 are numerical parameters.

This problem has the advantage of being both realistic from the physical point of view (the Woods-Saxon potential is very popular in nuclear physics) and convenient from the mathematical point of view. The latter comes from the fact that it can be solved analytically (see for instance Gy. Bencze, Commentationes Physico-Mathematicae, *31* (1966)) and thus the exact solution is available to be used as reference to check for the accuracy of any numerical scheme.

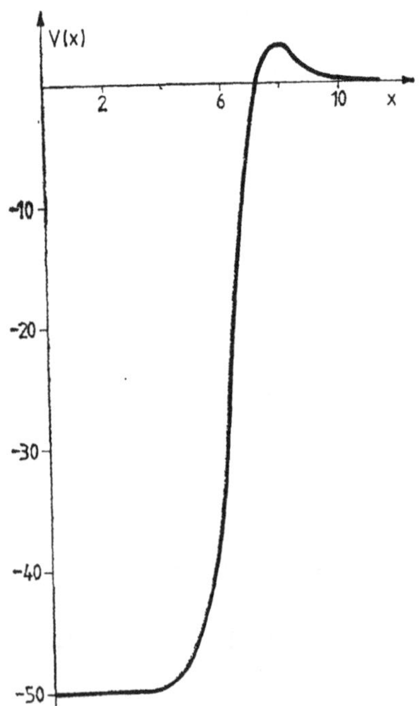

Fig. 4. — Potential $V(x)$ for the test equation (5.1.1)

We will be interested in the numerical computation of the eigenvalues $E_0, E_1, E_2, \ldots$ only, and to do it a shooting procedure will be used. The errors in eigenvalues will just reflect the errors in the methods to advance the solution.

The following parameters are used: $x_0 = 7$, $a_0 = .6$, $PP = -50$, $QQ = -PP/a_0$ and the shape of the resultant $V(x)$ is plotted on Fig. 4.

There are fourteen eigenvalues for this case. Their values which are exact in nine decimal figures are as follows

$$\begin{aligned} E_0 &= -49.457788728, & E_1 &= -48.148430420,\\ E_2 &= -46.290753954, & E_3 &= -43.968318432,\\ E_4 &= -41.232607772, & E_5 &= -38.122785097,\\ E_6 &= -34.672313206, & E_7 &= -30.912247488,\\ E_8 &= -26.873448916, & E_9 &= -22.588602258,\\ E_{10} &= -18.094688282, & E_{11} &= -13.436869040,\\ E_{12} &= -8.676081671, & E_{13} &= -3.908232481, \end{aligned} \tag{5.1.6}$$

and these are taken as reference eigenvalues.

5.2. ORGANIZATION OF THE PROGRAM

As explained in section 4.5, to search for the eigenvalues E_i, $i=0, 1, 2, \ldots$, a point x_{max} is first delimited to ensure that the approximation $V(x) = 0$ for $x > x_{max}$ does not affect the results in the desired precision. Once x_{max} is fixed the mathematical problem becomes

$$y'' + (E - V(x))y = 0,\ x \in [x_{min} = 0,\ x_{max}],\ E < 0 \tag{5.2.1}$$

with the boundary conditions

$$y(0) = 0,\ P(E) = 0$$

where

$$P(E) \equiv y'(x_{max}) + (-E)^{1/2}y(x_{max}) \tag{5.2.2 a}$$

cf. eq. (4.5.17), or

$$P(E) \equiv y(x_{max}) - \exp((-E)^{1/2}h)y(x_{max} + h), \tag{5.2.2 b}$$

see eq. (4.5.20).

Since it is well known that all eigenvalues are larger than $E_{min} = \min_{x \in [x_{min}, x_{max}]} V(x)$, only the energy interval $[E_{min}, E_{max} = 0]$ must be effectively explored. For the quoted potential we can take $E_{min} = -50$.

The computation of the eigenvalues by a shooting procedure basically requires the possibility of calculating $P(E)$ at any E. This essentially means to write a program which solves eq. (5.2.1) with the initial conditions $y(0) = 0$ and arbitrary $y'(0) \neq 0$. Once $y(x_{max})$ and $y'(x_{max})$ or $y(x_{max})$ and $y(x_{max} + h)$ were generated, the value of $P(E)$ is completely determined either as in eq. (5.5.2 a) which is the choice in case of *one-step methods*, or as in eq. (5.5.2 b) in case of *multistep methods*.

In practice some test values $E^0 = E_{min} < E^1 < E^2 < E^3 \ldots$ are taken in the energy domain and the values of $P(E^0)$, $P(E^1), P(E^2), \ldots$ are successively calculated until it is detected that the signs of $P(E^{i-1})$ and of $P(E^i)$ are different. The latter situation guarantees that the first eigenvalue lies in the interval (E^{i-1}, E^i) and this is further located with the desired accuracy by computing the root of $P(E)$ via some standard rootfinding procedure. Once this is found the execution goes forward to calculate and compare $P(E^i), P(E^{i+1})$ etc. with the aim of searching for the next eigenvalue, and so on.

In short, the program consists of two main blocks, one in which the given differential equation is integrated at any E to construct $P(E)$, and another in which the root of $P(E)$ is evaluated.

Six numerical methods are actually programmed to generate $P(E)$ namely (standard) Numerov method (algorithm (1.13.26)), the Numerov method in summed form (eqs (1.11.46 a) and (1.11.46 b)), the de Vogelaere method (eqs (1.12.13 a), (1.12.13 b) and (1.12.13 c)), the standard Runge-Kutta method (that is with parameters written in array (2.2.22)), CPM(0) (i.e. algorithm (3.3.25) with identifications (3.4.3), (3.4.7) and (3.4.8)) and CPM(1) (where the first-order corrections (3.4.46), (3.4.47), (3.4.50) and (3.4.51) are added to the previous CPM(0)).

These methods were chosen so as to produce compact experimental evidence over the relative merits of various classes of methods discussed in this book; note actually that the first three methods are multistep methods, the fourth is a genuine one-step method, and the last two methods belong to the class of piecewise perturbation methods.

The program is written in FORTRAN-4 in simple precision arithmetic for a computer IBM 370/135. Since the world length is here of about seven decimal figures in the mantissa and the maximal values of the potential and of the eigenvalues are of the order of tens, the maximal accuracy to be expected on the quoted computer is limited at the fifth decimal place in each eigenvalue. For this reason a tolerance of $8 \cdot 10^{-5}$ is used to locate each eigenvalue and the cut-off value $x_{max} = 10$ is found sufficient for such accuracy. As for the energy domain $[E_{min} = -50, E_{max} = 0]$ we choose equidistant test points $E^i = E_{min} + (i - 1)h_{en}$ where the energy step h_{en} is equal to 1.

5.3. THE PROGRAM

As stated, the eigenvalue problem (5.2.1) with the boundary conditions (5.2.2 a) or (5.2.2 b) will be computed by means of six distinct numerical methods. These are labelled by the integer L as follows

L = 1 standard Numerov method
L = 2 Numerov method in summed form
L = 3 de Vogelaere method
L = 4 standard Runge-Kutta method
L = 5 CPM(0)
L = 6 CPM(1)

The program described below consists of six separate modules, namely the main program and five subprograms: POT, PRELIM, NUMSOL, P and ROOT.

In the main program the entrance data are communicated, namely the reference eigenvalues given in eq. (5.1.6), ends of the space and energy domain, that is x_{min}, x_{max}, E_{min} and E_{max}, space and energy stepsizes h and h_{en}, numerical method to be used, and the parameters of the potential: a_0, x_0, PP and QQ. The main program orders the solution of the problem which is effectively carried out by subroutines PRELIM and NUMSOL. The obtained and the reference eigenvalues are finally compared and the result is printed. (The errors are printed in 10^{-4} units.) In list 1, the main program is presented for the search of the eigenvalues through all six methods at $h = 1/16$.

Subroutine POT. This is a FUNCTION-subprogram (see list 2) which calculates the value of $V(x)$ as given by eqs (5.1.4) and (5.1.5). The parameters of the potential are here communicated from the main program through the block COMMON/PARPOT/.

It is instructive to notice here a programming trick aimed to speed up the computation of $V(x)$. If eqs (5.1.4) and (5.1.5) would be used directly, then the evaluation of $V(x)$ would require the execution of EXP function twice, once to determine B and next in eq. (5.1.4). The statement 4 in POT avoids one exponentiation to generate B simply as a sum, B = 1 + EXPON, which is much faster (by about two orders of magnitude) than if the exponential term were here computed anew.

As a matter of fact, most of the programming tricks used in the whole program described here are based on two general criteria.

1. When a program implies repeated calculation of one and the same quantity the computation effort is significantly decreased if that quantity

```
      COMMON /PARPOT/ AO,XO,PP,QQ
      DOUBLE PRECISION EXEV(14)
      DIMENSION COMPEV(14)
      DATA EXEV/-49.457788728,-48.148430420,-46.290753954,-43.968318432,
     1-41.232607772,-38.122785097,-34.672313206,-30.912247488,-26.873448
     1916,-22.588602258,-18.094688282,-13.436869040,-8.676081671,-3.9082
     132481/
      AO=0.6
      XO=7.
      PP=-50.
      QQ=-PP/AO
      XMIN=0.
      XMAX=10
      EMIN=-50
      EMAX=0
      H=1./16
      HEN=1
      DO 1 L=1,6
      PRINT 10,L,H
   10 FORMAT(///,55X,'METHOD',I2,2X,'STEPSIZE',F8.5,/)
      CALL PRELIM(L,XMIN,XMAX,H,NSTEP)
      CALL NUMSOL(L,EMIN,EMAX,HEN,H,NSTEP,IMAX,CCMPEV)
      CO 2 I=1,IMAX
      IREF=I-1
      ERR=10000*(EXEV(I)-COMPEV(I))
      PRINT 20,IREF,EXEV(I),IREF,CCMPEV(I),IREF,ERR
   20 FORMAT(25X,'EXACT EV.(',I2,') = ',F9.5,5X,'COMPUTED EV.(',I2,') = '
     1,F9.5,5X,'ERROR(',I2,') = ',F8.1)
    2 CONTINUE
    1 CONTINUE
      END
```

List 1. Main program.

```
      FUNOTION POT(X)
      COMMON /PARPOT/ AO,XO,PP,QQ
      EXPON=EXP((X-XO)/AO)
      B=1+EXPON
      POT=PP/B+QQ*EXPON/(B*B)
      RETURN
      END
```

List 2. Function POT.

is calculated separately and stored in the core memory for further re-use whenever it is needed. This attitude is motivated by the fact that the activation of data from the core memory is one of the fastest operations in computers, anyway much faster than any arithmetic operation.

2. When an expression containing arithmetic operations is to be programmed we must first care to write it in a form in which the number of multiplications and of divisions is reduced as much as possible. The reason of this is that, from the four arithmetic operations, the multiplication and the division take much more computation time than the addition or the subtraction.

These points in mind we are now going to construct the subprograms PRELIM (list 3) and P (list 4) which collectively allow the integration of eq. (5.2.1) from x_{min} to x_{max} (or to $x_{max} + h$, according to the method used) at any E, thus producing the value of $P(E)$.
Each of the six numerical methods is analyzed separately.

Standard Numerov method

Its algorithm is *

$$y_{k+1} = \left(1 - \frac{1}{12} h^2 g_{k+1}\right)^{-1} \left[\left(2 + \frac{5}{6} h^2 g_k\right) y_k - \right.$$

$$\left. - \left(1 - \frac{1}{12} h^2 g_{k-1}\right) y_{k-1}\right], \tag{5.3.1}$$

$$g_i = V_i - E, \quad i = k-1, k, k+1, \quad k = 1, 2, 3, \ldots$$

The first starting value is known, $y_0 = y(x_{min} = 0) = 0$, while for the second starting value we can take $y_1 = \alpha$ where α is an arbitrary nonvanishing number. (Remember that, since eq. (5.2.1) is homogeneous and also $y(0) = 0$, the value of α affects only the *scale* of the solution, not its form.)

At each step k of the iteration process y_{k+1} is determined in terms of the known y_{k-1}, y_k and h but the evaluation of the expression on the right in eq. (5.3.1) actually requires the computation of g_i, $i =$

* In the following we omit to put the bar over y.

```
      SUBROUTINE PRELIM(L,XMIN,XMAX,H,NSTEP)
      COMMON/PTN/V(1500),V1(1500),V2(1500)
      NSTEP=(XMAX-XMIN)/H
      X=XMIN
      GO TO (11,11,13,14,15,16),L
C
C     NUMEROV AND NUMEROV IN SUMMED FCRM
C
   11 N=NSTEP+1
      HH=H*H/12.
      DO 1 K=1,N
      V(K)=HH*POT(X)
      X=X+H
    1 CONTINUE
      V(N+1)=0
      RETURN
C
C     DE VOGELAERE
C
   13 N=NSTEP+1
      HH=H*H/3.
      V1(1)=HH*POT(XMIN-H)
      DO 2 K=1,N
      V(K)=HH*POT(X)
      X=X+H
    2 CONTINUE
      RETURN
C
C     RUNGE - KUTTA
C
   14 N=NSTEP+1
      HH=H*H/2.
      H1=0.5*H
      DO 3 K=1,NSTEP
      V(K)=HH*POT(X)
      V1(K)=HH*POT(X+H1)
      X=X+H
    3 CONTINUE
      V(N)=HH*POT(X)
      RETURN
```

```
C
C     C P M ( 0 )
C
   15 HH=H*H
      VI1=POT(X)
      DO 4 K=1,NSTEP
      VI2=POT(X+0.2*H)
      VI3=POT(X+0.4*H)
      VI4=POT(X+0.6*H)
      VI5=POT(X+0.8*H)
      X=X+H
      VI6=POT(X)
      V(K)=HH*(19*(VI1+VI6)+75*(VI2+VI5)+50*(VI3+VI4))/288.
      VI1=VI6
    4 CONTINUE
      RETURN
C
C     C P M ( 1 )
C
   16 VI1=POT(X)
      HH=H*H
      DO 5 K=1,NSTEP
      VI2=POT(X+0.2*H)
      VI3=POT(X+0.4*H)
      VI4=POT(X+0.6*H)
      VI5=POT(X+0.8*H)
      X=X+H
      VI6=POT(X)
      V(K)=HH*(19*(VI1+VI6)+75*(VI2+VI5)+50*(VI3+VI4))/288.
      V1(K)=HH*(111*(VI6-VI1)+425*(VI5-VI2)-150*(VI4-VI3))/1344.
      V2(K)=HH*25*(5*(VI1+VI6)+6*(VI2+VI5)-11*(VI3+VI4))/1008.
      VI1=VI6
    5 CONTINUE
      RETURN
      END
```

List 3. Subroutine PRELIM.

```
      FUNCTION P(L,H,NSTEP,E)
      COMMON/PTN/V(1500),V1(1500),V2(1500)
      GO TO (10,20,30,40,50,60),L
C
C     STANDARD NUMEROV
C
   10 EH=H*H*E/12.
      NMAX=NSTEP+2
      D0=EH-V(1)
      D1=EH-V(2)
      Y0=0
      Y1=H
      DO 11 K=3,NMAX
      D2=EH-V(K)
      Y2=((2-10*D1)*Y1-(1+D0)*Y0)/(1+D2)
      D0=D1
      D1=D2
      Y0=Y1
      Y1=Y2
   11 CONTINUE
      P=Y1-Y0*EXP(-H*SQRT(-E))
      RETURN
C
C     NUMEROV IN SUMMED FORM
C
   20 EH=H*H*E/12.
      Z1=0
      S1=(1-(V(2)-EH))*H
      NMAX=NSTEP+2
      DO 21 K=2,NMAX
      Z0=Z1
      S0=S1
      D1=EH-V(K)
      Z1=Z0+S0
      S1=S0-12*D1*Z1/(1+D1)
   21 CONTINUE
      Y1=Z1/(1+D1)
      Y0=Z0/(1-(V(NMAX-1)-EH))
      P=Y1-Y0*EXP(-H*SQRT(-E))
      RETURN
```

```
C     DE VOGELAERE
C
   30 EH=H*H*E/3.
      YO=O
      ZO=H
      YPREV=-H
      FPREV=(V1(1)-EH)*YPREV
      FO=(V(1)-EH)*YO
      CO 31 K=1,NSTEP,2
      YNEXT=YO+ZO+2*FO-O.5*FPREV
      FNEXT=(V(K+1)-EH)*YNEXT
      Y1=YO+2*(ZO+2*FNEXT+FO)
      F1=(V(K+2)-EH)*Y]
      Z1=ZO+FO+4*FNEXT+F1
      YPREV=YNEXT
      FPREV=FNEXT
      YO=Y1
      FO=F1
      ZO=Z1
   31 CONTINUE
      P=ZO/H+YO*SQRT(-E)
      RETURN
C
C     STANDARD RUNGE - KLTTA
C
   40 EH=H*H*E/2.
      YO=O
      ZO=H
      CO 41 K=1,NSTEP
      FI=(V(K)-EH)*YO
      ZI=ZO+FI
      YI=YO+O.5*ZO
      FII=(V1(K)-EH)*YI
      ZII=ZO+FII
      YII=YO+O.5*ZI
      FIII=(V1(K)-EH)*YII
      ZIII=ZO+2*FIII
      YIII=YO+ZII
      FIIII=(V(K+1)-EH)*YIII
      Y1=YO+(ZO+ZIII+2*(ZI+ZII))/6.
      Z1=ZO+(FI+FIIII+2*(FII+FIII))/3.
      YO=Y1
      ZO=Z1
   41 CONTINUE
      P=ZO/H+YO*SQRT(-E)
      RETURN
```

List 4. Function P.

```
C
C      C P M ( 0 )
C
   50  EH=H*H*E
       Y0=0
       Z0=H
       TZ=0.4

       DO 51 K=1,NSTEP
       Z=V(K)-EH
       U=ABS(Z)
       IF(U.LT.TZ) GO TO 52
       SQU=SQRT(U)
       IF(Z.GE.TZ) GO TO 53
       OSI=COS(SQU)
       ETA0=SIN(SQU)/SQL
       GO TO 54
   53  EXPAR=EXP(SQU)
       EXPINV=1./EXPAR
       CSI=(EXPAR+EXPINV)*0.5
       ETA0=(CSI-EXPINV)/SQU
       GO TO 54
   52  CSI=(40320+Z*(20160+Z*(1680+Z*(56+Z))))/40320.
       ETA0=(5040+Z*(840+Z*(42+Z)))/5040.
   54  Y1=CSI*Y0+ETA0*Z0
       Z1=Z*ETA0*Y0+CSI*Z0
       Y0=Y1
       Z0=Z1
   51  CONTINUE
       P=Z0/H+Y0*SQRT(-E)
       RETURN
```

```
C
C      C P M ( 1 )
C
   60  EH=H*H*E
       Y0=0
       Z0=H
       TZ=0.4
       DO 61 K=1,NSTEP
       Z=V(K)-EH
       U=ABS(Z)
       IF(U.LT.TZ) GO TC 62
       SQU=SQRT(U)
       IF(Z.GE.TZ) GO TC 63
       CSI=COS(SQU)
       ETA0=SIN(SQU)/SQU
       ETA1=(CSI-ETA0)/Z
       ETA2=(ETA0-3*ETA1)/Z
       GO TO 64
   63  EXPAR=EXP(SQU)
       EXPINV=1./EXPAR
       CSI=(EXPAR+EXPINV)*0.5
       ETA0=(CSI-EXPINV)/SCU
       ETA1=(CSI-ETA0)/Z
       ETA2=(ETA0-3*ETA1)/Z
       GO TO 64
   62  CSI=(40320+Z*(20160+Z*(1680+Z*(56+Z))))/40320.
       ETA0=(5040+Z*(840+Z*(42+Z)))/5040.
       ETA1=(15120+Z*(1512+Z*(54+Z)))/45360.
       ETA2=(33264+Z*(2376+Z*(66+Z)))/498960.
   64  CORR1=V1(K)*ETA1
       CORR2=V2(K)*ETA2
       Y1=(CSI-CORR1)*Y0+(ETA0-CORR2)*Z0
       Z1=(ETA0+CORR2)*Z*Y0+(CSI+CORR1)*Z0
       Y0=Y1
       Z0=Z1
   61  CONTINUE
       P=Z0/H+Y0*SQRT(-E)
       RETURN
       END
```

List 4. (continued)

$= k - 1, k, k + 1$ as an intermediary stage. During the shooting procedure the iteration scheme (5.3.1) is repeated at each new E and thus the values of $V_0, V_1, V_2, \ldots$ are used repeatedly many times. On using criterion 1, the computation will be greatly facilitated if these values are calculated once, in a preliminary run, and stored.

Further time reduction is obtained by examining formula (5.3.1) in terms of criterion 2. In fact, if formula (5.3.1) would be used just at it stands, the execution will require 13 slow operations (i.e. multiplications + divisions) at each step k. This number is substantially reduced if the factor $\frac{1}{12} h^2$, which is a scaling factor of the g, is adequately used as follows.

First, not the true values of the potential at the mesh points are calculated in the preliminary run and stored in the FORTRAN vector V, but the *scaled values*

$$\mathrm{V(K)} = \frac{1}{12} h^2 V(x_{\min} + (\mathrm{K} - 1)h), \quad \mathrm{K} = 1, 2, 3, \ldots \tag{5.3.2}$$

Second, in the current iteration at given E a quantity $\mathrm{EH} = \frac{1}{12} h^2 E$ must be computed *before* the very loop over $\mathrm{K} = k + 2$. If so is done and if in each step K ($\mathrm{K} = 3, 4, 5, \ldots$) of the loop the quantities $\mathrm{D0 = EH - V(K-2)}$, $\mathrm{D1 = EH - V(K-1)}$, $\mathrm{D2 = EH - V(K)}$, $\mathrm{Y0} = y_{K-2}$, $\mathrm{Y1} = y_{K-1}$ and $\mathrm{Y2} = y_K$ are used, the iteration formula (5.3.1) becomes simply

$$\mathrm{Y2} = [(2 - 10 \cdot \mathrm{D1})\mathrm{Y1} - (1 + \mathrm{D0})\mathrm{Y0}]/(1 + \mathrm{D2}) \tag{5.3.3}$$

which consists of only 4 slow arithmetic operations.

To summarize, the computation of $P(E)$ becomes substantially faster when the corresponding block is programmed as a set of *two* distinct subprograms. The first of these accounts for the computation of the scaled values of the potential, eq. (5.3.2), and it is to be executed once, at the beginning of the whole computation at that h. The second is a subprogram in which the solution is advanced step by step through the recurrence formula of the algorithm in form (5.3.3).

Since such conclusion remains valid for all subsequent methods the SUBROUTINE PRELIM collects the generation of the scaled potential values needed by the six methods while the very recurrence formulas are collected in FUNCTION subprogram P. The values of the scaled quantities generated in PRELIM are communicated in P through the COMMON/PTN/ block.

In PRELIM instruction 3 calculates the total number of steps NSTEP of the interval $[x_{min}, x_{max}]$ while the GO TO statement (instruction 5) branches the execution according to the value of L. The generation of the scaled values of the potential at the mesh points for the Numerov method is accounted for in instructions 6—13. Note that although the total number of mesh points on the quoted interval is N=NSTEP + 1 the Numerov method will equally require $V(N + 1) = \frac{1}{12} h^2 V(x_{max} + h)$ whereas the determination (5.2.2 b) of $P(E)$ is appropriate for this two-step method. Instruction 12 assigns simply $V(N + 1) = 0$, which is consistent with the previous approximation, namely that we can take $V(x) = 0$ for $x > x_{max}$.

In P the execution is branched via a similar GO TO statement (instruction 3) and the set of instructions 4—19 is active for the standard Numerov scheme. A value $\alpha = h$ is actually assigned to start the computation and the whole sequence is programmed as explained before. The only slight modification (an improvement, according to criterion 1) is that D0 and D1 are not computed anew at each step K since these are exactly D1 and D2 of the previous step and are so identified in instructions 13 and 14. The value of P is calculated in instruction 18.

Numerov method in summed form

Its algorithm is

$$z_{k+1} = z_k + S_k \tag{5.3.4 a}$$

$$S_{k+1} = S_k + h^2 G_{k+1}, \quad k = 0, 1, 2, \ldots \tag{5.3.4 b}$$

where

$$z(x) = y(x) - \frac{1}{12} h^2 g(x) y(x) = \left(1 - \frac{1}{12} h^2 g(x)\right) y(x) \tag{5.3.5 a}$$

$$G(x; z) = g(x) z(x) \Big/ \left(1 - \frac{1}{12} h^2 g(x)\right), \quad g(x) = V(x) - E. \tag{5.3.5 b}$$

The analysis presented before for the standard Numerov scheme can be repeated easily and the result remains the same. In fact the same FORTRAN vector V(K), eq. (5.3.2) is generated (instructions 6—13 in PRELIM) and the same EH is first computed in P to be used in the loop over K.

The initial values in the loop are

$$z_0 = \left[1 - \frac{1}{12} h^2 (V(x_{\min}) - E)\right] y_0 = 0 \tag{5.3.6 a}$$

and

$$S_0 = z_1 - z_0 = z_1 = \left[1 - \frac{1}{12} h^2 (V(x_1) - E)\right] y_1 =$$

$$= [1 - (\mathrm{V}(2) - \mathrm{EH})]\, h \tag{5.3.6 b}$$

(again $y_1 = \alpha \equiv h$ is taken.)
In each step of the loop over K (K = 2, 3, 4, . . ., NMAX = NSTEP+2) the quantities D1 = EH − V(K), S0 = S_{K-2}, Z0 = z_{K-2}, are defined to produce S1 = S_{K-1} and Z1 = z_{K-1} through the recurrence formulas

$$\mathrm{Z1} = \mathrm{Z0} + \mathrm{S0} \tag{5.3.7 a}$$

$$\mathrm{S1} = \mathrm{S0} - 12 \cdot \mathrm{D1} \cdot \mathrm{Z1}/(1 + \mathrm{D1}) \tag{5.3.7 b}$$

which consist of a total of only 3 slow arithmetic operations; for the corresponding sequence of cards see instructions 20—34 of subprogram P.

Once the loop over K is terminated the numerical values of the original solution Y0(= $y(x_{\max})$) and Y1(= $y(x_{\max} + h)$) are generated in cards 31 and 32 and the value of P is computed in card 33 by formula (5.2.2 b).

It is interesting to remark that the computation through this method is expected to be slightly faster than through the previous method (the number of slow operations is 3 and 4, respectively), an advantage which should be added to the fact that, as the theory predicts, this method is also more robust than the other with respect to the round-off errors. These expectations will be actually confirmed by the runs but we must insist that the advantage with respect to the time consumed is only casual. In fact, it is due to the very problem investigated here, more exactly that only the eigenvalues are required. If we would also need the values of the solution, an extra division

$$\mathrm{Y1} = \mathrm{Z1}/(1 + \mathrm{D1}) \tag{5.3.7 c}$$

is to be added in each step K and thus the equality between the number of slow operations of the two methods is restored in full.

de Vogelaere method

Given y_{2k-1}, y_{2k} and y'_{2k} the solution is here advanced by three equations as follows

$$y_{2k+1} = y_{2k} + hy'_{2k} + \frac{h^2}{6}(4g_{2k}y_{2k} - g_{2k-1}y_{2k-1}) \tag{5.3.8 a}$$

$$y_{2k+2} = y_{2k} + 2hy'_{2k} + \frac{h^2}{3}(4g_{2k+1}y_{2k+1} + 2g_{2k}y_{2k}) \tag{5.3.8 b}$$

$$y'_{2k+2} = y'_{2k} + \frac{h}{3}(g_{2k}y_{2k} + 4g_{2k+1}y_{2k+1} + g_{2k+2}y_{2k+2}), \tag{5.3.8 c}$$

$$k = 0, 1, 2, \ldots$$

with $g(x) = V(x) - E$.

A substantial reduction of the number of slow operations is achieved if we work with the quantity $z(x) = hy'(x)$ instead of the derivative alone. In fact, if we multiply the last equation by h, the algorithm becomes more uniform, viz. given y_{2k-1}, y_{2k} and z_{2k}, the values of y_{2k+1}, y_{2k+2} and z_{2k+2} are determined as follows

$$y_{2k+1} = y_{2k} + z_{2k} + 2F_{2k} - 0.5F_{2k-1} \tag{5.3.9 a}$$

$$y_{2k+2} = y_{2k} + 2(z_{2k} + 2F_{2k+1} + F_{2k}) \tag{5.3.9 b}$$

$$z_{2k+2} = z_{2k} + F_{2k} + 4F_{2k+1} + F_{2k+2} \tag{5.3.9 c}$$

where

$$F(x) = \frac{h^2}{3} g(x) y(x) = \left(\frac{h^2}{3} V(x) - \frac{h^2}{3} E \right) y(x). \tag{5.3.10}$$

This structure of the recurrence relations suggests a strategy similar to that adopted in case of the previous methods.

To be specific, the following scaled values of the potential at the mesh points are computed in PRELIM,

$$\mathrm{V(K)} = \frac{1}{3} h^2 V(x_{\min} + (\mathrm{K} - 1)h), \quad \mathrm{K} = 1, 2, \ldots, \mathrm{NSTEP} + 1 \tag{5.3.11}$$

and stored, while in subprogram P the scaled energy $\mathrm{EH} = \frac{1}{3} h^2 E$ will be generated before the iteration loop.

To start the loop the values of y_0, z_0 and y_{-1} must be furnished. Two of them are $y_0 = 0$ (because so is the initial value) and arbitrary $z_0 = hy'_0$; $y'_0 = 1$ is assigned for convenience, so that $z_0 = h$; y_{-1} results from eq. (1.12.18) as $y_{-1} = -hy'_0 = -h$. Since the value of $\frac{h^2}{3} g(x_{\min} - h) y_{-1}$ is also needed to start the loop, the quantity $\frac{h^2}{3} V(-h)$ is also generated in PRELIM and stored at V1(1).

Clearly the iteration loop in P assumes only *odd* values of K($= 2k + 1$, $k = 0, 1, 2, \ldots$, NSTEP/2-1), i.e. K $= 1, 3, 5, \ldots$, NSTEP-1; note that NSTEP must be an *even* integer here. At each such K the computer knows YPREV ($= y_{2k-1}$), Y0($= y_{2k}$), Z0($= z_{2k}$), FPREV ($= F_{2k-1}$) and F0($= F_{2k}$), to compute YNEXT ($= y_{2k+1}$), Y1($=y_{2k+2}$) and Z1 ($= z_{2k+2}$) by the following sequence of instructions consistent with eqs (5.3.9 a-c) and (5.3.11)

$$\begin{aligned}
&\mathrm{YNEXT = Y0 + Z0 + 2 \cdot F0 - 0.5 \cdot FPREV} \\
&\mathrm{FNEXT = (V(K + 1) - EH)Y0} \\
&\mathrm{Y1 = Y0 + 2(Z0 + 2 \cdot FNEXT + F0)} \\
&\mathrm{F1 = (V(K + 2) - EH)Y1} \\
&\mathrm{Z1 = Z0 + F0 + 4 \cdot FNEXT + F1}
\end{aligned} \tag{5.3.12}$$

We see that execution of package (5.3.12) involves 7 slow arithmetic operations to advance the solution over a pair of *two* adjacent steps of the partition. That is to say that the de Vogelaere scheme uses $7/2 = 3.5$ slow operations per step and thus it is expected to be approximately as fast as each of the two schemes described before.

Once the loop is terminated the value of $P(E)$ is generated as in eq. (5.2.2 a).

Standard Runge-Kutta method

Since algorithm (2.2.23 a, b) is devised for differential equations of the first order or for systems of such equations, eq. (5.2.1) is first written in the latter form, viz.

$$\mathbf{y}' = \mathbf{f}(x, \mathbf{y}) \tag{5.3.13 a}$$

where $\mathbf{y}$ and $\mathbf{f}$ are two-dimensional column vectors,

$$\mathbf{y}(x) = \begin{pmatrix} y^1(x) \equiv y(x) \\ y^2(x) \equiv y'(x) \end{pmatrix}, \qquad \mathbf{f}(x, \mathbf{y}) = \begin{pmatrix} f^1(x, y^1, y^2) \equiv y^2 = y' \\ f^2(x, y^1, y^2) \equiv g(x)y^1 = g(x)y \end{pmatrix} \tag{5.3.13 b}$$

with $g(x) = V(x) - E$.

The execution of the algorithm first requires the calculation of the coefficients k_1, k_2, k_3 and k_4; of course each of these is a two-dimensional vector here. Thus, given y_k and y_k' we have in order

$$k_1^1 \equiv f^1(x_k, y_k^1, y_k^2) = y_k' \tag{5.3.14 a}$$

$$k_1^2 \equiv f^2(x_k, y_k^1, y_k^2) = g(x_k)y_k \equiv g_k y_k \tag{5.3.14 b}$$

$$k_2^1 \equiv f^1\left(x_k + \frac{1}{2}h,\ y_k^1 + \frac{1}{2}hk_1^1,\ y_k^2 + \frac{1}{2}hk_1^2\right) = y_k' + \frac{1}{2}hk_1^2 \tag{5.3.15 a}$$

$$k_2^2 \equiv f^2\left(x_k + \frac{1}{2}h,\ y_k^1 + \frac{1}{2}hk_1^1,\ y_k^2 + \frac{1}{2}hk_1^2\right)$$

$$= g\left(x_k + \frac{1}{2}h\right)\left(y_k + \frac{1}{2}hk_1^1\right) \equiv g_{k+\frac{1}{2}}\left(y_k + \frac{1}{2}hk_1^1\right) \tag{5.3.15 b}$$

$$k_3^1 = y_k' + \frac{1}{2} hk_2^2 \qquad (5.3.16\text{ a})$$

$$k_3^2 = g_{k+\frac{1}{2}}\left(y_k + \frac{1}{2} hk_2^1\right) \qquad (5.3.16\text{ b})$$

$$k_4^1 = y_k' + hk_3^2 \qquad (5.3.17\text{ a})$$

$$k_4^2 = g_{k+1}(y_k + hk_3^1). \qquad (5.3.17\text{ b})$$

The solution is advanced by means of eq. (2.2.23 a), viz.

$$y_{k+1} = y_k + \frac{h}{6}[k_1^1 + k_4^1 + 2(k_2^1 + k_3^1)] \qquad (5.3.18\text{ a})$$

$$y_{k+1}' = y_k' + \frac{h}{6}[k_1^2 + k_4^2 + 2(k_2^2 + k_3^2)]. \qquad (5.3.18\text{ b})$$

The direct use of these formulas should be however avoided since it requires too many slow arithmetic operations. Substantial gain is actually achieved by modifying them in a convenient way. This way is very much like the one used in the case of the de Vogelaere method and it consists of working with some new scaled quantities, viz. $z(x) = hy'(x)$ and $G(x) = \frac{1}{2} h^2 g(x)$.

Thus, given Y0($= y_k$) and Z0($= hy_k' \equiv hk_1^1$), eq. (5.3.14 b) is multiplied by the scaling factor $\frac{1}{2} h^2$ to yield

$$\text{FI}\left(\equiv \frac{1}{2} h^2 k_1^2\right) = G_k y_k. \qquad (5.3.19)$$

Formulas (5.3.15 a) and (5.3.15 b) are multiplied by h and by $\frac{1}{2} h^2$, respectively and written as a set of three formulas, that is

$$\text{ZI}(\equiv hk_2^1) = \text{Z0} + \text{FI} \qquad (5.3.20\text{ a})$$

$$\text{YI}\left(\equiv y_k + \frac{1}{2} hk_1^1\right) = \text{Y0} + 0.5 \cdot \text{Z0} \qquad (5.3.20\text{ b})$$

and

$$\text{FII}\left(\equiv \frac{1}{2}h^2k_2^2\right) = G_{k+\frac{1}{2}} \cdot \text{YI}. \tag{5.3.20 c}$$

Similar procedures are applied to the pair (5.3.16 a)—(5.3.16b)

$$\text{ZII}\,(\equiv hk_3^1) = \text{Z0} + \text{FII} \tag{5.3.21 a}$$

$$\text{YII}\left(\equiv y_k + \frac{1}{2}hk_2^1\right) = \text{Y0} + 0.5\cdot\text{ZI} \tag{5.3.21 b}$$

$$\text{FIII}\left(\equiv \frac{1}{2}h^2k_3^2\right) = G_{k+\frac{1}{2}}\cdot\text{YII} \tag{5.3.21 c}$$

and to the pair (5.3.17 a)—(5.3.17 b)

$$\text{ZIII}\,(\equiv hk_4^1) = \text{Z0} + 2\cdot\text{FIII} \tag{5.3.22 a}$$

$$\text{YIII}\,(\equiv y_k + hk_3^1) = \text{Y0} + \text{ZII} \tag{5.3.22 b}$$

$$\text{FIIII}\left(\equiv \frac{1}{2}h^2k_4^2\right) = G_{k+1}\cdot\text{YIII}. \tag{5.3.22 c}$$

With the new coefficients the formulas (5.3.18 a) and (5.3.18 b) become in order

$$\text{Y1}(\equiv y_{k+1}) = \text{Y0} + [\text{Z0} + \text{ZIII} + 2(\text{ZI} + \text{ZII})]/6 \tag{5.3.23}$$

a nd

$$\text{Z1}(\equiv hy'_{k+1}) = \text{Z0} + [\text{FI} + \text{FIIII} + 2(\text{FII} + \text{FIII})]/3. \tag{5.3.24}$$

In the program the scaled values of the potential at the mesh points and at the mid-points are correspondingly calculated in PRELIM (cards 22—31) and stored in two FORTRAN vectors V and V1, viz.

$$\text{V(K)} = \frac{1}{2}h^2V(x_{\min} + (\text{K} - 1)h),\ \text{K} = 1, 2, \ldots, \text{N} = \text{NSTEP}+1 \tag{5.3.25 a}$$

and

$$\mathrm{V1(K)} = \frac{1}{2} h^2 V\left(x_{\min} + \left(\mathrm{K} - \frac{1}{2}\right) h\right), \quad \mathrm{K} = 1, 2, \ldots, \mathrm{NSTEP}. \tag{5.3.25 b}$$

In subprogram P (cards 55—75), the scaled energy $\mathrm{EH} = \frac{1}{2} h^2 E$ is computed before the very iteration loop. At each step K (= 1, 2, ..., NSTEP) of the loop the computation is performed by means of formulas (5.3.19), (5.3.20—22 a, b, c) and (5.3.23—24) which contain a total of 11 slow operations. If this number is compared to the number of slow operations for the previous methods (these were 4, 3 and 3.5, respectively) one may expect that the standard Runge-Kutta method is approximately three times slower than each of these.

CPM(0)

The general algorithm of the piecewise perturbation methods is of the form

$$\begin{pmatrix} y(x_{k+1}) \\ y'(x_{k+1}) \end{pmatrix} = \begin{pmatrix} u(h) & v(h) \\ u'(h) & v'(h) \end{pmatrix} \begin{pmatrix} y(x_k) \\ y'(x_k) \end{pmatrix}, \quad k = 0, 1, 2, \ldots \tag{5.3.26}$$

see eq. (3.3.25).

In case of CMP(0) a constant F is introduced, eq. (3.4.3), which for $g(x) = V(x) - E$ is $F = \bar{V} - E$ with

$$\bar{V} = \frac{1}{h} \int_{x_k}^{x_{k+1}} V(x')\, \mathrm{d}x' \tag{5.3.27}$$

and u, v, u' and v' are the reference propagators given in eqs (3.4.7) and (3.4.8) in terms of the basic functions $\xi(F, h)$ and $\eta_0(F, h)$.

Several slow operations are saved if the barred functions $\bar{\xi}$ and $\bar{\eta}_0$ of a single compressed argument $Z = Fh^2$ are used instead of the original ξ and η_0 and also when the algorithm is adjusted to work with the scaled function $z(x) = hy'(x)$. In fact eq. (5.3.26) becomes

$$y(x_{k+1}) = \bar{\xi}(Z)y(x_k) + \bar{\eta}_0(Z)z(x_k), \tag{5.3.28 a}$$

$$z(x_{k+1}) = Z\bar{\eta}_0(Z)y(x_k) + \bar{\xi}(Z)z(x_k). \tag{5.3.28 b}$$

In PRELIM (cards 32—44) the scaled piecewise reference constants $h^2\bar{V}_k$ are generated and stored in the FORTRAN vector V

$$\mathrm{V(K)} = \frac{h^2}{288}[19(V_1 + V_6) + 75(V_2 + V_5) + 50(V_3 + V_4)],$$

$$\mathrm{K} = 1, 2, 3, \ldots, \mathrm{NSTEP} \tag{5.3.29 a}$$

where

$$V_i = V\left(x_K + \frac{1}{5}(i-1)\,h\right), \quad x_K = x_{\min} + (\mathrm{K}-1)h,$$

$$i = 1, 2, 3, 4, 5, 6 \tag{5.3.29 b}$$

see eq. (3.4.37), while in P (cards 80—102) the scaled energy EH = $= h^2E$ is generated before the loop.

At each K in the loop the main body of the computation refers to the evaluation of the functions $\bar{\xi}(Z)$ and $\bar{\eta}_0(Z)$; these are named in the program as CSI and ETA0 respectively.

The value of the argument $Z = h^2(\bar{V}_k - E) = \mathrm{V(K)} - \mathrm{EH}$ is first computed (instruction 81) and compared to the threshold value TZ = 0.4 previously introduced by instruction 79. Three regimes in Z are such delimited, namely Z < − TZ, − TZ ⩽ Z ⩽ TZ, and Z > TZ.

Z < − TZ. $\bar{\xi}(Z)$ and $\bar{\eta}_0(Z)$ are computed according to the upper determination in eqs (3.4.15) and (3.4.16) and this implies the use of the FORTRAN functions SQRT, SIN and COS.

− TZ ⩽ Z ⩽ TZ. Power series are used in this range,

$$\bar{\xi}(Z) = 1 + \frac{1}{2!}Z + \frac{1}{4!}Z^2 + \frac{1}{6!}Z^3 + \ldots \tag{5.3.30 a}$$

and

$$\bar{\eta}_0(Z) = Z + \frac{1}{3!}Z^2 + \frac{1}{5!}Z^3 + \ldots, \tag{5.3.30 b}$$

see eq. (3.4.19). It can be checked easily that five terms in $\bar{\xi}$ and four in $\bar{\eta}_0$ are sufficient to calculate these functions with seven exact figures (remember that the program is devised for a computer IBM 370/135 in simple precision). To minimize the number of slow operations in such an approach it is convenient to write the truncated series as sets of nested multiplications, that is by using the identity

$$a_0Z^0 + a_1Z^1 + a_2Z^2 + \ldots + a_pZ^p =$$

$$= a_p\left(\frac{a_0}{a_p} + Z\left(\frac{a_1}{a_p} + Z\left(\frac{a_2}{a_p} + \ldots + Z\left(\frac{a_{p-1}}{a_p} + Z\underbrace{\Big)\Big)\ldots\Big)}_{p \text{ times}}\right.\right.\right. \tag{5.3.31}$$

In case of $\bar{\xi}$ we have $p = 4$ and $a_0 = 1$, $a_1 = 1/2$, $a_2 = 1/24$, $a_3 = 1/720$ and $a_4 = 1/40320$ and thus $a_0/a_4 = 40320$, $a_1/a_4 = 20160$, $a_2/a_4 = 1680$ and $a_3/a_4 = 56$ so that altogether

$$\text{CSI} = (4320 + \text{Z}(20160 + \text{Z}(1680 + \text{Z}(56 + \text{Z}))))/40320. \tag{5.3.32 a}$$

Likewise, for $\bar{\eta}_0$ we have

$$\text{ETA0} = (5040 + \text{Z}(840 + \text{Z}(42 + \text{Z})))/5040. \tag{5.3.32 b}$$

$Z > TZ$. $\bar{\xi}(Z)$ and $\bar{\eta}_0(Z)$ are computed by the lower determination in eqs (3.4.15) and (3.4.16). Direct programming of these forms will require the FORTRAN functions SQRT, SINH and COSH but we prefer a slight modification which turns out to be faster. It consists of writing hyperbolic sinus and cosinus in terms of the exponential functions, viz.

$$\sinh(u) = \frac{1}{2}(\exp(u) - \exp(-u)), \; \cosh(u) = \frac{1}{2}(\exp(u) + \exp(-u)).$$

In fact, on using

$$\text{SQU} = \text{SQRT(Z)} \tag{5.3.33 a}$$

$$\text{EXPAR} = \text{EXP(SQU)} \tag{5.3.33 b}$$

$$\text{EXPINV}(\equiv \text{EXP}(-\text{SQU})) = 1/\text{EXPAR} \tag{5.3.33 c}$$

the values of $\bar{\xi}(Z)$ and $\bar{\eta}_0(Z)$ are computed as

$$\text{CSI} = (\text{EXPAR} + \text{EXPINV})/2 \tag{5.3.34 a}$$

$$\text{ETA0} = (\text{CSI} - \text{EXPINV})/\text{SQU}. \tag{5.3.34 b}$$

We thus see that with this procedure the computation of the pair SINH and COSH is replaced by execution of EXP function once, plus three slow arithmetic operations and this is actually a significant gain in computation time. Once the evaluation of $\bar{\xi}(Z)$ and of $\bar{\eta}_0(Z)$ is terminated the solution is advanced through eqs (5.3.28 a) and (5.3.28 b).

The execution time per step for this method depends on the regime in Z. It is the largest in the trigonometric range (where the functions SQRT, SIN and COS are active) and the smallest in the series range, where only arithmetic operations are employed. In the latter regime 12 slow operations are encountered (seven in eqs (5.3.32 a) and (5.3.32 b) and five in eqs (5.3.28 a) and (5.3.28 b)). As a matter of fact, the trigonometric and the exponential branches are active only when h is large enough to secure that $|Z| >$ TZ while the series branch becomes increasingly active when h is decreased. For this reason, in the limit of small h, that is when the series branch becomes the only active one, the computation with CPM(0) is expected to be approximately as fast as with the standard Runge-Kutta method. It is also important to have in mind the direct dependence of the number of slow operations in the series branch on the value assigned to TZ. (For instance, if we take TZ $=10^{-2}$, three terms in $\bar{\xi}(Z)$ and in $\bar{\eta}_0(Z)$ are only needed to reach the same seven figures accuracy and thus the number of slow operations in their nested multiplication representation is reduced to a total of 4.) This circumstance may offer good opportunity to further improve the efficiency of the CPM(0).

CPM(1)

The algorithm remains of the form (5.3.26) but now the propagators are sums of the zeroth-order contributions and the first-order corrections.

We recall here again the expressions of the latter,

$$u_1(h) = -v_1'(h) = -\frac{1}{4}(f_1^* + f_2^* h)\,\eta_1(F, h) \tag{5.3.35 a}$$

cf. eqs (3.4.46) and (3.4.51), and

$$u_1'(h)/F = -v_1(h) = \frac{1}{12} f_2^* \eta_2(F, h), \tag{5.3.35 b}$$

cf. eqs (3.4.47) and (3.4.50).

We thus see that the computation of the four corrections essentially implies the evaluation of two distinct quantities only. Another remark to be adequately exploited is that f_1^* and f_2^* are *independent of E* although they are formed through eqs (3.4.36 b) and (3.4.36 c) in which $f(x) = V(x) - E$.

Altogether if we write the first-order corrections in terms of the barred functions $\bar{\eta}_s$ these can be structured as follows

$$u_1(h) = -v_1'(h) = -\left[\frac{1}{4} h^3(f_1^* + f_2^* h)\right] \bar{\eta}_1(Z) \tag{5.3.36 a}$$

and

$$u_1'(h)/F = -v_1(h) = \left[\frac{1}{12} h^5 f_2^*\right] \bar{\eta}_2(Z), \tag{5.3.36 b}$$

i.e. as products of *two* factors in which one depends on the potential only.

For this reason the following *three* FORTRAN vectors are filled up in PRELIM (see cards 45—60)

- vector V, which collects the values of the scaled constants $h^2\bar{V}_K$ as in eq. (5.3.29 a, b),
- vector V1, which collects the front factors in the last member of eqs (5.3.36 a),

$$\mathrm{V1(K)}\left(= \frac{1}{4} h^3(f_1^* + f_2^* h)\right) = \frac{h^2}{1344}[111(V_6 - V_1) + $$
$$+ 425(V_5 - V_2) - 150(V_4 - V_3)] \tag{5.3.37}$$

— vector V2, which collects the front factors in the last member of eq. (5.3.36 b) divided by h, that is

$$\mathrm{V2(K)}\left(=\frac{1}{12}h^4 f_2^*\right)=\frac{25h^2}{1008}[5(V_1+V_6)+6(V_2+V_5)-11(V_3+V_4)], \tag{5.3.38}$$

K = 1, 2, 3, ..., NSTEP.

In subprogram P the algorithm of CPM(1) is written as

$$y(x_{k+1}) = (\bar{\xi}(Z) - \mathrm{CORR1})y(x_k) + (\bar{\eta}_0(Z) - \mathrm{CORR2})z(x_k) \tag{5.3.39 a}$$

$$z(x_{k+1}) = Z(\bar{\eta}_0(Z) + \mathrm{CORR2})y(x_k) + (\bar{\xi}(Z) + \mathrm{CORR1})z(x_k) \tag{5.3.39b}$$

for any $k\,(= K - 1) = 0, 1, 2, \ldots$, with

$$\mathrm{CORR1} = \mathrm{V1(K)}\bar{\eta}_1(Z) \tag{5.3.40 a}$$

$$\mathrm{CORR2} = \mathrm{V2(K)}\bar{\eta}_2(Z) \tag{5.3.40 b}$$

To calculate $\bar{\eta}_1(Z)$ and $\bar{\eta}_2(Z)$ at each K the same procedure as in case of CPM(0) is adopted, that is a threshold value TZ = 0.4 is first assigned to distinguish three regimes in Z.

When either (i) $Z < -\mathrm{TZ}$ or (ii) $Z > \mathrm{TZ}$ the computation is carried out by using the very definitions (3.4.18 a) and (3.4.18 b)

$$\bar{\eta}_1(Z) = [\bar{\xi}(Z) - \bar{\eta}_0(Z)]/Z \tag{5.3.41 a}$$

and

$$\bar{\eta}_2(Z) = [\bar{\eta}_0(Z) - 3\bar{\eta}_1(Z)]/Z. \tag{5.3.41 b}$$

While (iii) $-\mathrm{TZ} \leqslant Z \leqslant \mathrm{TZ}$, series (3.4.19) are used with four terms included,

$$\bar{\eta}_1(Z) = (15120 + Z(1512 + Z(54 + Z)))/45360 \tag{5.3.42 a}$$

$$\bar{\eta}_2(Z) = (33264 + Z(2376 + Z(66 + Z)))/498960. \tag{5.3.42 b}$$

As for the time consumed in the execution at each K, this depends on the regime encountered. When cases (i) or (ii) are active, five slow arithmetic operations are added (those encountered to compute eqs (5.3.41 a), (5.3.41 b), (5.3.40 a) and (5.3.40 b)) to the CPM(0) algorithm and since the latter involves the computation of the functions SQRT, SIN and COS or of SQRT and EXP, according to the case, the quoted addition is very small indeed, almost negligible. The situation is significantly changed when regime (iii) is active (this happens for sufficiently small h) when 8 such operations (eqs (5.3.42 a), (5.3.42 b), (5.3.40 a) and (5.3.40 b)) are added over the 12 operations required by the CPM(0). The computation through CPM(1) is thus expected to be slower by a factor $20/12 \simeq 1.7$ than with CPM(0).

Subroutine NUMSOL

This subroutine (list 5) explores the mesh points E^i of the energy interval $[E_{min}, E_{max}]$ to locate the positions of the eigenvalues. It actually orders the computation of $P(E^i)$ and compares its sign with that of $P(E^{i-1})$. Whenever these signs are different subprogram ROOT is called to localize the eigenvalue with the desired accuracy.

FUNCTION *subprogram* ROOT

This subprogram (list 6) calculates the root of function P(E) through successive bisections until the width of the interval where the root is lying is smaller than $\text{TOL} = 8 \cdot 10^{-5}$, to finally identify the position of the root by linear interpolation (card 19).

5.4. DISCUSSION OF THE RESULTS

The program presented above was carried out on a computer IBM 370/135 with various values of h and the results are collected in Tables 17—22. Stepsizes of the form $h = 2^{-n}$, $n = 0, 1, 2, 3, 4, 5, 6$ and 7 were tried for all methods but in each of the six tables results are reported from only the stepsizes which yielded correctly the total number, 14, of eigenvalues of the spectrum. The output from the main program of list 1 (that is with $h = 1/16$) is just photographed as list 7.

We are now going to briefly examine these results in several respects.

```
      SUBROUTINE NUMSOL(L,EMIN,EMAX,HEN,H,NSTEP,IMAX,COMPEV)
      DIMENSION COMPEV(14)
      IMAX=0
      E=EMIN
      P1=P(L,H,NSTEP,E)
    1 ENEXT=E+HEN
      IF(ENEXT.GT.EMAX) ENEXT=EMAX
      P2=P(L,H,NSTEP,ENEXT)
      PROD=P1*P2
C
C     NOTICE.  THE FOLLOWING SEQUENCE IS ACTUALLY WRITTEN UNDER THE ASSUMPTION
C     THAT NONE OF E AND ENEXT IS AN EIGENVALUE, THAT IS P1.NE.0 AND P2.NE.0.
C
      IF(PROD.GT.0.) GO TO 2
      IMAX=IMAX+1
      COMPEV(IMAX)=ROOT(E,ENEXT,P1,P2,L,H,NSTEP)
    2 P1=P2
      E=ENEXT
      IF(E.EQ.EMAX) RETURN
      GO TO 1
      END
```

List 5. Subroutine NUMSOL.

```
      FUNCTION ROOT(EA,EB,PA,PB,L,H,NSTEP)
      E1=EA
      E2=EB
      P1=PA
      P2=PB
      TOL=8.E-5
    2 ETEST=0.5*(E1+E2)
      PTEST=P(L,H,NSTEP,ETEST)
      IF(PTEST.EQ.0.) GO TO 4
      PROD=P1*PTEST
      IF(PROD.GT.0.) GO TO 1
      E2=ETEST
      P2=PTEST
      GO TO 3
    1 E1=ETEST
      P1=PTEST
    3 DEV=ABS(E2-E1)
      IF(DEV.GT.TOL) GO TO 2
      ROOT=(E1*P2-E2*P1)/(P2-P1)
      RETURN
    4 ROOT=ETEST
      RETURN
      END
```

List 6. Function ROOT.

Table 17

Absolute errors, in units 10^{-4}, in the eigenvalues of the test problem (5.1.1 —5) produced by the standard Numerov method at various values of h

i	h				
	1/8	1/16	1/32	1/64	1/128
0	.2	−.3	−.9	−1.5	−57.8
1	.2	.3	1.4	11.4	−27.9
2	−.2	−.2	−.8	3.5	−33.3
3	1.2	−1.5	−3.8	−20.0	−37.1
4	4.7	−.8	−3.5	−14.3	−87.1
5	11.9	−.2	−4.0	−10.4	−36.6
6	25.9	.5	−3.8	−21.5	−80.3
7	49.7	2.3	−3.1	−11.1	−77.4
8	88.5	4.7	−2.1	−9.0	−24.4
9	146.5	8.2	−3.4	−18.3	−35.1
10	227.8	13.4	−3.5	−13.7	−21.8
11	336.6	19.8	−1.9	−13.4	−45.1
12	471.9	28.2	−.8	−9.3	−39.9
13	622.5	37.3	.4	−7.1	−61.9
CPU time (seconds)	7.30	13.7	27.5	54.3	109.

Table 18

Ibid. for the Numerov method in summed form

i	h				
	1/8	1/16	1/32	1/64	1/128
0	0	−.2	0	−.2	−.2
1	.2	0	−.2	−.2	−.2
2	−.2	0	−.2	−.2	−.5
3	1.7	0	−.2	−.2	−.2
4	4.7	.3	−.3	−.2	−.2
5	11.9	.8	−.2	−.2	−.3
6	26.1	1.5	0	−.2	−.3
7	49.9	2.9	−.2	−.3	−.3
8	88.7	5.3	.2	−.3	−.3
9	146.3	9.0	.3	−.2	−.3
10	228.0	13.9	.6	−.2	−.3
11	336.6	20.6	1.0	−.1	−.3
12	472.0	28.8	1.6	−.1	−.3
13	622.5	38.0	2.1	0	−.2
CPU time (seconds)	5.70	11.2	22.2	44.8	91

Table 19

Ibid. for the de Vogelaere method

i	h				
	1/8	1/16	1/32	1/64	1/128
0	−.5	−.2	−.3	−.3	−.3
1	−1.2	−.2	−.2	−.2	0
2	−9.5	−.8	−.2	−.3	−.2
3	−38.8	−2.7	−.3	−.5	−.3
4	−144.6	−7.6	−.6	−.3	−.3
5	−274.8	−18.3	−1.2	−.2	−.2
6	−568.4	−38.8	−3.1	−.5	−.3
7	−1048.3	−73.7	−4.9	−.8	−.5
8	−1763.5	−128.6	−8.4	−.9	−.2
9	−2745.8	−209.7	−13.7	−1.2	−.3
10	−3992.0	−322.6	−21.1	−1.7	−.5
11	−5441.6	−470.1	−31.0	−2.4	−.6
12	−6948.2	−650.1	−43.2	−3.1	−.6
13	−8214.1	−844.9	−56.5	−3.9	−.7
CPU time (seconds)	5.90	11.3	22.2	44.7	93.

Table 20

Ibid. for the Runge-Kutta method

i	h					
	1/4	1/8	1/16	1/32	1/64	1/128
0	−.3	−.2	−.2	−.2	−.3	−.2
1	−2.7	0	.2	0	−.2	.2
2	−20.8	−1.4	−.2	.2	0	0
3	−86.2	−6.4	−.9	−.2	−.3	−.2
4	−251.0	−18.2	−1.2	−.2	−.3	0
5	−581.8	−44.6	−3.1	−.2	−.2	.2
6	−1149.4	−94.0	−6.4	−.5	−.3	−.3
7	−2006.8	−177.3	−12.2	−1.1	−.5	−.2
8	−3162.1	−306.4	−21.2	−1.7	−.3	−.2
9	−4550.0	−493.2	−34.8	−2.4	−.5	−.2
10	−6000.1	−747.2	−53.6	−3.7	−.6	−.3
11	−7221.7	−1072.2	−78.4	−5.3	−.7	−.3
12	−7814.0	−1458.3	−108.8	−7.4	−.8	−.4
13	−7328.9	−1862.8	−142.1	−9.7	−1.0	−.4
CPU time (seconds)	8.52	17.0	33.3	66.1	133.	269.

Table 21

Ibid. for the CPM(0)

i	h							
	1	1/2	1/4	1/8	1/16	1/32	1/64	1/128
0	−316.0	−78.9	−20.1	−5.2	−1.5	−.6	−.6	−.3
1	−905.0	−229.9	−57.8	−14.6	−3.8	−1.1	−.3	−.2
2	−1558.4	−410.0	−103.6	−26.1	−6.6	−1.7	−.6	−.2
3	−1982.9	−599.7	−152.7	−38.6	−9.9	−2.7	−1.1	−.5
4	−2642.5	−781.7	−200.7	−50.8	−13.0	−3.4	−.9	−.2
5	−5493.0	−940.2	−244.1	−61.6	−15.6	−4.0	−1.2	−.3
6	−2550.5	−1059.6	−279.4	−70.8	−18.3	−4.9	−1.7	−.5
7	1983.6	−1124.4	−302.3	−77.1	−19.8	−5.0	−1.5	−.8
8	−676.4	−1130.2	−309.0	−79.0	−20.1	−5.2	−1.7	−.5
9	−5901.5	−1026.8	−295.0	−76.0	−19.4	−5.0	−1.5	−.6
10	−9745.9	−617.2	−256.0	−66.5	−17.1	−4.6	−1.5	−.6
11	−5999.6	−960.0	−186.9	−49.6	−12.9	−3.5	−1.2	−.6
12	8238.3	167.7	−81.3	−23.4	−6.3	−1.9	−.7	−.4
13	8115.1	1563.9	68.3	14.1	3.1	.5	−.1	−.2
CPU time (seconds)	4.38	8.29	15.0	23.2	31.0	62.3	125.	254.

Table 22

Ibid. for the CPM(1)

i	h							
	1	1/2	1/4	1/8	1/16	1/32	1/64	1/128
0	−23.2	−2.0	−.5	−.2	−.5	−.2	−.3	−.2
1	−122.4	−8.9	−.9	−.3	0	0	−.2	0
2	−322.7	−27.3	−2.1	−.2	−.2	−.2	−.2	0
3	−612.6	−60.0	−4.4	−.9	−.3	−.3	−.3	−.5
4	−1063.1	−107.0	−8.1	−.6	−.2	−.2	−.2	−.2
5	−1725.3	−167.1	−13.0	−.9	−.2	−.2	−.3	−.2
6	−1969.5	−236.5	−19.4	−1.7	−.6	−.5	−.3	−.6
7	−904.4	−309.0	−26.9	−2.3	−.6	−.3	−.5	0
8	−85.3	−377.3	−34.9	−2.7	−.6	−.3	−.5	−.2
9	−1482.7	−427.9	−43.0	−3.4	−.6	−.3	−.5	−.3
10	−4341.6	−452.7	−50.4	−3.8	−.6	−.5	−.4	−.3
11	−6282.3	−486.2	−55.6	−4.4	−.6	−.4	−.4	−.2
12	−3102.3	−441.2	−57.5	−4.6	−.6	−.4	−.4	−.3
13	4396.3	−218.6	−54.4	−4.5	−.6	−.3	−.4	−.3
CPU time (seconds)	5.24	10.0	18.7	31.3	51.8	104.	209.	419.

```
                                   METHOD 1   STEPSIZE 0.06250

EXACT EV.( 0) = -49.45778     COMPUTED EV.( 0) = -49.45775     ERROR( 0) =     -0.3
EXACT EV.( 1) = -48.14842     COMPUTED EV.( 1) = -48.14845     ERROR( 1) =      0.3
EXACT EV.( 2) = -46.29074     COMPUTED EV.( 2) = -46.29073     ERROR( 2) =     -0.2
EXACT EV.( 3) = -43.96831     COMPUTED EV.( 3) = -43.96815     ERROR( 3) =     -1.5
EXACT EV.( 4) = -41.23260     COMPUTED EV.( 4) = -41.23255     ERROR( 4) =     -0.8
EXACT EV.( 5) = -38.12277     COMPUTED EV.( 5) = -38.12276     ERROR( 5) =     -0.2
EXACT EV.( 6) = -34.67230     COMPUTED EV.( 6) = -34.67235     ERROR( 6) =      0.5
EXACT EV.( 7) = -30.91225     COMPUTED EV.( 7) = -30.91248     ERROR( 7) =      2.3
EXACT EV.( 8) = -26.87344     COMPUTED EV.( 8) = -26.87392     ERROR( 8) =      4.7
EXACT EV.( 9) = -22.58859     COMPUTED EV.( 9) = -22.58942     ERROR( 9) =      8.2
EXACT EV.(10) = -18.09468     COMPUTED EV.(10) = -18.09602     ERROR(10) =     13.4
EXACT EV.(11) = -13.43687     COMPUTED EV.(11) = -13.43884     ERROR(11) =     19.8
EXACT EV.(12) =  -8.67608     COMPUTED EV.(12) =  -8.67890     ERROR(12) =     28.2
EXACT EV.(13) =  -3.90823     COMPUTED EV.(13) =  -3.91196     ERROR(13) =     37.3

                                   METHOD 2   STEPSIZE 0.06250

EXACT EV.( 0) = -49.45778     COMPUTED EV.( 0) = -49.45776     ERROR( 0) =     -0.2
EXACT EV.( 1) = -48.14842     COMPUTED EV.( 1) = -48.14842     ERROR( 1) =      0.0
EXACT EV.( 2) = -46.29074     COMPUTED EV.( 2) = -46.29074     ERROR( 2) =      0.0
EXACT EV.( 3) = -43.96831     COMPUTED EV.( 3) = -43.96831     ERROR( 3) =      0.0
EXACT EV.( 4) = -41.23260     COMPUTED EV.( 4) = -41.23264     ERROR( 4) =      0.3
EXACT EV.( 5) = -38.12277     COMPUTED EV.( 5) = -38.12285     ERROR( 5) =      0.8
EXACT EV.( 6) = -34.67230     COMPUTED EV.( 6) = -34.67245     ERROR( 6) =      1.5
EXACT EV.( 7) = -30.91225     COMPUTED EV.( 7) = -30.91254     ERROR( 7) =      2.9
EXACT EV.( 8) = -26.87344     COMPUTED EV.( 8) = -26.87398     ERROR( 8) =      5.3
EXACT EV.( 9) = -22.58859     COMPUTED EV.( 9) = -22.58949     ERROR( 9) =      9.0
EXACT EV.(10) = -18.09468     COMPUTED EV.(10) = -18.09607     ERROR(10) =     13.9
EXACT EV.(11) = -13.43687     COMPUTED EV.(11) = -13.43893     ERROR(11) =     20.6
EXACT EV.(12) =  -8.67608     COMPUTED EV.(12) =  -8.67896     ERROR(12) =     28.8
EXACT EV.(13) =  -3.90823     COMPUTED EV.(13) =  -3.91203     ERROR(13) =     38.0

                                   METHOD 3   STEPSIZE 0.06250

EXACT EV.( 0) = -49.45778     COMPUTED EV.( 0) = -49.45776     ERROR( 0) =     -0.2
EXACT EV.( 1) = -48.14842     COMPUTED EV.( 1) = -48.14841     ERROR( 1) =     -0.2
EXACT EV.( 2) = -46.29074     COMPUTED EV.( 2) = -46.29066     ERROR( 2) =     -0.8
EXACT EV.( 3) = -43.96831     COMPUTED EV.( 3) = -43.96803     ERROR( 3) =     -2.7
EXACT EV.( 4) = -41.23260     COMPUTED EV.( 4) = -41.23184     ERROR( 4) =     -7.6
EXACT EV.( 5) = -38.12277     COMPUTED EV.( 5) = -38.12094     ERROR( 5) =    -18.3
EXACT EV.( 6) = -34.67230     COMPUTED EV.( 6) = -34.66843     ERROR( 6) =    -38.8
EXACT EV.( 7) = -30.91225     COMPUTED EV.( 7) = -30.90488     ERROR( 7) =    -73.7
EXACT EV.( 8) = -26.87344     COMPUTED EV.( 8) = -26.86058     ERROR( 8) =   -128.6
EXACT EV.( 9) = -22.58859     COMPUTED EV.( 9) = -22.56763     ERROR( 9) =   -209.7
EXACT EV.(10) = -18.09468     COMPUTED EV.(10) = -18.06242     ERROR(10) =   -322.6
EXACT EV.(11) = -13.43687     COMPUTED EV.(11) = -13.38986     ERROR(11) =   -470.1
EXACT EV.(12) =  -8.67608     COMPUTED EV.(12) =  -8.61108     ERROR(12) =   -650.1
```

```
                                   METHOD 4   STEPSIZE 0.06250

EXACT EV.( 0) = -49.45778     COMPUTED EV.( 0) = -49.45776     ERROR( 0) =     -0.2
EXACT EV.( 1) = -48.14842     COMPUTED EV.( 1) = -48.14844     ERROR( 1) =      0.2
EXACT EV.( 2) = -46.29074     COMPUTED EV.( 2) = -46.29073     ERROR( 2) =     -0.2
EXACT EV.( 3) = -43.96831     COMPUTED EV.( 3) = -43.96822     ERROR( 3) =     -0.9
EXACT EV.( 4) = -41.23260     COMPUTED EV.( 4) = -41.23248     ERROR( 4) =     -1.2
EXACT EV.( 5) = -38.12277     COMPUTED EV.( 5) = -38.12247     ERROR( 5) =     -3.1
EXACT EV.( 6) = -34.67230     COMPUTED EV.( 6) = -34.67166     ERROR( 6) =     -6.4
EXACT EV.( 7) = -30.91225     COMPUTED EV.( 7) = -30.91103     ERROR( 7) =    -12.2
EXACT EV.( 8) = -26.87344     COMPUTED EV.( 8) = -26.87132     ERROR( 8) =    -21.2
EXACT EV.( 9) = -22.58859     COMPUTED EV.( 9) = -22.58511     ERROR( 9) =    -34.8
EXACT EV.(10) = -18.09468     COMPUTED EV.(10) = -18.08932     ERROR(10) =    -53.6
EXACT EV.(11) = -13.43687     COMPUTED EV.(11) = -13.42903     ERROR(11) =    -78.4
EXACT EV.(12) =  -8.67608     COMPUTED EV.(12) =  -8.66520     ERROR(12) =   -108.8
EXACT EV.(13) =  -3.90823     COMPUTED EV.(13) =  -3.89402     ERROR(13) =   -142.1

                                   METHOD 5   STEPSIZE 0.06250

EXACT EV.( 0) = -49.45778     COMPUTED EV.( 0) = -49.45763     ERROR( 0) =     -1.5
EXACT EV.( 1) = -48.14842     COMPUTED EV.( 1) = -48.14804     ERROR( 1) =     -3.8
EXACT EV.( 2) = -46.29074     COMPUTED EV.( 2) = -46.29008     ERROR( 2) =     -6.6
EXACT EV.( 3) = -43.96831     COMPUTED EV.( 3) = -43.96732     ERROR( 3) =     -9.9
EXACT EV.( 4) = -41.23260     COMPUTED EV.( 4) = -41.23131     ERROR( 4) =    -13.0
EXACT EV.( 5) = -38.12277     COMPUTED EV.( 5) = -38.12122     ERROR( 5) =    -15.6
EXACT EV.( 6) = -34.67230     COMPUTED EV.( 6) = -34.67047     ERROR( 6) =    -18.3
EXACT EV.( 7) = -30.91225     COMPUTED EV.( 7) = -30.91026     ERROR( 7) =    -19.8
EXACT EV.( 8) = -26.87344     COMPUTED EV.( 8) = -26.87143     ERROR( 8) =    -20.1
EXACT EV.( 9) = -22.58859     COMPUTED EV.( 9) = -22.58665     ERROR( 9) =    -19.4
EXACT EV.(10) = -18.09468     COMPUTED EV.(10) = -18.09297     ERROR(10) =    -17.1
EXACT EV.(11) = -13.43687     COMPUTED EV.(11) = -13.43558     ERROR(11) =    -12.9
EXACT EV.(12) =  -8.67608     COMPUTED EV.(12) =  -8.67545     ERROR(12) =     -6.3
EXACT EV.(13) =  -3.90823     COMPUTED EV.(13) =  -3.90854     ERROR(13) =      3.1

                                   METHOD 6   STEPSIZE 0.06250

EXACT EV.( 0) = -49.45778     COMPUTED EV.( 0) = -49.45773     ERROR( 0) =     -0.5
EXACT EV.( 1) = -48.14842     COMPUTED EV.( 1) = -48.14842     ERROR( 1) =      0.0
EXACT EV.( 2) = -46.29074     COMPUTED EV.( 2) = -46.29073     ERROR( 2) =     -0.2
EXACT EV.( 3) = -43.96831     COMPUTED EV.( 3) = -43.96828     ERROR( 3) =     -0.3
EXACT EV.( 4) = -41.23260     COMPUTED EV.( 4) = -41.23259     ERROR( 4) =     -0.2
EXACT EV.( 5) = -38.12277     COMPUTED EV.( 5) = -38.12276     ERROR( 5) =     -0.2
EXACT EV.( 6) = -34.67230     COMPUTED EV.( 6) = -34.67224     ERROR( 6) =     -0.6
EXACT EV.( 7) = -30.91225     COMPUTED EV.( 7) = -30.91219     ERROR( 7) =     -0.6
EXACT EV.( 8) = -26.87344     COMPUTED EV.( 8) = -26.87338     ERROR( 8) =     -0.6
EXACT EV.( 9) = -22.58859     COMPUTED EV.( 9) = -22.58853     ERROR( 9) =     -0.6
EXACT EV.(10) = -18.09468     COMPUTED EV.(10) = -18.09462     ERROR(10) =     -0.6
EXACT EV.(11) = -13.43687     COMPUTED EV.(11) = -13.43681     ERROR(11) =     -0.6
EXACT EV.(12) =  -8.67608     COMPUTED EV.(12) =  -8.67602     ERROR(12) =     -0.6
EXACT EV.(13) =  -3.90823     COMPUTED EV.(13) =  -3.90817     ERROR(13) =     -0.6
```

List 7.

A. *Experimental reflection of the type of approximation used to generate the algorithms*

The standard Numerov method (in the following abbreviated as N1), its summed form (N2), the de Vogelaere (V) and Runge-Kutta (RK) methods are all devised on the approximation that the *solution* is well described by piecewise polynomials and hence the quality of the numerical solution produced by them will directly reflect the degree of validity of such approximation.

For these methods the meaning of the condition that h is sufficiently small is that such h must secure good piecewise polynomial approximation of the solution, see also section 3.1. It follows that when the solution of the differential equation to be solved is flat then h — which is sufficiently small in the above sense — is actually rather big, anyway bigger than in the case when the solution happens to exhibit narrowly oscillating behaviour.

In contrast, the basic approximation to derive piecewise perturbation methods refers to the *coefficients of equation* and these are approximated by piecewise polynomials. Hence, the condition that h is sufficiently small is here dictated by the quality of the approximation of the *coefficients* of the equation, not of the shape of the solution as before.

In our test-problem the shape of the solution strongly depends on E. In fact, the eigenfunction y_i which corresponds to the eigenvalue E_i has just i zeros on (x_{min}, x_{max}) so that, while the solution attached to E_0 is actually very flat, the eigenfunction attached to E_{13} will have thirteen nœuds. It is then expected that at one and the same h the accuracy is the best for E_0 to deteriorate rapidly with i in the case of the classical methods N1, N2, V and RK, but the accuracies will be distributed more or less uniformly over the whole spectrum for the piecewise perturbation methods CPM(0) and CPM(1).

These behaviours are nicely illustrated in tables 17—22. Compare, for illustration, N1 and CPM(0) at $h = 1/8$. The accuracy of N1 is the highest when $i = 0$ and the lowest when $i = 13$ so that the claim that $h = 1/8$ is "sufficiently small" is clearly much better motivated in the first case than in the second. The situation is different for CPM(0) where the accuracies are seen to be distributed uniformly vs. i and thus the same claim remains justified over the whole spectrum.

B. *Experimental reflection of the theoretical predictions*

Theoretical predictions say that methods N1, N2, V, RK and CPM(1) are of *fourth* order while CPM(0) is of *second* order. This is equivalent to saying that, in the limit of small stepsizes, the ratio of the errors resultant from the runs at h and at $h/2$ is expected to be $2^4 = 16$ for the fourth order methods and $2^2 = 4$ for the second-order method.

This prediction is well confirmed and for illustration we take the case of the errors in E_{10} at $h = 1/8$ and $h = 1/16$. For methods N1, N2, V and RK we have in order: $227.8/13.4 = 17$, $228.0/13.9 \simeq 16.4$, $-3992.0/-322.6 \simeq 12.4$ and $-747.2/-53.6 \simeq 13.9$, while for CPM(0) we obtain $-66.5/-17.1 \simeq 3.88$. Confirmation at this pair of step-widths is less indicated for CPM(1) since the errors in the results at $h = 1/16$ are here of the order of the round-off error, but it becomes obvious if the pairs of data at $h = 1/4$ and at $1/8$ are compared.

Theoretical predictions also say that the accumulation of the round-off errors is more drastic for N1 (the accumulated round-off error is here predicted to increase as h^{-2} when h is decreased) than for the other methods where an increase as h^{-1} is expected.

The experimental evidence partially confirms such predictions. In fact, examination of the results at $h = 1/64$ and at $h = 1/128$ clearly indicate dramatic increase of these errors in case of N1, in some cases just by the predicted $2^2 = 4$ factor. They also confirm that the level of the round-off errors is much higher with this method than with any other of the six methods. Yet, an increase of a factor 2 when h is halved, as the theory predicts for the other methods, is not detected; only some fluctuations in the same range are actually observed. The latter indicates that the quoted values of h are still too large to cause the activation of the h^{-1} law of the growth of the round-off errors *.

C. *Comparison of methods for efficiency*

When the program was written particular care was paid to optimize each of the six methods for efficiency.

---

* The fact that the eigenvalues are particularly "resistant" to the round-off errors (anyway much more resistant than the eigenfunctions) is well known among persons with some computational experience. Qualitatively the reason is that the influence of round-off errors in such as only the *scale* of the function P(E) is distorted; the positions of the roots of P(E) remain almost unchanged.

We also found that methods N1, N2 and V are expected to be almost equally fast (remember that they involve in order 4, 3 and 3.5 slow arithmetic operations at each step) while method RK which involves 11 such operations is expected to be about three times slower than the former. As for the piecewise perturbation methods CPM(0) and CPM(1) the computation effort per step depends on the value of Z and this is slower in the trigonometric and the exponential ranges than in the series range. If only the latter is active CPM(0) with its 12 slow operations per step is expected to be about as fast as RK while CPM(1) which involves 20 such operations will be 1.7 times slower than RK.

Although estimates of such sort are manifestly of approximate nature (remember that they take into account only the *slow* arithmetic operations) they are usually rather realistic. In fact they are well confirmed by the experimental results, compare for instance the CPU times at $h = 1/128$.

The values of CPU times mentioned in tables 17—22 show that for methods N1, N2, V and RK times are in all cases approximately doubled when h is halved, and this is a normal result. The same experimental conclusion remains valid also in case of CPM(0) and of CPM(1) but only when $h \leqslant 1/16$. This is also a normal result since it reflects directly the fact that when $h > 1/16$ *all* three regimes in Z are active at different frequencies, depending on h. Specifically, if $h = 1$ the computation meets almost exclusively the slow trigonometric and exponential ranges. The frequency of activation of the faster series range increases progressively when h is decreased and this range becomes the *only active* one when $h \leqslant 1/16$, thus explaining the quoted dependence of the CPU times for CPM(0) and CPM(1).

It is also interesting to note that this program uses *equal* stepsizes and thus it somehow favours the multistep methods N1 and N2 for which this is the only possible stepsize regime. Expressed in other words, in this program methods N1 and N2 work at their best while the other four methods were prevented of taking advantage of their capability of being used at variable stepsize. Yet, the results collected in tables 17—22 are sufficient to conclude that, as a rule, if the computation of the low lying eigenvalues is only required then *all* methods are more or less equivalent for efficiency. A real difference in terms of efficiency occurs only at higher eigenvalues and this is clearly in favour of the piecewise perturbation methods.

Subject index

FSC
www.fsc.org
MIX
Papier | Fördert
gute Waldnutzung
FSC® C083411

Zeitfracht Medien GmbH
Ferdinand-Jühlke-Straße 7
99095 Erfurt, Deutschland
produktsicherheit@kolibri360.de